Earthquakes - From Tectonics to Buildings

Edited by Walter Salazar

Published in London, United Kingdom

IntechOpen

Supporting open minds since 2005

Earthquakes - From Tectonics to Buildings
http://dx.doi.org/10.5772/intechopen.87816
Edited by Walter Salazar

Part of IntechOpen Book Series: Earthquakes, Volume 3
Book Series Editor: Valentina Svalova

Contributors
Ewald Brückl, Peter Carniel, Stefan Mertl, Rita Meurers, Rohollah Rostami, Slobodan B. Mickovski, Subhamoy Bhattacharya, Nicholas Hytiris, Yuriy Gatinsky, Tatiana Prokhorova, Enzo Mantovani, Caterina Tamburelli, Daniele Babbucci, Marcello Viti, Nicola Cenni, Mahmood Ahmad, Xiaowei Tang, Pauline Galea, Daniela Farrugia, Sebastiano D'Amico, Hiroshi Kawase, Eri Ito, Kenichi Nakano, Shigeki Senna, Rob Westaway, Feezan Ahmad, Marijana Hadzima-Nyarko, Ahsan Nawaz, Asim Farooq

First published in London, United Kingdom, 2021 by IntechOpen
IntechOpen is the global imprint of INTECHOPEN LIMITED, registered in England and Wales, registration number: 11086078, 5 Princes Gate Court, London, SW7 2QJ, United Kingdom
Printed in Croatia

British Library Cataloguing-in-Publication Data
A catalogue record for this book is available from the British Library

Additional hard and PDF copies can be obtained from orders@intechopen.com

Earthquakes - From Tectonics to Buildings
Edited by Walter Salazar
p. cm.
Print ISBN 978-1-83962-424-7
Online ISBN 978-1-83962-429-2
eBook (PDF) ISBN 978-1-83962-430-8
ISSN 2631-9152

IntechOpen Book Series

Earthquakes

Volume 3

Dr. Walter Salazar is a structural civil engineer who obtained a doctoral degree in Engineering Seismology from the Interdisciplinary Graduate School of Science and Engineering, Tokyo Institute of Technology, Japan, in 2004. Dr. Salazar has been active in site-effects and seismic hazard research, producing several peer-reviewed maps for El Salvador, Jamaica, and the Eastern Caribbean. He has published sixty articles in peer-reviewed journals, books, and international conferences. In 2011, he received a Distinguished Salvadoran National Award. He is a peer reviewer for several scientific journals. Currently, Dr. Salazar is a Professor of Structural Engineering at the Catholic University of El Salvador.

Editor of Volume 3:
Walter Salazar, Ph.D.
Faculty of Engineering and Architecture,
Catholic University of El Salvador UNICAES, Santa Ana, El Salvador

Book Series Editor: Valentina Svalova
Institute of Environmental Geoscience RAS, Russia

Scope of the Series

The book series addresses the multi-disciplinary topic of earthquake hazards and risks, one of the fastest growing, relevant and applied fields of research and study practiced within the geosciences and environment. It also addresses principles, concepts and paradigms of the earthquake connected disciplines, as well as operational terms, materials, tools, techniques and methods including processes, procedures and implications.
This book series aims to equip professionals and others with a formal understanding of earthquake hazards and risk topics, to clarify the similarities or differences in fundamental concepts and principles in the discipline, to explain the relevance and application of primary tools and practices in earthquake risk study, to direct geologists, engineers, architects, planners, teachers, students, and others interested in the earthquake discipline to authoritative and vetted sources, and last but not least, to capture the wide range of expanding disciplinary activities under a single umbrella of earthquake hazard, disaster and risk concept.

Contents

Preface

The unprecedented Covid-19 pandemic of the last year has threatened the global economy and situated all of us in a critical psychological challenge to survive such an immense impact on our lives and loved ones. Employees have had to transition to working from home and students have had to adjust to full-time remote learning. Teachers have learned to communicate via telematics tools with students and peers. We prepared this book during a time of unique challenges that heightened the importance of publishing sound, expertly reviewed research. This book is the result of the tireless effort of twenty-seven distinguished scholars from eight different countries. It is confirmation that researchers will continue their important work even during times of crisis.

Earthquakes - From Tectonics to Buildings is organized into three sections containing eight chapters. Section 1, "Tectonics and Seismicity," includes Chapters 1–4. In Chapter 1, Gatinsky and Prokhorova explain the intracontinental seismic activity in South Siberia and the Russian far east. The seismogenic zones' geodynamic activity connects, besides plate interaction, with deep lithosphere structure and anomalies of the different geophysical fields and blocks' kinematics. In Chapter 2, Mantovani, Tamburelli, Babbucci, Viti, and Cenni explain the short-term development of Italy's Periadriatic zone's ongoing tectonic processes influence in the spatio-temporal distribution of major shocks, emphasizing the future challenges of the seismic hazard evaluation in the region. In Chapter 3 Brückl, Carniel, Mertl, and Meurers present a beautiful example of how citizens and authorities can be involved in seismological data acquisition in Vienna, Austria. They develop innovative tools for ground motion visualization and epicentral locations. In Chapter 4, Westaway presents a compendium of the induced seismicity from nearby oil production at Newdigate, Surrey, England. He proposes a model incorporating poroelastic and fault asperities' effects into Mohr-Coulomb failure calculations.

Section 2, "Site Effects Evaluation," includes Chapters 5–7. In Chapter 5 Ito, Nakano, Senna, and Kawase propose a new method to fill the gap between observed S-wave spectral amplification factors and calculated theoretical 1D transfer functions based on a robust, strong motion database and current velocity models in Japan. Chapter 6 by Farrugia, Galea, and D'Amico focuses on unique profile soil patterns characterized by buried low-velocity layers in the Maltese Islands (Central Mediterranean), studying the dispersion of Rayleigh waves and their implication in earthquakes ground motion estimation. In Chapter 7, Ahmad, Tang, Ahmad, Hadzima-Nyarko, Nawaz, and Farooq present a critical review of the factors of soil liquefaction by applying interpretive structural modeling (ISM) and the MICMAC approach.

Section 3, "Building Foundations," includes Chapter 8 by Rostami, Mickovski, Hytiris, and Bhattacharya. In this chapter, the authors explore the pile foundations' dynamic behavior in liquefiable soils based on robust analysis methods. Additionally, they discuss remediation techniques for earthquake resistance.

Walter Salazar, Ph.D.
Professor,
Faculty of Engineering and Architecture,
Catholic University of El Salvador UNICAES,
Santa Ana, El Salvador

Section 1

Tectonics and Seismicity

Chapter 1

Tentative Intracontinental Seismic Activity in South Siberia and Russian Far East

Yuriy Gatinsky and Tatiana Prokhorova

Abstract

Overwhelming majority of minor, strong and major earthquakes in south Siberia and Russian Far East coincide with relatively narrow intracontinental zones on the boundaries of blocks and lithosphere plates. The geodynamic activity of these zones connects, besides the plate interaction, with deep lithosphere structure and anomalies of the different geophysical fields as well as with blocks' kinematics. Authors' located areas of the seismic centers origin and the possible manifestations of the high seismicity based on the distribution of the maximal volumes of releasing seismic energy. We established these areas, with certain care, in the northeast Altai and adjacent part of the west Sayany, in the west of the east Sayany, around the Baikal Lake and in northwest Transbaikalia, in the east of Transbaikalia between the Vitim River and upper stream of the Aldan River, and in the north of the Sakhalin Island. The majority of minor and strong, rarely major, earthquakes took place in these areas. Deep and near surface structural peculiarities influence on these areas' geodynamics and allow establishing possible levels of seismic energy releasing. We draw areas of intensive seismic energy releasing with its calculating for each from investigated regions. They gravitate towards interblock zones, which separate crust blocks and the North Eurasian Lithosphere Plate. The fulfilled investigation allows establishing specific areas of the increased seismicity in south Siberia, Russian Far East and adjacent territories.

Keywords: seismic active zone, strong and major earthquakes, seismic energy, area of origin of seismic centers, active fault, block, interblock zone, lithosphere plate, deep structure, geophysical field

1. Introduction

In 2013–2018, authors analyzed at the relatively small scale the structure and seismicity of intracontinental interblock zones within the Asiatic part of Russia and adjacent countries together with calculation of the releasing seismic energy and construction of their dissipation graphs as well as the deep seismic sections [1, 2]. The results of this analyzing allowed selecting the most active seismic zones, including overwhelming majority of minor and strong earthquakes' epicenters, and examined in detail the lithosphere deep structure under them and not far of them with establishing the possible connection of these zones with the geodynamic activity and anomalies of different geophysical fields. With special attention, we considered kinematics of blocks and main lithosphere plates as well as the connection with them active processes in interblock zones. All blocks considered in our paper are delineated by active faults, data on which we taken from the works of [3, 4], including results of author's field investigations in Transbaikalia in 2008 [5]. The modern block structure of central Asia was considered and validated in some previous author's works [1, 6–8].

In 2018–2020, we investigated in the more large scale the influence of the deep and near surface lithosphere structure on the interblock zones geodynamic activity with possible levels of seismic energy releasing. Within the most active from these zones, we established areas with seismic energy maximal releasing. They situate in the northwest Altai together with the adjacent part of the west Sayany, in the west of the east Sayany, around the Baikal Lake and in northwest Transbaikalia, in the east of Transbaikalia between the Vitim River and upper stream of the Aldan River and in the north of the Sakhalin Island.

2. Areas of seismic centers origin in South Siberia and Russian Far East

Based on the earthquake epicenters distribution and maximal volumes of seismic energy releasing we located, with certain care, five specific areas of origin of seismic centers (OSC) with seismic energy volumes calculation in them. Active faults limit the majority of these areas. The fulfilled investigation of these intracontinental areas allows showing increased seismicity distribution in south Siberia, Far East and adjacent territories. Below we consider geodynamics and seismicity of each from five selected OSC areas.

3. OSC area in the Altai

The area includes seismic active zones selecting the Altai Block from the North Eurasian Plate (NEP), West Mongolia and Sayany blocks within the south part of the Altai Republic, west part of the Tuva Republic in Russia and adjacent territory of northwest Mongolia (**Figure 1**). Its length is 356 km, width – 87-243 km and the total seismic energy volume achieves $1.70833 \cdot 10^{16}$J (here and below we calculate cumulative energy volume during the course of 1966–2017 using for its calculation data from NEIC2019 catalog). Left- and right-lateral northwest slips with compression component predominate within the area as well as near latitudinal and northwest thrusts to the north and northwest. Maximal magnitudes (M) of earthquakes reach 5–8. The greatest seismic activity coincides with the boundary of the Altai Block and NEP, where the energy releasing increases up to 10^{11}J. According to seismic cross section [1] right-lateral slips and thrusts to the northeast develop here in the depth 10-34 km. East on the boundary between West Mongolia and Sayany blocks left-lateral slips predominate in the depth of 8-32 km. Tectonic strain axis's have the northeast direction.

The heat flow (HF) value changes from 25 to 28 $\mu W/m^{-2}$ within NEP near Novosibirsk, up to 55–96 to the south in the Altai Mountains [10]. Within the Mining Altai in east Kazakhstan HF is equal 29-61 $\mu W/m^{-2}$. East on the boundary between West Mongolia and Sayany blocks HF value increases up to 80 $\mu W/m^{-2}$ and larger. The International Terrestrial Reference Frame (ITRF2014) vectors of horizontal displacement have mainly east and southeast direction in stations near Novosibirsk (88.7–92.2°) with velocities 23.9–27.1 mm/y. However, the Altai Station north of Ust-Kamenogorsk fixes the displacement to 350° NW with the velocity 50 mm/y. The large gradient in the crust structure takes place at the boundary between Altai and Junggar blocks south from investigated region, where crust thickness decreases southeast from 55 to 60 down to 35-40 km [11]. Data on HF and the Global Navigation Satellite System (GNSS) velocities as well on S-wave velocities below give the representation of the crust and lithosphere activity, which can have the indirect connection with the seismicity level.

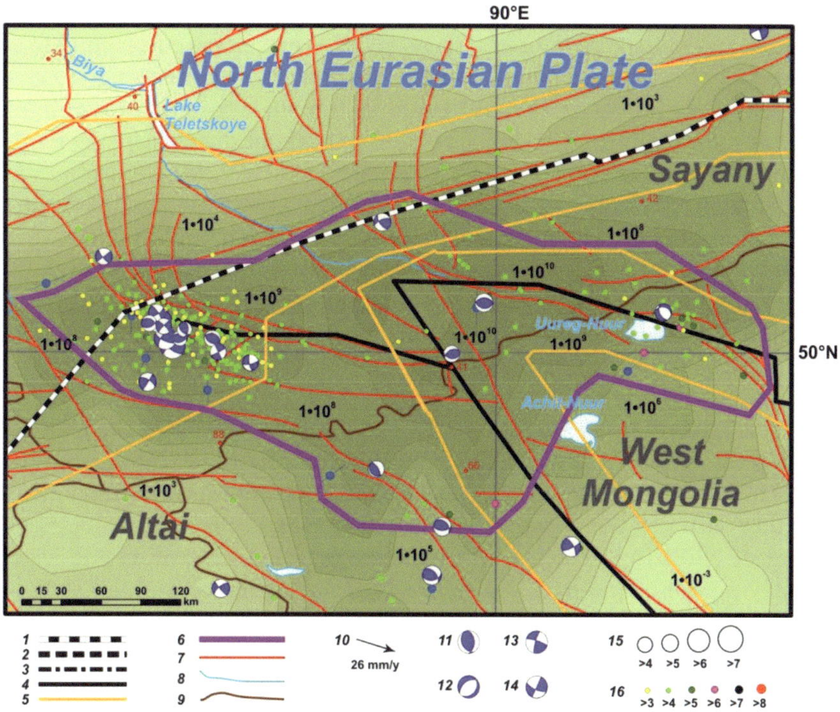

Figure 1.
The OSC area situated in the northeast altai and west Sayany within seismic active zones selects the Altai block from NEP, Sayany and West Mongolia blocks. The summary legend for ***Figures 1, 2 and 3-5*** *includes boundaries of: 1 – Lithosphere plate, 2 – The same supposed, 3 – Transitional zone, 4 – Block, 5 – Interblock zone, 6 – OSC area (see the part 2). Other symbols: 7 – Active fault, 8 – River, 9 – State frontier, 10 – ITRF2014 (see the part 3) vectors of horizontal displacement, 11–14 – Mechanism in hypocenters after the centroid moment tensor (CMT2019) according to NEIC2018 seismological data (11 – Compression, 12 – Extension, 13 – Left-lateral slip, 14 – Right-lateral slip; the predominant orientation of segments in mechanism corresponds to a displacement direction), 15 – The magnitude scale of earthquake epicenters by CMT, 16 – The magnitude scale of earthquake epicenters, for which CMT2019 data are absent (according to NEIC seismological data). Thin lines with dark-blue circles correspond to tectonic stress axis's [9]. Green color fields, divided by isohypsis, show volumes of seismic energy releasing. Increasing color intensity corresponds to increasing energy volume on* 10^1*J or* 10^{-1} *J. the minimal value is* 10^{-7}*J, the maximal –* 10^{14}*J. The black numerals in the scheme correspond to energy volume values in joules. Dotes with red figures near them correspond to values of the heat flow in* $\mu W/m^{-2}$*. The total extensions of studied blocks see in the* ***Figure 6****. The authors of the chapter constructed this scheme as all others.*

The geodynamic and seismicity data analysis allow supposing future earthquakes intensity within the Altai OSC area. Events with M7–8 are possible in the west quarter of this area within the Mountain Altai at the left bank of the Argug River, which is the Biya River left source, southwest from the Belyashi settlement. The hypocenters' predominant depth is there about 10-20 km that allows waiting mainly destructive events in this part of the area. Earthquakes with M5–6 are more probable in future within the rest territory of the OSC area.

4. OSC area in the East Sayany

This area embraces seismic active interblock zones separating the Sayany Block from NEP and from the Hangay Block in the east part of the Tuva Republic, southwest of Buryat Republic and adjacent part of north Mongolia (**Figure 2**). Its length is 350 km, width – 100-216 km and the total seismic energy volume achieves $1.805.11{\cdot}10^{16}$J. The most intensive seismic energy releasing up to 10^{9-11}J characterizes the left bank of the Ka-Hem River in the upper stream and south from it in both sides of the Russia and Mongolia Frontier. The majority of epicenters correspond to events with M3–6 and M of some of them in the Mongolia territory increase up to 7–9.

In the OSC area's north parts near latitudinal thrusts to the north predominate with the left-lateral slip component. The stead left-lateral slips take place south between Sayany and Hangay blocks. Transverse north – northeast subsidiary faults develop inside the Sayany Block. The majority of tectonic stress axis's have the strike changing from near longitudinal in the west up to northeast in the east. The transition from transpression to transtension conditions takes place in south Siberia from west to east within interblock zones separating NEP from blocks, which situate south. Thrusts predominate on Altai and west Sayany, where instrumentally measuring crust shortening reaches 2-4 mm/y [12]. They changes on slips with the thrust component in the east Sayany and further east on left-lateral slips with the extension component in the Tunka depression approaching to the Lake Baikal. More easts within the Baikal Rift System the extension completely predominates. The tectonic stress tensors' analysis confirms this changing as well as our field investigation together with V.A. San'kov near the Lake Baikal and in Transbaikalia [5, 13].

The steady NEP displacement east characterizes regarded territory with velocities 25-26 mm/y and the azimuth 89° NE on the Novosibirsk Station and 104.5° SE on the Badary Station north of the east Sayany. Permanent azimuth deflections east and southeast correspond to the plate small clockwise rotation. The maximal difference between measuring and a model vector of horizontal displacement reaches 68° in the NEP – Sayany interblock zone, where active slips predominate. The Novosibirsk Station records the uplift at 8.0 mm/y. The lesser uplift with velocities 2.0–2.5 mm/y characterizes the east Sayany. High magnetic anomalies up to +50 - +100*n* coincide with interblock zones between NEP, Altai and Sayany blocks as well as with faults within them (*http://projects.gtk.fi/WDMAM/project/perugia/index.html*).

The negative gravitational anomalies at the Bouguer Reduction up to −50 – -150mGal, characterize the great part of central Asia, including Hangay and Sayany

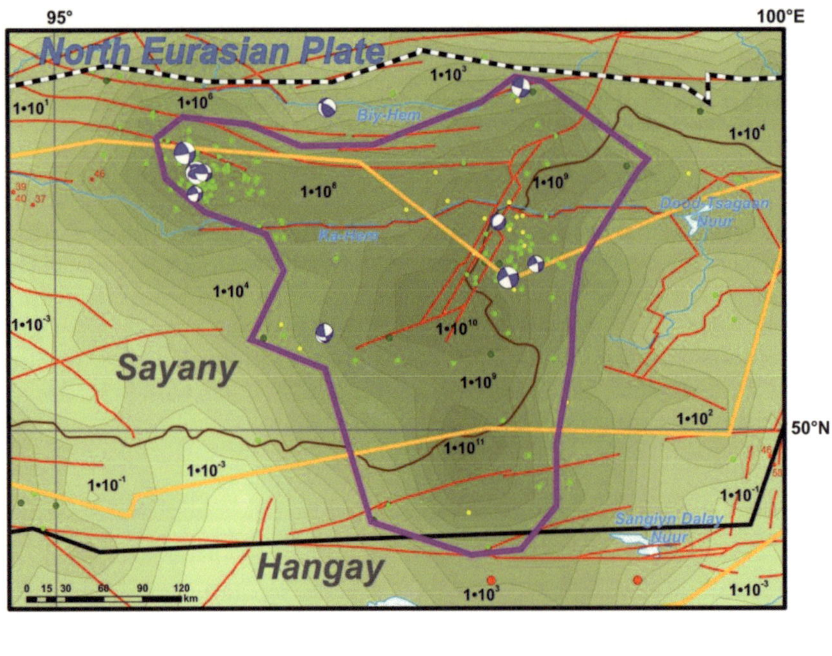

Figure 2.
The OSC area in east Sayany takes place within seismic active zones separating the Sayany block from NEP and the Hangay block. See the summary legend for the ***Figure 1****.*

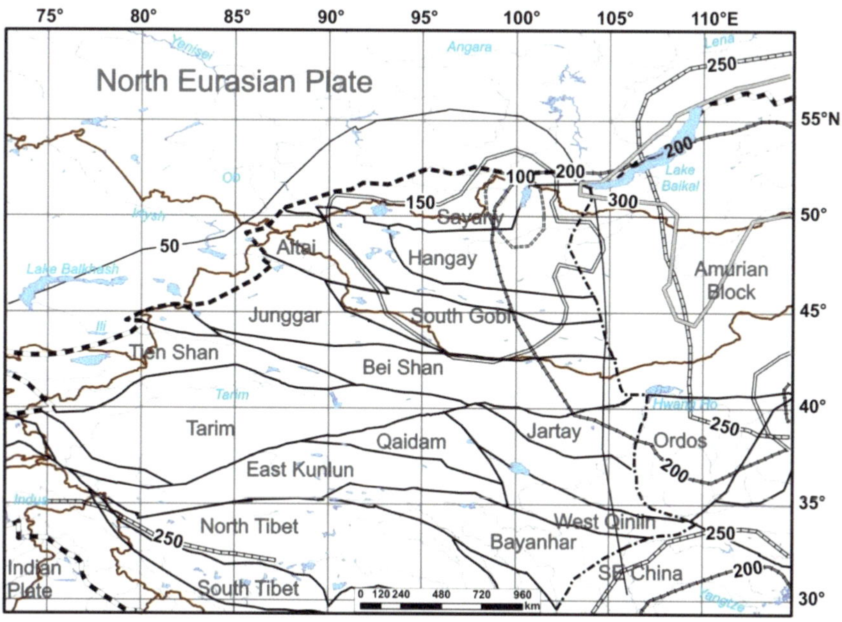

Figure 3.
Central Asian mantle anomalies. For block, transitional zone and plate boundaries see summary legend for the ***Figure 1****. Gray lines of different types limit projections of S-waves' velocity slowing-down up to ≤4.2–4.25 km/sec-1 on 50, 100, 150, 200, 250, and 300 km depth levels after seismic tomography data. Authors of the paper drew these lines after data contained in the work [16]. Note that both parts of OSC areas with possible earthquakes' M to 7–9 (see* ***Figures 1, 2****) coincide with near isometric outline (counter), under which S-wave velocities slow down at the depth 150 km. That allows supposing the connection of the increase seismicity there with the possible mantle plume rising.*

blocks. In the Junggar–Altai interblock zone they reach even −200mGal [14]. High negative values of these anomalies correspond to significant increasing of the crust thickness within the above-mentioned blocks. HF values increase in the interblock zone between Sayany – NEP up to 47-150 $\mu W/m^{-2}$ and between Hangay – Sayany blocks within the Hubsugul Rift up to 80–140. The increasing HF values often coincide with areas, which situate above supposed mantle plumes [15]. One of those plumes can situate after seismic tomography data at the boundary between Hangay and Sayany blocks in east Tuva in depth 100-150 km (see **Figure 3**). There the lithosphere thickness decreases down to 70-50 km and the temperature at the depth 50 km can reach 1000–1200° C after relation of isotopes 3He/4He [17].

The increased level of seismicity can characterize in the future two parts of the OSC area with possible earthquakes M to 7–9. The first of them takes place on the left bank of the Ka-Hem River in its upper stream between the Lake Tere-Hol in the west and the Russia – Mongolia frontier in the east (see **Figure 2**). The second part situates south from it in both sides of the Russia and Mongolia Frontier. The hypocenters' concentration takes places at the depth 9-1 km, so we can wait mainly destructive events in this part of the area. Earthquakes with M4.99–5.99 are more probable in future within the rest territory of the OSC area.

5. The OSC area in the Baikal system and northwest Transbaikalia

The OSC area includes territory between NEP and the Amurian Block (or the small plate in other interpretation [18]) within the Baikal Rift System and its northeast continuation (**Figure 4**). The area length is 950 km, its width – 11-230 km and the total releasing energy volume reaches 1.79362·10^{15}J. Normal northeast faults predominate there with the left-lateral slip component. Tectonic stress axis's have stead northeast direction. The releasing energy level oscillates from 10^4 to 10^{12}J, M changes from 3 to 6 reaching 6–7 for some certain events. The zone of active extension stands out within all Baikal Rift System after mechanism earthquake solutions. Releasing energy maximal volumes coincides with this zone. The extension processes do not get out from the crust (10-33 km) according to analysis of the hypocenters development after CMT2017 data. Separate faults in adjacent parts of the Amurian Block are left-lateral slips, rarely thrusts to southwest.

Normal faults and left-lateral slips with extension component prevail near southwest ending the Baikal Rift System at the depth of 10-16 km within the interblock zone dividing NEP and the Amurian Block. Yu. Gatinsky and G. Vladova together with V. San'kov discovered during 2008 field itineraries in the Barguzin Depression distinct changing slips, causing seismic dislocations, by later normal faults [5]. These faults cross from the Paleozoic granite to the yang alluvium corresponding to the time progress of the transtension process. The maximal increasing the seismic energy dissipation up to 1.4·10^{15}J occurs above normal faults restricting the Baikal Trough in southwest [1]. Just there near the Kultuk settlement the earthquake with M6.3 took place in 2008. The earthquake near destroyed the settlement and some communication ways. West we observed the displacement of modern streamlets' thalwegs along NW slips with amplitudes up to 10 m. NEP displaces on 105.6° SE with the velocity 25.9 mm/y according to measuring in the Irkutsk Station into absolute coordinates of ITRF2014. Vectors within the Amurian Block have azimuths of 106.9° -108.0° SE on the Ulan-Bator Station with the velocity 31.2 mm/y and up to 121° SE and more on some Chinese stations with velocities 26-35 mm/y. HF values reach 96-140 μW/m^{-2} in the Lake Baikal Trough in comparison with 36–79 at its sides within NEP and the Amurian Block.

Note, that in the east part of Central Asia horizontal displacements predominate into transtension conditions with opening numerous rifts in the Baikal System,

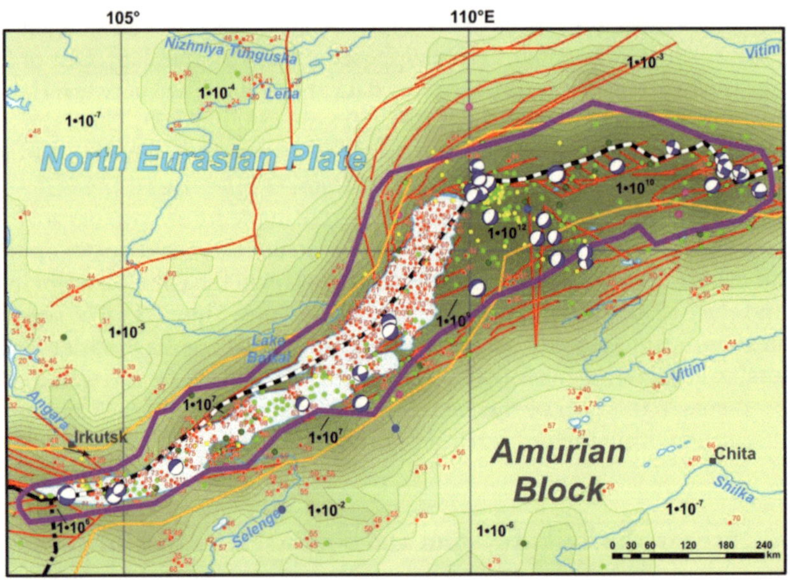

Figure 4.
The OSC area within the Lake Baikal and northwest Transbaikalia takes place in seismic active zones separating NEP and the Amurian block. See summary legend for the ***Figure 1****.*

around the Ordos Block and in boundaries of some other blocks. The development of those extension processes has different interpretation: squeezing out east some blocks including the Amurian Block under the influence of the Hindustan – Asia collision [12], or the upper mantle flow generated by the deep submergence of the Pacific slab under Asia [19–21]; mantle plume raising under north Mongolia and the Baikal region (see **Figure 3**) resulted in the crust extension and rift formation [7, 17]. The increase anisotropy exists in the majority regions of central Asia with coinciding north-northeast direction of splitting *Pn*-waves high velocities with the axes of maximal compression and GPS vectors [22], that shows the complete deformation coupling within crust and lithosphere mantle. The authors of paper [23] arrived at the same conclusion for the region of Baikal, west and central Mongolia.

Data on geodynamics and seismicity of the regarded OSC area allow predicting maximal intensive seismic events with M up to 7–9 within northwest third of the area territory: to the north in the Stanovoi Upland at both banks of the Upper Angara River and farther south east of the Lake Baikal in the north part of the Barguzin Range. New earthquakes with M up to 6–7 can be in the other part of the OSC area. The main hypocenters concentrations of preceding events were at the depth 10-16 km, so within regarded area disastrous earthquakes will be the most probably, most likely as near Kultuk in 2008.

6. The OSC area in the northeast Transbaikalia

The area situates in the territory of Zabaikalian Krai, south Yakutsk, and northwest of the Amurian District between the Vitim River and upper stream of the Aldan River (**Figure 5**). Its length is 464 km, width – 64-108 km, and the total seismic energy reaches $4.64153 \cdot 10^{16}$J, that is nearly equally to the energy of the OSC area of the north Pamir - $4.50343 \cdot 10^{16}$J [24]. The energy level oscillates on the regarded area between 10^4 - 10^{12}J and M in epicenters changes from 3 up to 6. M of separate events reaches 6–7. Northeast normal faults with the left-lateral slip component predominate in the west of the OSC area and northwest left-lateral slips develop in the east. Tectonic stress axis's have the stead northeast direction.

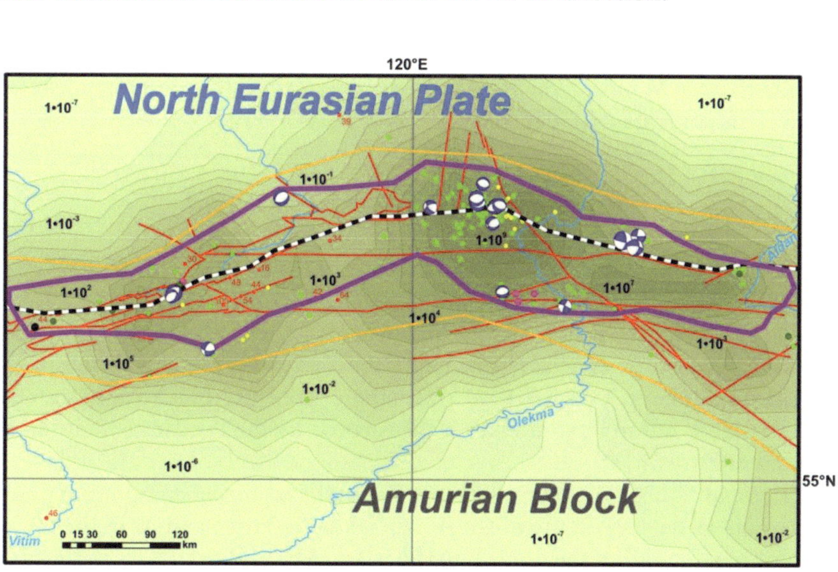

Figure 5.
The OSC area in northeast Transbaikalia settles down within the seismic active interblock zone between NEP and the Amurian block. See summary legend for the ***Figure 1****.*

According measuring at the Yakutsk Station NEP moves on 121° SE with velocity 20.0 mm/y, but simultaneously displaces south relatively the stable Eurasia in the NNR_NUVEL_1A System. Azimuths of the vast Amurian Block change from 106° SE on the Ulan-Bator Station up to 110° near Beijing with velocities 29-32 mm/y. This block turns counter-clockwise after SLR and GPS data 2000–2008 coinciding with right-lateral displacing along the large-scale Tanlu Slip System on the block east boundary. The same turn fixed by results of the Belgic-Russia tidal gravitation profile along 50° N latitude [25, 26]. The alternation of wide northeast strips with +100 - -100*n* takes place in the magnetic field above the Amurian Block. They apparently can reflect the early Precambrian block fundament structures strike. After seismic tomographic data, the reduced lithosphere thickness characterizes the north part of the block (100-105 km in comparison with 200 km under NEP) [27]. The low values of the rock density and electric resistance apparently correspond to lithosphere deep stratification on boundaries of 10, 20, 40 and 70-90 km [28].

Low *Pn*-waves velocities characterize the upper mantle under the transitional zone between continent and ocean under Amurian and neighboring Japanese-Korean blocks [22]. Some researches suppose existing of the strong upper mantle substance flow on the southeast direction based on the analysis of lengthening deformation axis's and anisotropy in border fields of NEP and Amurian Block. This deep flow causes the NEP clockwise turn confirmed by ITRF2014 vectors [23]. The above-mentioned simultaneous counter-clockwise turn of the Amurian Block apparently results in opening the Baikal Rift. At the same time, we cannot exclude the influence of plume tectonic processes on geodynamics of the Amurian Block – NEP zone. After seismic tomography data isometric counters of S-waves' velocity slowing-down projections (see **Figure 3**) up to ≤4.2–4.25 km/sec^{-1} take places on 50, 100, 150, 200, 250, and 300 km depth levels [16]. Note that the majority of the high heat-flow meanings up to 64-100 μW/m^{-2} coincide with S-waves velocity slowing-down projections [8]. These data allow supposing ascent of heating and non-consolidated mantle material to the earth's surface with increasing seismicity level.

Speaking on the future seismic events it is needed to distinguished the east part of the regarded OSC area on Olekma River booth banks in the region of BAM railway stations of Khani and Yuktali as well as east up to left bank of the Upper Aldan River. There it is possible to wait for new earthquakes with M up to 7–8. In the other area territory M most likely will be not higher than 6. The light hypocenters concentrations coincide with depths of 10–11 and 21-26 km.

7. The OSC area in the north half of the Sakhalin Island

We select this area within the interblock zone dividing Japanese-Korean and Okhotsk blocks (**Figure 6**). Its length is 530 km, width – 154 km, the total seismic energy releasing after instrumental measuring data (NEIC2017) reaches $1.15535 \cdot 10^{16}$J. The energy level in the central part of the area is about 10^{11}J. Epicenters with M4–5 are numerous, with 6–8 are only single. This interblock zone stretches through all island, but in its south half is "absorbed" by the Pacific subduction zone dipping under Eurasia [1].

Longitudinal right-lateral slips with north-northeast thrust component predominate in the up-to-date structure of the north Sakhalin. Hypocenters in the crust (10-40 km) correspond to the dip slip in the island east and to thrusts in the west. The intensive earthquakes with M7.2 in Moneron (1971) and Neftegorsk (1995) occurred in northeast Sakhalin within its coastal part. Their hypocenters depth reaches 30-60 km. North Sakhalin GPS vectors direction is about 130° SE with

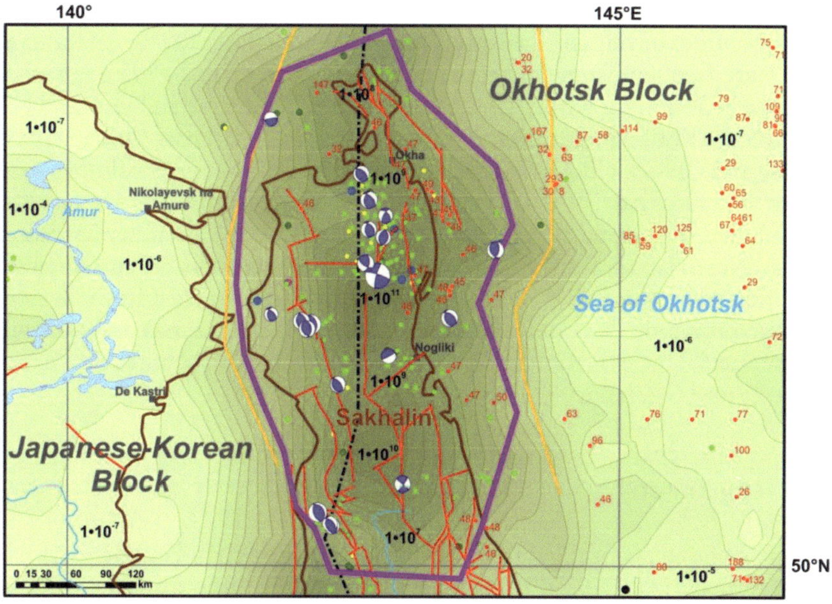

Figure 6.
The OSC area in the north half of the Sakhalin Island situates within the seismic active interblock zone dividing Japanese-Korean and Okhotsk blocks. See summary legend for ***Figure 1****.*

velocities 18.1–22.1 mm/y. After measuring in the ITRF2014 System on Khabarovsk and Seoul stations the Japanese-Korean Block moves on 115–120° SE with velocities 22-25 mm/y. This block also turns clockwise to southeast with the velocity of 0.9–1.5 mm/y according with measuring in the Sikhote-Alin geodynamic net [29]. This turn can evoke up-to-date extension and crust non-consolidation in northeast China within the Songliao Depression [30] differ from main lithosphere plates displacing by their kinematics. The Okhotsk Block displaces on 149° SE with the velocity 22.9 mm/y, North Japanese – on 132° SE with the velocity 25.0 mm/y. In the central part of the Honshu Island, the uplift fixes with the velocity 3.7 mm/y after ITRF2014. The Pacific Plate moves relatively Eurasia on 288°NW with the velocity 75 mm/y in the NNR_NUVEL_1A System.

The HF average value in Sakhalin Island is 76 $\mu W/m^{-2}$ increasing within adjacent aqyatories up to 123-200 $\mu W/m^{-2}$ [31, 32]. After the same authors magnetic field anomalies within island do not exceed 0-200*n* sharply increase in the Okhotsk Sea up to 1000-1200*n*. Gravitational anomalies on Sakhalin in the Bouguer reduction come to −30 - +50mGal [15]. The crust thickness under island reaches 30-35 km, the lithosphere thickness – 52 km [31]. Completing the examination of the north Sakhalin OSC area we can expect the probably origin of earthquakes with M up to 7–8 in the center of the island north part within the Neftegorsk Region, as well as west and southwest from it. For the more south regions of OSC area and the extreme north of the island we can, with certain care, suppose, that M of future seismic events will be not higher than 5–6. Hypocenters concentrations situate at the depth 5-55 km.

8. Conclusion

We selected five OSC areas in intracontinental interblock zones, in which the most intensive seismic events can take place. The OSC areas with maximal seismic activity divide the Altai Block and NEP, Hangay and Sayany blocks,

Amurian Block and NEP within republic's Altai, Tuva, Buryat, Irkutsk District and Mongolia. Volumes of releasing seismic energy reach in them 10^{7-12}J, earthquakes with M4–6 are frequent, some infrequent events have M up to 7–9. The seismicity intensity is a little weaker within OSC areas dividing the Amurian Block and NEP in Transbaikalia Krai, Amurian District, Mongolia, and northeast China, as well as Japanese and Okhotsk blocks in Sakhalin, where seismic energy volumes reach 10^{6-10}J, earthquakes with M5–7 are frequent, infrequent events have M up to 8.

We excluded from our analysis continental margin zones, dividing Japanese-Korean and North Japan, Kuril-Kamchatka and Okhotsk blocks because they situate in connection with west Pacific subduction zones as distinguished from investigated above intracontinental OSC areas. These continental margin zones represent subjects of the special investigation and partly were examined in some earlier papers of Yu. Gatinsky and G. Vladova [33, 34].

Acknowledgements

This investigation is fulfilled with assistance of the Russian Foundation for Basic Research (Projects No. 13-05-00109 and 18-05-00160). Authors during many years discussed a matter of central Asia up-to-date geodynamics including active seismic zones of south Siberia and Far East with Professor S.I. Sherman, Academician A.F. Grachev, and Academician D.V. Rundquist, who were the great connoisseurs of these regions geology. His papers and monographs represent the essential contribution to understanding and solving problem touched in our paper.

Author details

Yuriy Gatinsky* and Tatiana Prokhorova
Institute of Earthquake Prediction Theory and Mathematical Geophysics, Russian Academy of Sciences, Moscow, Russia

*Address all correspondence to: gatinsky@gmail.com

References

[1] Gatinsky Yu. G., Prokhorova T. V. Seismic active zones in South Siberia, Russian Far East, and adjacent countries. Russian Journal of Earth Science. 2015; 15: ES3003. DOI: 10.2205/2015ES00 0554

[2] Gatinsky Yu. G., Prokhorova T. V., Rundquist D. V. The 102-103° E Geodivider in the Modern Lithosphere Structure of Central Asia. Geodynamics & Tectonophysics. 2018; 9 (3): 989-1006. https://doi.org//10.5800/GT-2018-9-3-0380.

[3] Trifonov, V. G., Soboleva O. V., Trifonov R. V., Vostrikov G. A. Recent Geodynamics of the Alpine-Himalayan Collision Belt. In: Leonov Yu.G., editor. Transactions of the Geological Institute RAS, 54. 2002. 225 p. (in Russian).

[4] Sherman S. I., Sorokin A. P., Sorokina A. T., Gorbunova E. A., Bormotov V. A. New Data on the Active Faults and Zones of Modern Lithosphere Destruction in the Amur Region. Doklady Earth Sciences. 2011; 439(2): 1146-1151. DOI:10.1134/S1028334X11080186

[5] Gatinsky Yu. G., Prokhorova T. V., Rundquist D. V., Vladova G. L. Zones of catastrophic earthquakes of Central Asia: Geodynamics and seismic energy. Russian Journal of Earth Science. 2009; 11(1): ES1001. DOI:10.220 5/2009ES000326

[6] Gatinsky Yu. G., Rundquist, D. V. Geodynamics of Eurasia: Plate Tectonics and Block Tectonics. Geotectonics. 2004; 38 (1): 1-16.

[7] Gatinsky Yu., Rundquist D, Vladova G., Prokhorova T. Up-to-date geodynamics and seismicity of Central Asia. International Journal of Geosciences. 2011; 2 (1): 1-12. DOI:10.4236/ijg.2011.21001

[8] Gatinsky Yu. G., Prokhorova T. V. Superficial and Deep Structure of Central Asia as Example of Continental Lithosphere Heterogeneity. Universal Journal of Geoscience. 2014; 2(2): 43-52. DOI:10.13189/ujg.2014.0202020380.

[9] Heidbach O., Rajabi M., Reiter K., Ziegler M., WSM Team. World Stress Map Database Release 2016. GFZ Data Services. 2016. DOI:10.5880/WSM.2016.001

[10] Duchkov A.D., Zheleznyak M.N., Ayunov D.E., Veselov O.V., Sokolova L.S., Kazantsev S.A., GornovP.Yu.,DobretsovN.N.,BoldyrevI.I., Pchelnikov D.V., Dobretsov A.N. Geothermal Atlas of Siberia and Far East (2009-2012). 2012. Available from: http://maps.nrcgit.ru/geoterm/ (last accessed 11.06.2019).

[11] Crustal Structure of China from Deep Seismic Sounding Profile. 2001. https://earthquake.usgs.gov/data/crust/china.php

[12] San'kov V. A., Lukhnev A. I., Melnikova V. I., et al. Present-day Tectonic deformations of the Southern Mounting Frame of the Siberian Platform from GPS Geodesy Data. In: Proceedings International Seminar on the use of Space Techniques for Asia-Pacific Regional Crustal Movements Studies, Irkutsk, August 2002. 2003; Moscow, GEOS: 118-126.

[13] San'kov, V. A., Lukhnev A. V., Radziminovich N. A., et al. A Quantitative estimate of modern deformations of the Earth's crust in the Mongolian block (Based on GPS-geodesy and seismotectonic data). Doklady Earth Sciences. 2005; 403: 946-949.

[14] Bonvalot S., Balmino G., Briais A., Kuhn M., Peyrefitte A., Vales N., Biancale R., Gabalda G.,

Moreaux G., Reinquin F., Sarrailh M. World Gravity Map. Scale 1:50000000. BGI-CGMW-CNES-IRD, Paris. 2012.

[15] Huang J., Zhao D. Seismic imaging of the crust and upper mantle under Beijing and surrounding regions. Physics of the Earth and Planetary Interiors. 2009; 173(3-4): 330-348. https://doi.org/10.1016/j.pepi.2009.01.015.

[16] Kozhevnikov V. M., Yanovskaya T. B. S-wave velocities distribution in the lithosphere of the Asian continent after data of surface Rayleigh waves. In: Levi K. G. and Sherman S. I. , editors. Actual problems of modern geodynamics of central Asia. 2005. Novosibirsk, Siberian Branch RAS: 46-64 (in Russian).

[17] Grachev A. F. Modern Volcanism, Mantle Plumes, and Their Connection with the Stress Intensity in the Lithosphere. In: Grachev A. F., editor. Neotectonics, Geodynamics and Seismicity of Northern Eurasia. 2000; Probel, Moscow: 245-266 (in Russian with English Summary).

[18] Zonenshain, L. P., Savostin L. A. Geodynamics of the Baikal Rift Zone and Plate Tectonics of Asia Tectonophysics. 1981; 176(1-2): 1-45. DOI:10.1016/0040-1951(81)90251-1

[19] Rasskasov S. V., Chuvashova I. S., Yasnygina T. A., Saranina E. V., Fefelov N. N., Brandt I. S., Brandt S. B. Slab and over-slab later Cenozoic melting in zones of convergent boundaries of Asia and in east Hangay of central Mongolia. Proceedings of the Irkutsk University, Earth Sciences Series. 2008; 5(1): 236-249 (in Russian with English Summary).

[20] Barruol G., Deschamps A., DeverchereJ.,MordvinovaV.V.,UlzibatM., Perrot J. Upper mantle flow beneath and around the Hangay dome, Central Mongolia. Earth and Planetary Science Letters. 2008; 1-2: 221-233. DOI:10.1016/j.epsl.2008.07.027

[21] Parfeevets A. V., San'kov V. A. Late Cenozoic tectonic stress fields of the Mongolian microplate. Comptes Rendus Geoscience. 2012; 344(3-**4**): 227-238. DOI:10.1016/j.crte.2011.09.009

[22] Pei S., Zhao J., Sun Y., Xu Z., Wang S., Liu H., Rowe C. A., Toksoz M. N., Gao X. Upper Mantle Seismic Velocities and Anisotropy in China Determined Through Pn and Sn Tomography. Journal of Geophysical Research. 2007; 112: B05312, DOI:10.1029/2006JB004409

[23] San'kov V. A., Lukhnev A. V., Parfeevets A. V., Miroshnichenko A. I., Ashurkov S. V. Coupling of the Crustal and Upper Mantle Deformations in the Mongolia–Siberian Mobile Area. Doklady Earth Sciences. 2011; 436(1): 159-164, DOI:10.1134/S1028334X11010302

[24] Gatinsky Yu. G., Prokhorova T. V., Rundquist D. V. Zones of the Origin of Seismic Centers in the Pamir–Tien Shan Sector of High Asia. Doklady Earth Sciences. 2017; 475(2): 887-890, DOI:10.1134/S1028334X17080049

[25] Timofeev V. Yu., Ardyukov D. G., Gornov P. Y., et al. GPS and tidal method for geodynamic study in Siberia and in Far East of Russia. Proceedings of the APSC Symposium: Space Geodesy and Dynamic Planet, (16-18 October 2006, Korea). 2006: 196-209, CAPSG Korea.

[26] Timofeev V. Yu., Gornov P. Y., Ardyukov D. G., et al. Plate motion modeling (by example of the Amurian Plate). In: Problems of Far East and east Siberia seismicity and up-to-date geodynamics. Khabarovsk. 2010: 65-67 (in Russian).

[27] Ashurkov S. V., San'kov V. A., Miroshnichenko A. I., Lukhnev A. V.,

Sorokin M. P., Serov M. A., Byzov L. M. GPS geodetic constrains of the Amurian Plate. Russian Geology and Geophysics. 2011; 52(2): 239-249. DOI:10.1016/j.rgg.2010.12.017

[28] Shevchenko B. F., Kaplun V. B. The dip geodynamic model of the Eurasian and Amurian lithosphere plates connecting region. Lithosphera. 2007; 4: 3-20 (in Russian).

[29] Shestakov N., Gerasimenko M., Takahashi H., et al. Present tectonics of the southeast of Russia as seen from GPS observations. Geophysical Journal International. 2011; 184(2): 529-540. DOI:10.1111/j.1365-246X.2010.04871.x

[30] Khain V. E. Tectonics of Continents and Oceans (Year 2000). Scientific World, Moscow. 2001: 606 p. (in Russian with English Summary).

[31] Rodnikov A. G., Zabarinskaya L. P., Sergeeva N. A. Deep structure of seismic dangerous regions of the Earth (Sakhalin Island). Bulletin of the RAS Earth Science Department. 2014; 6(NZ1001): 1-8. DOI:10.2205/2014NZ000121 (in Russian).

[32] Lomtev V. L., Litvinova L. V. On the structure of the north Sakhalin submarine margin. Geology and minerals of the World Ocean. 2013; 1: 93-103 (in Russian).

[33] Gatinsky Yu. G., Vladova G. L., Rozhkova V. V. Seismicity and Metallogeny of Convergent Plate Boundaries in Subduction Zones. Doklady Earth Sciences. 2000; 371A: 583-587.

[34] Gatinsky Yu. G., Vladova G. L. Subduction zones of SE Asia: main types, seismicity and mineralization. In: Proceedings of the VAG International Symposium 2008, 7-9 November, Hanoi, Vietnam. Special issue of Journal of Geology: International Year of Planet Earth. 2008: 9-16. http://idm.gov.vn/nguon_luc/Xuat_ban/2008/b31-32/b9.htm

Chapter 2

Tectonics and Seismicity in the periAdriatic Zones: Implications for Seismic Hazard in Italy

Enzo Mantovani, Caterina Tamburelli, Daniele Babbucci, Marcello Viti and Nicola Cenni

Abstract

The recognition of the seismic zones most prone to next major earthquakes in Italy would considerably help the choice of the most efficient prevention plan. This work describes an attempt to gain reliable information about that problem by exploiting the knowledge about the short-term development of the ongoing tectonic processes in the study area and its influence on the spatio-temporal distribution of major shocks. In the periAdriatic zones, such distribution is connected with the progressive northward displacement of the Adria plate, that is controlled by the progressive activation of the decoupling fault systems in the surrounding belts (Dinarides, Apennines and Eastern Southern Alps). The reliability of this hypothesis is evaluated by analysing the seismic histories of the periAdriatic zones. The regularity patterns that are tentatively recognised in such histories are used to identify the most probable location of next major shocks. Further insights into the present seismic hazard in the Southern Apennines and Calabria are tentatively inferred from tectonic connections between these regions and other periAdriatic zones, suggested by the seismic histories in the last 2–4 centuries and the geodynamic/tectonic context in the central Mediterranean area.

Keywords: seismic hazard, tectonics, Apennines, Italy, deterministic approach

1. Introduction

It is well known that seismic activity in the periAdriatic zones (**Figure 1**) is related to the interaction of the Adriatic plate (Adria hereafter) with the surrounding belts (**Figure 2**).

Stressed by the convergence of the confining structures, Adria tries to move roughly northward [10–12, 14–16].

This gradual displacement is allowed by the activation of the decoupling fault systems located along the lateral boundaries of Adria (Dinarides and Apennines) and in the northern front of that plate, in the Eastern Southern Alps. The central and southern Dinarides and the Eastern Southern Alps are characterised by thrust faults while a dextral transpressional regime prevails in the northern

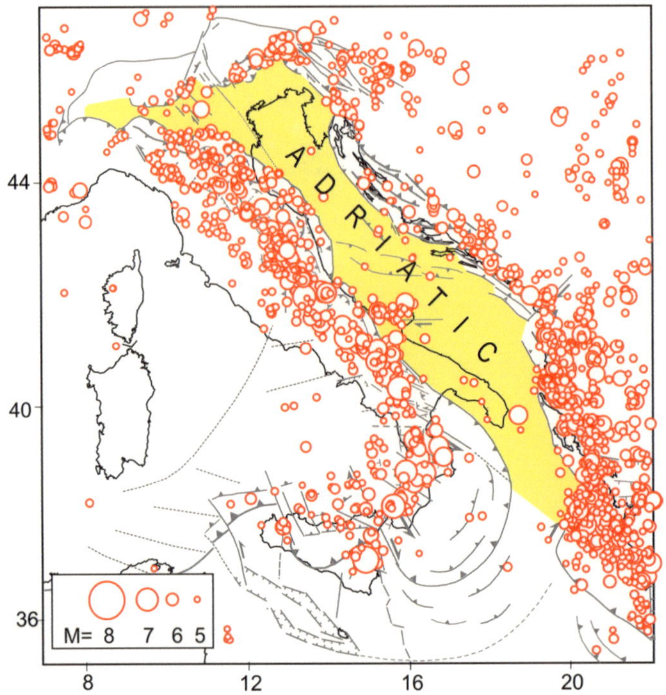

Figure 1.
Major earthquakes (red circles, M ≥ 5) since 1000 A. D in the periAdriatic zones [1–9].

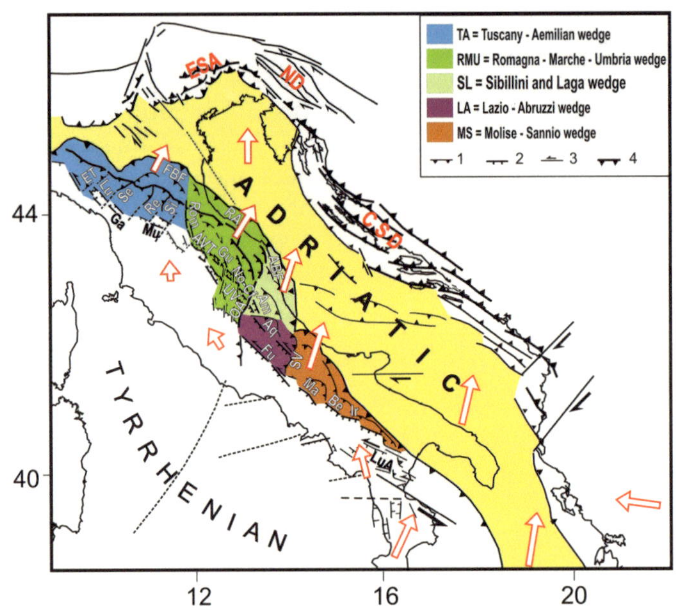

Figure 2.
Tectonic sketch of the Adriatic region (e.g., [10–12]). The main wedges in the eastern sector of the Apennine belt are evidenced by colours (inset). See text for explanations. The proposed kinematic pattern with respect to Eurasia [13, 14] is indicated by red empty arrows. 1) Compressional, 2) extensional 3) transcurrent features, 4) Outer fronts of Neogenic belts. Am = Amatrice fault system, Aq = L'Aquila fault system, AVT = Alta Valtiberina trough, CSD = Central-Southern Dinarides; ESA = Eastern Southern Alps; ET = Enza-Taro thrust, FBF = Ferrara buried folds, Fu = Fucino fault system; Ga = Garfagnana, Lu = Lunigiana, LuA = Lucanian Apennines, Ma, Be, Ir = Matese, Benevento and Irpinia fault systems, ND = Northern Dinarides, No-Cf-Gu = Norcia-Colfiorito-Gubbio fault system; OV = Olevano-Antrodoco-Sibillini transversal thrust, RA = Rimini-Ancona thrust front, Re = Reno thrust, Rom = Romagna fault system, Se = Secchia thrust, Si = Sillaro thrust, SV = Sangro-Volturno oblique thrust, UV = Umbra valley.

Dinarides e.g., [17–21]. The decoupling mechanisms along the western Adria boundary (Apennines) are more complex [10–12, 16], due to the presence of a shallow crustal structure (eastern sector of the chain, coloured in **Figure 2**), that is moving independently from Adria and the western (Tyrrhenian) sector of the belt. This tectonic/kinematic context has been determined by the fact that in the most recent evolution (Quaternary) the outer chain, stressed by Adria, has undergone longitudinal shortening, accommodated by major deformations:

- Strong uplift, recognised in various sectors of the chain [22–25].

- Formation of arcs, as the Campania-Lucania and the Matese-Benevento in the Southern Apennines (see [12] and references therein), the Gran Sasso in the Central Apennines [26, 27] and the Emilian and Ferrara buried folds in the Northern Apennines [28]. This deformation is also suggested by the transition from a cylindrical to a non-cylindrical (arcs) geometry of the orogenic accretion e.g., [28].

- The zones of interaction between the main belt sectors are characterised by transversal/oblique thrusts, as the Olevano-Antrodoco-Sibillini Mts. and the Sangro-Volturno [26, 29, 30].

- Roughly NE ward extrusion of major wedges, with particular regard to the Molise-Sannio (MS) and the Romagna-Marche-Umbria (RMU). This process is compatible with a large amount of geological evidence reported by Viti et al. [12] and references therein).

The divergence between the MS and RMU escaping wedges with respect to the inner less deformed belt has caused the formation of extensional and transtensional fault systems along the axial part of the chain, where a number of troughs has developed. Roughly NW-SE sinistral transtensional and transpressional faults developed in the Lucania Apennines e.g. [31–34]. Extensional faults are recognised in the Irpinia, Benevento and Matese zones, along the inner side of the MS wedge e.g., [35, 36]. The L'Aquila and Fucino transtensional fault systems allow the relative motion between the Lazio-Abruzzi (LA) wedge and the inner belt [26, 37–40]. The decoupling of the RMU wedge from the inner belt is accommodated by the Norcia-Colfiorito-Gubbio-Alta Valtiberina extensional and transtensional fault system and the parallel Umbra Valley trough (e.g., [41–43] and references therein). The simultaneous development of uplift and extensional features in the belt cannot easily be explained without assuming belt-parallel compression as driving force.

The occurrence of several major earthquakes in the Romagna Apennines reveals the presence of an important roughly S-N fault system (Rom in **Figure 2**, [44] and references therein, [42, 45]). This discontinuity allows the RMU wedge to decouple from the Tuscany-Emilia Apennines sector that is not parallel to the Adria plate.

Another evidence consistent with the longitudinal compressional regime in the northernmost Apennines is the presence of transverse thrust faults, as the Sillaro, Reno, Secchia and Enza-Taro faults [23, 46, 47].

The kinematic field that is suggested by the Quaternary deformation pattern in the Apennine belt [11, 48, 49] is compatible with the present displacement field, inferred from geodetic GPS data e.g., [11, 50, 51], which indicates that the outer Adriatic sector of the Apennine chain is moving faster (4–5 mm/y) and more northward with respect to the inner belt (about 1 mm/y).

2. Short-term kinematics of Adria and spatio-temporal distribution of seismicity in the surrounding belts

In the short-term the northward displacement of Adria does not develop continuously over time. Each seismic decoupling along the Adria lateral boundaries (Dinaric and Apennine belts) triggers the acceleration of the involved Adriatic sector e.g., [52–54]. These local accelerations induce an increase of stress at the other still blocked Adria boundaries, where consequently the probability of earthquake occurrence gets higher. When such stressed zones are then affected by major shocks, the acceleration involves more northern zones of Adria up to reach the thrust front of Adria in the Eastern Southern Alps.

In order to check the above seismotectonic interpretation, we have divided the periAdriatic boundary zones in a number of sectors (**Figure 3**). The eastern lateral boundary of Adria includes the Central-Southern Dinarides (CSD in **Figure 4**), and the Northern Dinarides (ND). The western lateral boundary (Apennine belt) is divided in more sectors, being characterised by a more complex tectonic setting, as discussed in the previous section. The Southern Apennines (SA) are mainly characterised by extensional faulting. In the Central Apennines (CA) transtensional decoupling fault systems (L'Aquila and Fucino) prevail. The Northern Apennines are divided in various sectors, due to their complex tectonic setting, with particular regard to the Romagna-Marche-Umbria wedge (RMU). The southern part of the western RMU boundary (Norcia-Colfiorito-Gubbio fault system, RMUWB), is mainly characterised by extensional faults. Considering the peculiar seismotectonic role of the Northern RMU wedge, its boundaries, i.e. the Rimini-Ancona thrust front (RA), the Romagna fault (Rom) and the Alta Valtiberina trough (AVT), are taken as three independent zones. The Emilia Apennines (EM) is the belt sector that lies just north of the RMU wedge. The last sector in **Figure 4** (ESA-ND) is the zone where the Adria plate underthrusts the Eastern Southern Alps.

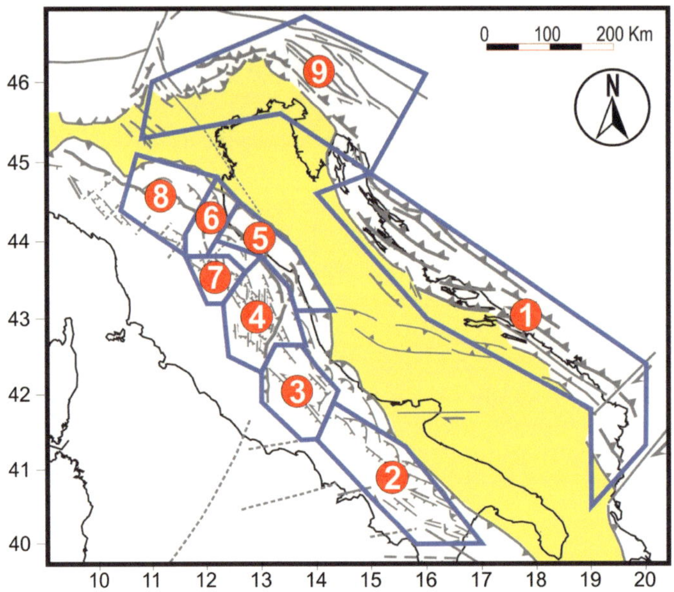

Figure 3.
Geometries of the periAdriatic boundary zones adopted for determining the seismicity patterns shown in ***Figure 4****. 1) Central-Southern Dinarides (CSD in* ***Figure 4****), 2) Southern Apennines (SA), 3) Central Apennines (CA), 4) Southern part of the western boundary of the RMU wedge (RMUWB), 5) Rimini-Ancona thrust front (RA), 6) Romagna fault system (Rom), 7) Alta Valtiberina trough (AVT), 8) Emilia Apennines (EM), 9) Eastern Southern Alps and Northern Dinarides (ESA-ND).*

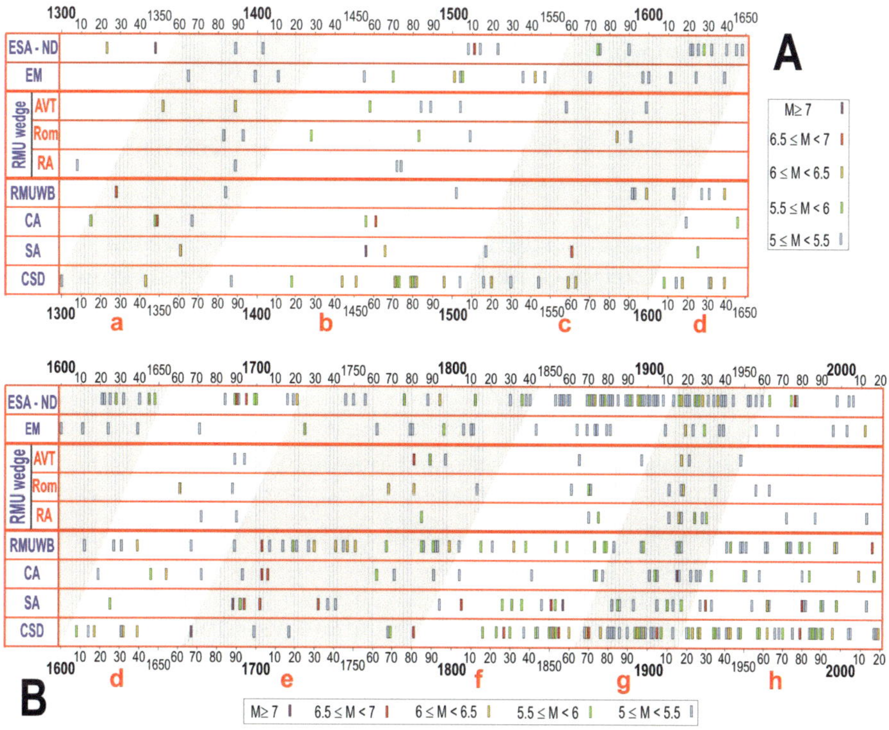

Figure 4.
(A, B) *Time patterns of seismic activity in the periAdriatic zones. AVT = Alta Valtiberina trough, CA = Central Apennines, CSD=Central-Southern Dinarides, ESA-ND = Eastern Southern Alps and Northern Dinarides, EM = Emilia Apennines, RA = Rimini-Ancona thrust front, RMUWB=Southern western boundary of the RMU wedge, Rom = Romagna fault system, SA = Southern Apennines. Earthquakes are indicated by bars with colours related to magnitude (scale in* **A** *and* **B***). Sources of seismicity data in the caption of* **Figure 1***. The grey and white bands tentatively include the events that are supposed to belong to the presumed migrating sequences, identified by letters (***a-h***). The geometries of the zones considered and the spatial distribution of major shocks in the various sequences are shown in the* **Figures** *3 and 5 respectively.*

In the seismicity time patterns shown in **Figure 4**, we tentatively recognise the following peculiar features:

- In the zones considered, most intense seismicity tends to concentrate in short periods (crises), separated by longer phases of lower activity.

- The crises tend to occur later and later as the zones involved are located more and more to the north, delineating a sort of migrating pattern (seismic sequence).

- A number of sequences may be recognised in the period considered, as tentatively evidenced by grey and white bands and letters (from **a** to **h**) in **Figure 4**.

- The time development of the proposed sequences tends to occur in two phases. During the first phase, seismicity mainly affects the southern and central Dinarides, the central and southern Apennines and the western extensional boundary of the RMU wedge (the Norcia-Colfiorito-Gubbio fault system). This phase generally involves several shocks of $M \geq 5.0$ in each zone and generally lasts some tens of years.

- In most cases, the second phase starts with a crisis in the Romagna decoupling fault system, followed (within 10–20 years) by the activation of the inner

(Alta Valtiberina trough) and outer (Rimini-Ancona thrust) boundaries of the northern RMU wedge.

- Then, seismic activity mostly involves the main fault systems in the Emilia Apennines, the Eastern Southern Alps and the Northern Dinarides, over periods of about one-two decades.

- Major earthquakes at the northern Adria front mostly occurred some years after the main seismic decouplings around the northern RMU wedge (**Figure 4**). This tendency is consistent with the hypothesis that the release of the RMU wedge favours the acceleration of the northern Adria domain [11, 42].

The spatial distribution of the shocks in the 8 sequences evidenced in **Figure 4** is shown in **Figure 5**.

In most of the proposed sequences seismic activity took place in all periAdriatic zones. Moreover, one could note that when a zone is characterised by low seismicity, in the following sequence such zone is often characterised by intense earthquakes. For example, in the sequence **c** the Central Apennines did not experience any event with M ≥ 5.0 while strong earthquakes (1646 M = 5.9, 1654 M = 6.3) hit that zone in the subsequent sequence. In the Southern Apennines, after a period of low activity from 1562 to 1687 (only one earthquake with M ≥ 5.0 in the sequences **c** and **d**), a phase of intense seismicity took place in the following sequence **e** (1688 M = 7.1, 1692 M = 5.9,

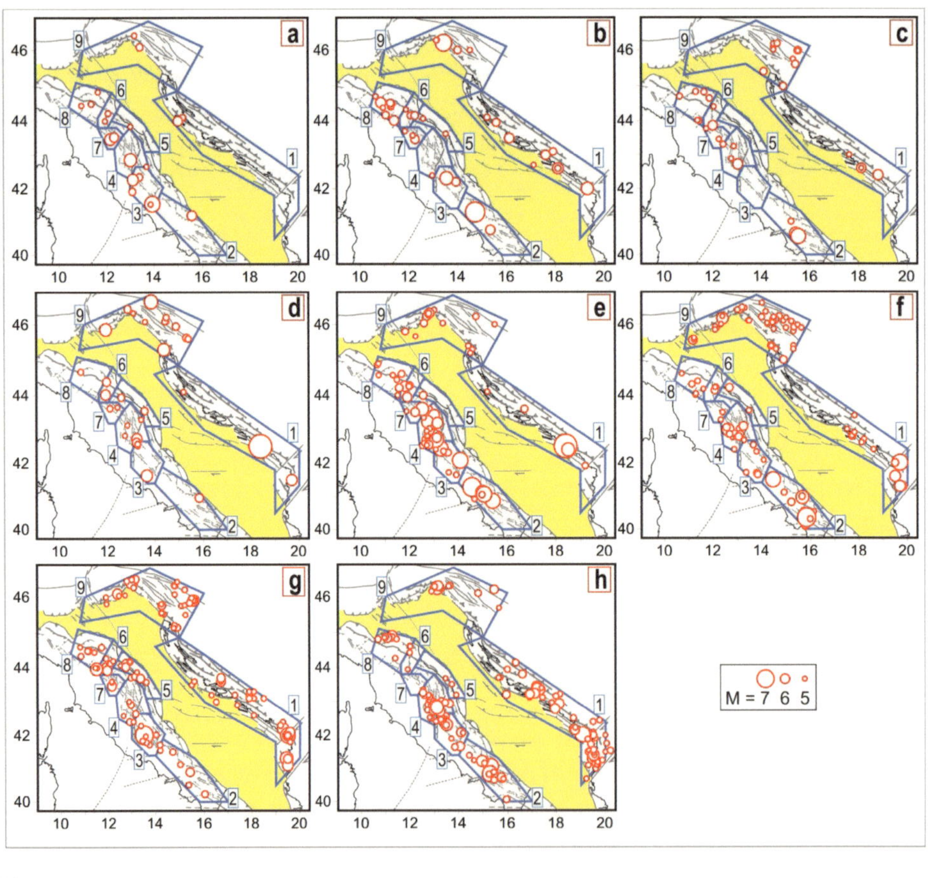

Figure 5.
Spatial distribution of major (M ≥ 5.0) earthquakes in the seismic sequences tentatively evidenced in ***Figure 4****. Numbers as in* ***Figure 3****.*

1694 M = 6.7). The strong 1915 Fucino earthquake (M = 7.1, sequence **g**) took place after a sequence (**f**) characterised by relatively low seismicity (few events with M ≤ 5.5).

3. Some remarks on the Apennine zones most prone to the next strong earthquakes

Taking into account the regularity patterns that we tentatively recognise in the seismic sequences so far developed (**a-g** in **Figure 4**), we try to gain insights into how the last, still ongoing, sequence (**h** in **Figure 4**) might develop in the next future. In this regard, it must be considered that the first phase of that sequence has so far involved several earthquakes in the Southern and Central Dinarides and the Southern and Central Apennines. The acceleration of southern Adria triggered by such seismic decouplings has presumably stressed and deformed the RMU wedge, increasing its tendency to separate from the inner belt. This hypothesis may explain why a number of major extensional shocks (1979 M = 5.8, 1984 M = 5.6, 1997 M = 6.0, 5.7, 5.6, 5.5, 2016 M = 6.2, 6.1, 6.6, [7]) occurred along the western border of the RMU wedge (Norcia-Colfiorito-Gubbio fault system). The NE ward acceleration of the southern RMU wedge may have emphasised stresses (and thus seismic hazard) at the northern boundaries of that wedge (Romagna fault, Alta Valtiberina trough and Rimini-Ancona thrust front). Thus, one could expect that the present seismic hazard in such zones is higher than in the other Apennine fault systems. This hypothesis is also suggested by the fact that the last significant earthquakes (M ≥ 5.3) in the above zones occurred about 100 years ago, i.e. a quiescence longer than the previous ones (**Figure 4**).

Another zone where tectonic load may currently be high is the Emilia Apennines and the related buried folds (**Figure 2**), since such structures, including the Mugello trough, have been stressed by the push of the RMU wedge during the last tens of years. The above hypothesis is consistent with the fact that intense earthquakes (2012, M = 6.1, 5.9) recently occurred in the Ferrara buried folds (lying outside the Emilian Apennines) and that moderate seismicity (M = 4.5) affected the Mugello trough on December 2019.

The kinematic field delineated by geodetic data [11, 50, 51] suggests that the separation between the inner and outer Apennine belts is developing at rates of about 3–4 mm/y, which implies that a displacement of about 30–40 cm has been accumulated since the last activations of the fault systems surrounding the northern RMU wedge (about 100 years). This displacement is comparable to the fault slip associated with a M = 5–6 earthquake e.g., [55].

4. Present seismic hazard in the Southern Apennines: further evidence from a seismotectonic correlation

Further information on the present seismic hazard in the Southern Apennines could be inferred from a correlation that has been recognised between the major earthquakes in that zone and the ones in the Southern Dinarides [56–61].

The possibility that intense seismic activity in the Southern Apennines may be influenced by the occurrence of major shocks in the Southern Dinarides has been first suggested by the fact that the strong April 1979 Montenegro event (M = 6.9) was followed by the strong November 1980 Irpinia earthquake (M = 6.8) in the Southern Apennines (**Figure 6**). The idea that the above correspondence may be a systematic phenomenon was then suggested by the fact that in the last two centuries similar correspondences occurred other times (**Figure 6B**). From the list of events given in

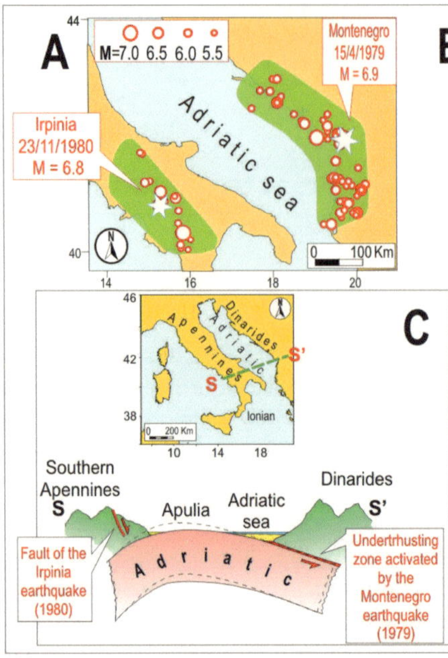

B

Southern Dinarides (M > 5.5)	Southern Apennines (M ≥ 5.5)
1823 (5.7)	1826 (5.7)
1827 (6.5), 1830 (5.6)	1831 (5.5)
1833 (6.5)	1836 (5.9)
1843 (5.6, 6.2),1848 (6.4), 1850 (5.6)	1851 (6.5, 5.5),1853 (5.6)
1851 (6.0, 6.6, 6.3, 6.1), 1852 (6.2), 1853 (5.7), 1855 (6.6, 6.8)	1857 (7.1)
1860 (6.2), 1864 (5.9)	
1869 (5.6, 6.0, 6.2), 1870 (6.5) 1876 (6.3, 5.6)	
1894 (5.9), 1895 (6.2),1896 (5.9, 6.2)	
1905 (6.6, 6.1), 1906 (6.4, 5.6), 1907 (5.7, 6.2)	1910 (5.8)
	1917 (5.5)
1920 (5.9, 5.7), 1921 (5.6), 1926 (5.9, 6.3), 1927 (6.0)	1930 (6.7)
1934 (5.6),1935 (5.6), 1940 (5.6), 1948 (5.6)	
1958 (5.6), 1959 (5.9, 6.3, 5.7), 1962 (6.2)	1962 (5.7, 6.2)
1979 (6.9, 5.7, 6.4)	1980 (6.8)
1982 (5.7), 1988 (5.7)	1990 (5.8)
1996 (6.0)	1998 (5.5)
2019 (6.2)	

Figure 6.
A) *Geometry of the zones implied in the presumed interrelation between Southern Dinarides-Albanides and Southern Apennines and location of the earthquakes given in the table. The stars indicate the locations of the 1979 and 1980 earthquakes in Montenegro and Irpinia.* ***B)*** *List of the major seismic events occurred since 1810. The events with M ≥ 6.0 are in red. Seismicity data as in* ***Figure 1****.* ***C)*** *Structural sketch, through a transversal section in the southern Adriatic area (S-S′), evidencing the vertical flexure of the Adriatic lithosphere overthrusted by the Dinaric belt, on one side, and plunged under the Apennine belt, on the other side (e.g., [62]). The vertical scale is exaggerated in order to make more evident the possible effect of a seismic slip (red arrow) along the subduction fault beneath the Dinaric belt. The dashed lines indicate the presumed profile of the Adriatic lithosphere before a seismic slip in the Southern Dinarides.*

this figure, one can note that in the period considered all the shocks with M ≥ 6.0 in the Southern Apennines have been preceded within few years (less than 5) by one or more earthquakes with M ≥ 6 in the Southern Dinarides. The above correspondence does not worsen significantly even if a lower threshold (M = 5.5) is considered, given that only one of the 15 Southern Apennine events failed to be preceded by comparable events in the Southern Dinarides. The above evidence may indicate that a fault in the Southern Apennines cannot easily activate without the contribution of a post-seismic perturbation triggered by one or more major shocks in the Southern Dinarides.

Since the probability that such a correspondence merely occurs by chance is very small [57, 58], it is plausible to suppose that a close tectonic connection exists between the two zones (**Figure 6C**). The occurrence of a major seismic slip at a thrust fault beneath the Southern Dinarides, such as the one that developed with the 1979 Montenegro event (estimated to be 1–2 metres, e.g. [63]), implies a comparable displacement of the adjacent Adria domain, which causes a reduction of vertical flexure in the southern Adriatic domain, as sketched in the section of **Figure 6**. Such process is expected to induce extensional strain in the Southern Apennines, which may favour the activation of the belt-parallel normal faults recognised in that zone, as for instance the one that generated the 1980 strong earthquake in the Irpinia zone e.g., [35, 64]. This hypothesis is confirmed by the results of numerical modelling of the strain perturbation that was presumably induced in the Irpinia zone by the 1979 Montenegro event [57–60]. Moreover, the strain rate induced by the Montenegro earthquake is expected to reach its maximum amplitude in the Southern Apennines about 1–2 years after the triggering event, a delay fairly consistent with the time interval that elapsed between the April 1979 Montenegro and November 1980 Irpinia shocks. The possible relationship between stress/strain rate increase and triggering of seismic activity has been pointed out in several works e.g., [52–54, 60, 65–69].

The fact that the above significant correlation can be recognised for the most recent, complete and reliable part of the seismic catalogue may imply that this phenomenon can represent a tool for recognising the periods when the probability of strong shocks in Southern Apennines is undergoing a significant increase. In this view, the fact that since 1979 no earthquakes with M ≥ 6.5 have occurred in the Southern Dinarides (**Figure 6**) could imply that at present the probability of major shocks in the Southern Apennines is relatively low. Some doubts about this prediction may be raised by the fact that a significant shock recently occurred in the Southern Dinarides (2019, M = 6.2). The previous seismic histories would suggest that such event is slightly weak for triggering significant seismicity in the Southern Apennines. However, possible uncertainties in the estimated magnitude could reflect on the reliability of the above prediction.

5. Present seismic hazard in Calabria

The analysis of the seismic histories of Calabria and the Hellenides sector lying between the Ionian islands and Albania, along with the geodynamic context in the central Mediterranean area, suggests a possible connection between these two zones [58, 59, 61]. This interpretation is consistent with the structural/tectonic setting sketched in the section of **Figure 7**, which implies that a seismic slip at the Hellenic thrust zone reduces the upward vertical flexure of the Adriatic lithosphere, so attenuating the resistance that the Calabrian wedge encounters in overthrusting such lithosphere. Since, this last process underlies the main genetic mechanism of Calabrian shocks, one can realise why an earthquake in the Hellenides may cause an increase of seismic hazard in Calabria.

The above interpretation and its implications on the interaction of the Calabrian and Hellenic seismic sources is consistent with the quantification of the effects of post-seismic relaxation induced by strong earthquakes in the Hellenides [58, 59, 61], which provides insights into the most probable delay between the presumed precursor and the induced event.

The possibility that the above phenomenon was systematic is supported by the comparison of the seismic histories of the two zones involved (**Table 1**), which indicates that all Calabrian seismic crises with M ≥ 6.0 have been preceded, within

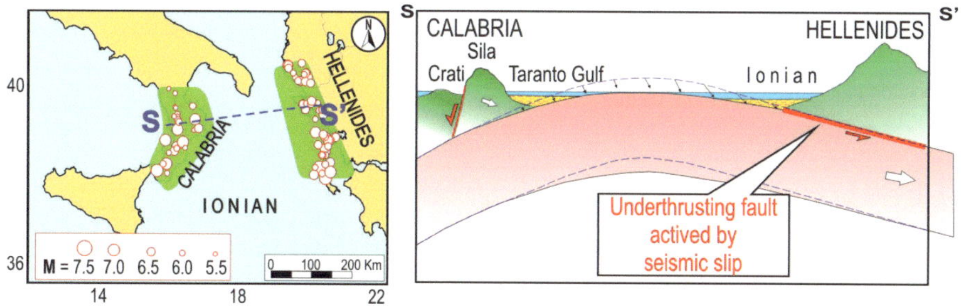

Figure 7.
*Geometry of the presumably interrelated Calabrian and Hellenic seismic zones and trace of the section (S-S') are shown in the map. Red circles indicate the epicentres of the earthquakes that have occurred in the two zones since 1600 a. D (**Table 1**). The section illustrates a tentative reconstruction (vertically exaggerated) of the reduction of vertical flexure of the Adriatic plate (dashed line) that may occur in response to a strong decoupling earthquake in the Hellenic thrust zone. This effect may favour the outward escape of the uplifted Calabrian wedge towards the Ionian domain. Seismicity data as in **Figure 1**.*

Hellenides (M ≥ 6.0)	Calabria (M ≥ 5.5)
1601 (6.3)	
1612 (6.3), 1613 (6.3)	
1625 (6.5)	**1626 (6.1)**
1630 (6.5), 1636 (7.2)	**1638 (6.8, 7.1)**
1638 (6.3)	1640 (5.8)
1650 (6.2), **1658 (6.7)**	**1659 (6.6)**
1666 (6.2), 1674 (6.3)	
1701 (6.6), 1704 (6.4)	1708 (5.6)
1709 (6.2), 1714 (6.3)	
1722 (6.3), 1723 (6.1, 6.3), **1732 (6.6)**	
1736 (6.0), 1741 (6.3), **1743 (6.9)**	1743 (5.9), 1744 (5.7), 1749 (5.8)
1759 (6.3), **1766 (6.6), 1767 (6.7)**	1767 (5.9)
1769 (6.8), 1772 (6.1)	
1773 (6.5)	**1783 (7.1, 6.7, 7.0)**
1783 (6.5, 6.6), 1786 (6.5)	**1791 (6.1)**
1809 (6.1), 1815 (6.3), **1820 (6.6)**	
1823 (6.3), **1825 (6.7)**	**1832 (6.7)**
1833 (6.5)	1835 (5.9), **1836 (6.2)**
1851 (6.8)	**1854 (6.3)**
1854 (6.0), 1858 (6.0, 6.2, 6.4), 1859 (6.0,6.2)	
1860 (6.4), 1862 (6.4, 6.2), 1865 (6.3), **1866 (6.6,** 6.2, 6.1. 6.4), **1867 (7.2), 1869** (6.0, **6.7)**	**1870 (6.2)**
1872 (6.0)	
1885 (6.0)	1886 (5.6), 1887 (5.6)
1893 (6.6)	**1894 (6.1)**
1895 (6.2, **6.5**, 6.2, 6.2), **1897 (6.6)**, 1912 (6.1)	**1905 (7.0), 1907 (6.0), 1908 (7.1)**, 1909 (5.5), 1913 (5.6)
1914 (6.0), 1915 (6.1, 6.3, 6.0),	
1920 (6.0, **6.5)**	1928 (5.9)
1930 (6.2)	
	1947 (5.7)
1948 (6.5, 6.5)	
1953 (6.0, **6.6, 7.0**, 6.2)	
1983 (6.7, 6.0)	
2003 (6.2)	
2014 (6.1, 6.1), **2015 (6.5)**	

Table 1.
*List of major Hellenic and Calabrian events, occurred since 1600 a.D. in the zones depicted in **Figure 7** (the shocks with M ≥ 6.0 in Calabria and M ≥ 6.5 in Hellenides are in **bold**). Seismicity data as in **Figure 1**.*

10 years, by at least one event with M ≥ 6.5 in the Hellenides. Even if lower magnitudes (M ≥ 5.5) are considered, the correspondence remains fairly significant, since only 3 (out of 29) Calabrian events have not been preceded by equivalent shocks in the Hellenides. The above evidence could imply that a major earthquake can hardly occur in Calabria without being preceded by significant seismic activity in the Hellenides [58, 59].

On the other hand, considering the opposite aspect of the presumed interrelation, one can note that only 11, out of 22, Hellenic seismic crises with M ≥ 6.5 were followed by a Calabrian earthquake with M ≥ 6.0. This indicates that the role of the Hellenic events as precursors of Calabrian shocks is affected by significant uncertainty. This problem mainly concerns the most recent time, given that since 1948 no Hellenic events with M ≥ 6.5 have been followed by an event in Calabria with M ≥ 5.5 (**Table 1**). Such long quiescence (73 years) is rather anomalous with respect to the previous behaviour, in particular with the fact that from 1626 to 1947 the average inter-event time between M ≥ 5.5 Calabrian shocks was about 16 years and was never longer than 41 years.

In order to find a possible explanation of the present long quiescence and of the fact that since the middle of the XX century the correspondence between Hellenic and Calabrian earthquakes has undergone a considerable worsening, we advance the hypothesis that such anomalous behaviour is an effect of the considerable increase of E-W compressional stress that developed in the Hellenic and Ionian zones in response to the large westward displacement of the Anatolian-Aegean system since 1939, when a very strong earthquake in Eastern Anatolia (M = 8) triggered the progressive activation of the entire North Anatolian fault system (NAF in **Figure 8**, e.g., [71]).

The peculiarity of the above seismic sequence in the NAF is the fact that it also involved the activation of the central NAF, which had been almost silent for a long time e.g., [72]. This rare event favoured a significant westward displacement (some metres) of the whole Anatolian wedge, causing a considerable increase of E-W compression in the zones stressed by the convergence of this block with the Africa-Adriatic domain (**Figure 8**). The least action principle suggests that the fast shortening required by such dynamics was mainly accommodated by the outward extrusion of the Peloponnesus and the central Aegean zones, i.e. the orogenic

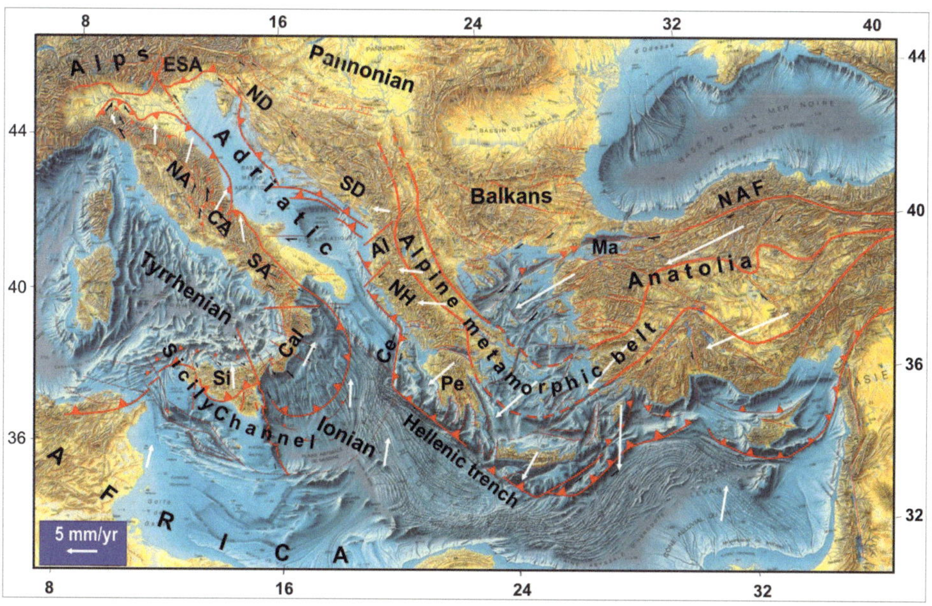

Figure 8.
Proposed plate/microplate configuration and kinematic pattern in the Central Mediterranean and Aegean-Anatolian region [14]. White arrows indicate the presumed velocity field with respect to Eurasia. Land and seafloor morphological features from Le Pichon and Biju-Duval [70]. Thick red lines delimitate for reference the inner part of the Alpine metamorphic belt. Al = Albanides; CA, NA, SA = Central, Northern and Southern Apennines, Cal = Calabrian Arc, Ce = Cephalonia fault system, ESA = Eastern Southern Alps, Ma = Marmara, NAF = North Anatolian fault system, ND = Northern Dinarides, NH = Northern Hellenides, Pe = Peloponnesus, SD = Southern Dinarides, Si = Sicily. Symbols as in ***Figure 2****.*

structures which were facing the thin and dense (low buoyancy) Ionian oceanic lithosphere e.g., [73]. The extrusion of the northern Hellenides (facing the thicker and more buoyant Adriatic continental domain) would have instead encountered much higher resistance. This hypothesis may explain why since about 1947 (when the effects of such strong perturbation might have reached the western Hellenic zone) most seismic activity has occurred in the Aegean zones lying south of the Cephalonia fault system and the North Aegean trough, while minor activity has instead occurred in the Northern Hellenides (**Figure 9**).

Since the activation of that Hellenic thrust zone is supposed to be a necessary condition for the occurrence of Calabrian earthquakes (**Figure** 7 and **Table 1**), the above evidence could explain why since 1947 no major events have occurred in Calabria. The same interpretation may help to understand why in the 1850–1908 time interval, characterised by very high seismic activity in the Hellenides sector, very strong earthquakes occurred in Calabria (**Table 1**).

The evidence and arguments described above suggest that the probability of strong earthquakes in Calabria will not undergo a significant increase until the

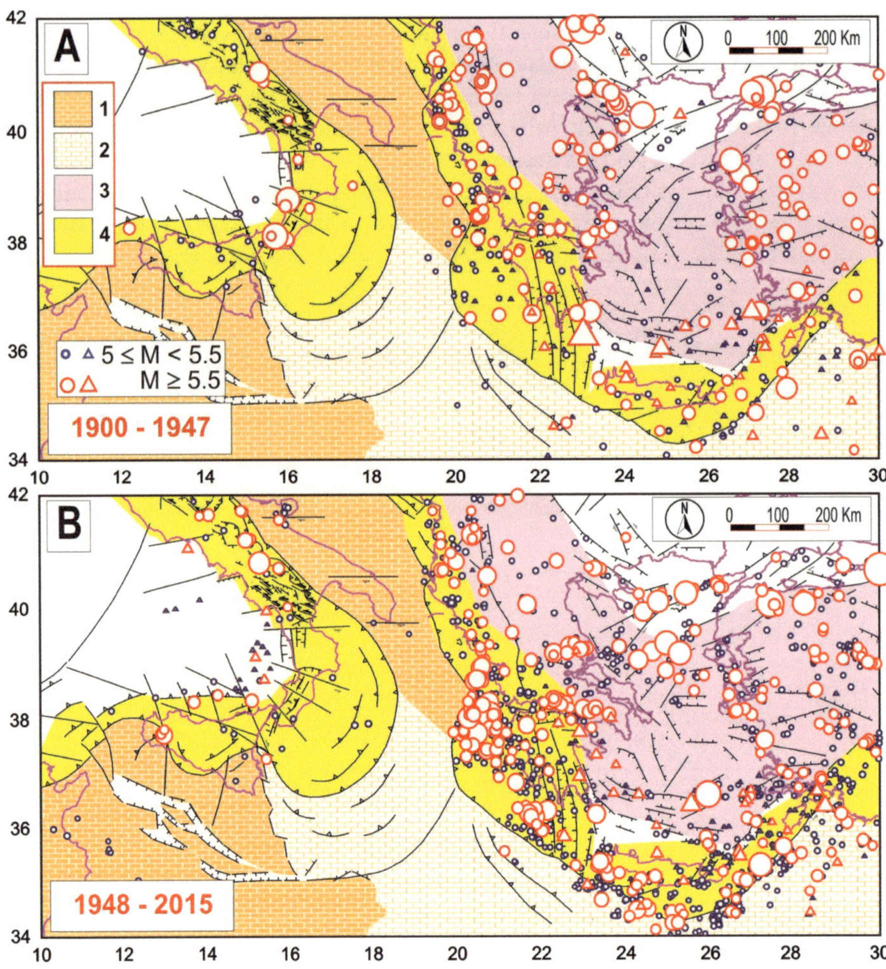

Figure 9.
*Distribution of major earthquakes occurred in two time intervals (**A** and **B**) which respectively preceded and followed the presumed arrival in the Aegean area of the effects of the large westward displacement of the Anatolian wedge, triggered by the strong 1939 earthquake (M = 8) in the easternmost north Anatolian fault system [74, 75]. 1) Africa-Adriatic domain 2) oceanic Ionian domain 3) Alpine metamorphic belt 4) orogenic belts. Circles and triangles respectively indicate focal depths lower and greater than 60 km. Seismicity data as in **Figure 1**.*

occurrence of major shocks in the Hellenides thrust zone. The fact that three earthquakes with M > 6 recently occurred in the Cephalonia zone (2014 and 2015) cannot easily be taken as a possible precursor of Calabrian shocks, since in the tectonic context created by the Anatolian westward displacement other very strong events (1953, M = 7.0, 6.6; 1983 M = 6.7) affected the Cephalonia fault without inducing significant seismic activity in Calabria.

6. Conclusions

It is advanced the hypothesis that the spatio-temporal distribution of major earthquakes in the periAdriatic zones (**Figure 4**) is closely connected with the progressive roughly northward displacement of the Adria plate. This motion is allowed by the seismic activations of the decoupling fault systems located along the lateral boundaries of Adria (Dinarides and Apennines) and the Eastern Southern Alps. This migrating pattern of earthquakes may tentatively be recognised in the period considered (1300–2020), delineating 7 already developed sequences and one partially developed migration. Taking into account the regularities that we tentatively recognise in the first 7 seismic sequences and the main features of the last ongoing one (which has already involved intense seismic crises in the Southern and Central Apennines and in the western boundary of the RMU wedge in the Northern Apennines) we suppose that the boundaries of the northern RMU wedge (Rimini-Ancona thrust, Romagna fault and Alta Valtiberina trough), along with the Emilia Apennines (stressed by the RMU wedge, **Figure 2**) are the zones most prone to the next strong earthquakes in the Apennine belt.

Further insights into the present seismic hazard in two major Italian seismic zones (Southern Apennines and Calabria) are tentatively inferred from the presumed tectonic connection of such regions with other periAdriatic zones. The first tectonic connection (suggested by seismic histories of about two centuries, **Figure 6**) provides that a strong earthquake (M ≥ 6.0) in the Southern Apennines cannot easily occur if not preceded (within 5 years) by a shock with M ≥ 6.5 or by more than one shock with M ≥ 6.0 in the Southern Dinarides. Even if weaker shocks are taken into account, the correlation remain significant, since almost all Southern Apennines shocks with M ≥ 5.5 (14 out of 15) have been preceded by seismic phases in the Southern Dinarides involving more than one shock (2–5) with M > 5.5.

Assuming that the presumed implications of the above correspondence can be taken as realistic for the next years, one could try to estimate the present seismic hazard in the Southern Apennines. To this purpose, one have to take into account the recent seismic activity in the Southern Dinarides, which only includes an event with M = 6.2 in 2019 (Albania). The fact that the magnitude of such shock was lower than 6.5 would imply a low probability for the occurrence of a Southern Apennine shock with M ≥ 6, while the occurrence of a weaker shock cannot easily be excluded.

The possible tectonic connection between Calabrian and Hellenic earthquakes (**Figure 7** and **Table 1**) is suggested by the seismic histories of these two zones for the period 1600–1947. However, in the subsequent time, this correspondence cannot be recognised, mainly due to the fact that no more earthquakes with M ≥ 5.5 have occurred in Calabria. We suggest that such quiescence is an effect of the considerable westward displacement that the whole Anatolian wedge has undergone due to the activation of the full NAF fault system. That event has caused a noticeable increase of E-W compression in the Ionian and Calabrian zones, so enhancing the resistance against the outward escape of the Calabria wedge. Since this process is the main genetic mechanism of seismicity in Calabria, the above effect can help to explain the recent seismic quiescence in that zone (**Table 1**). If this interpretation is realistic, one could suppose that in the present context seismic hazard in Calabria is not high, since in recent times no major earthquakes have occurred in the Hellenides sector lying north of the Cephalonia zone, i.e. the area that generated the main precursors of the strongest Calabria earthquakes. Recent seismicity only affected the Cephalonia fault system (2014, M = 6.1, 6.1 and 2015, M = 6.5), that in the ongoing stress regime have already involved other major shocks with no effects in Calabria.

The above considerations about the present seismic hazard in the Southern Apennines and Calabria reinforce our conviction that the Northern Apennine zones cited above should be taken as priority zones in a prevention plan in Italy.

Author details

Enzo Mantovani[1*], Caterina Tamburelli[1], Daniele Babbucci[1], Marcello Viti[1] and Nicola Cenni[2]

1 Dip.to di Scienze Fisiche, della Terra e dell'Ambiente, Univ. di Siena, 53100, Siena, Italy

2 Dip.to di Geoscienze, Univ. di Padova, 35131, Padova, Italy

*Address all correspondence to: enzo.mantovani@unisi.it

References

[1] Ekström G, Nettles M, Dziewonski AM. The global CMT project 2004-2010: Centroid-moment tensors for 13,017 earthquakes, Phys Earth Planet Inter (2012) 200-201:doi:10.1016/j.pepi.2012.04.002

[2] Godey S, Bossu R, Guilbert J, Mazet-Roux G The Euro-Mediterranean bulletin: A comprehensive seismological bulletin at regional scale. Seismological Research Letters (2006) 77:460-474

[3] Grünthal G, Wahlström R. The European-Mediterranean earthquake catalogue (EMEC) for the last millennium, J Seismology (2012) 16: DOI: 10.1007/s10950-012-9302-y

[4] ISIDe Working Group (INGV). Italian seismological instrumental and parametric Database (2010): http://iside.rm.ingv.it

[5] Makropoulos K, Kaviris G and Kouskouna V. An updated and extended earthquake catalogue for Greece and adjacent areas since 1900. Natural Hazards and Earth System Sciences (2012) 12:1425-1430.

[6] Papazachos BC and Papazachos CB. The earthquakes of Greece. Geophys Lab Publ (1989). University of Thessaloniki, Greece

[7] Rovida A, Locati M, Camassi R, Lolli B, Gasperini P. Catalogo Parametrico dei Terremoti Italiani (CPTI15), versione 2.0. Istituto Nazionale di Geofisica e Vulcanologia (INGV), (2019) https://doi.org/10.13127/CPTI/CPTI15.2

[8] Shebalin NV, Leydecker G, MokrushinaNG,TatevossianRE,ErtelevaOO, Vassiliev VYu. Earthquake catalogue for central and southeastern Europe, 342 BC - 1990 AD. European Commission, Final Report to Contract No ETNU-CT930087 Brussels (1998).

[9] Stucchi M, Rovida A, Gomez Capera AA, Alexandre P, Camelbeeck T, DemirciogluMB,GasperiniP,KouskounaV, Musson RMW, Radulian M, Sesetyan K, Vilanova S, Baumont D, Bungum H, Fäh D, Lenhardt W, Makropoulos K, Martinez Solares JM, Scotti O, Zivcic M, Albini P, Batllo J, Papaioannou C, Tatevossian R, Locati M, Meletti C, Viganò D, Giardini D. SHARE European earthquake catalogue (SHEEC) 1000-1899, J Seismology (2012) 17: doi:10.1007/s10950-012-9335-2.

[10] Mantovani E, Babbucci D, Tamburelli C, Viti M. A review on the driving mechanism of the Tyrrhenian–Apennines system: Implications for the present seismotectonic setting in the central-northern Apennines. Tectonophysics (2009) 476: doi:10.1016/j.tecto.2008.10.032.

[11] Mantovani E, Viti M, Babbucci D, Tamburelli C, Cenni N. How and why the present tectonic setting in the Apennine belt has developed. Journal of the Geological Society of London (2019) 176: https://doi.org/10.1144/jgs2018-175

[12] Viti M, Mantovani E, Babbucci D, Tamburelli C. Quaternary geodynamics and deformation pattern in the southern Apennines: Implications for seismic activity. Boll Soc Geol It (2006) 125:273-291.

[13] Mantovani E. Viti M. Babbucci D, Albarello D. Nubia-Eurasia Kinematics: An alternative interpretation from Mediterranean and North Atlantic evidence. Annals of Geophysics(2007) **50**:311-336.

[14] Viti M, Mantovani E, Babbucci D, Tamburelli C. Plate kinematics and geodynamics in the Central Mediterranean. Journal of Geodynamics (2011) 51: doi:10.1016/j.jog.2010.02.006.

[15] Mantovani E, Viti M, Babbucci D, Tamburelli C, Albarello D. "Geodynamic

connection between the indentation of Arabia and the Neogene tectonics of the Central-Eastern Mediterranean region" In: Dilek Y, Pavlides S, editors. Post-Collisional Tectonics and Magmatism in the Mediterranean Region and Asia. Geol. Soc. Am. (2006). p. 15-49.

[16] Mantovani E, Viti M, Babbucci D, Tamburelli C, Cenni N. Geodynamics of the central western Mediterranean region: Plausible and non-plausible driving forces. Marine and Petroleum Geology (2020) 113: https://doi.org/10.1016/j.marpetgeo.2019.104121

[17] Falcucci E, Poli ME, Galadini F, Scardia G, Paiero G, Zanferrari A. First evidence of active transpressive surface faulting at the front of the eastern southern Alps, northeastern Italy: Insight on the 1511 earthquake seismotectonics. Solid Earth (2018): https://doi.org/10.5194/se-9-911-2018

[18] Kastelic V, Carafa MMC. Fault slip rates for the active external Dinarides thrust-and-fold belt. Tectonics (2012) 31: doi:10.1029/2011TC003022

[19] Kuk V, Prelogovic E, Dragicevic I. Seismotectonically active zones in the Dinarides. Geol Croatica (2000) 53:295-303.

[20] Louvari E, Kiratzi AA, Papazachos BC, Katzidimitriou P. Fault-plane solutions determined by waveform modelling confirm tectonic collision in the eastern Adriatic. Pure App Geophysics (2001) 158:1613-1637.

[21] Moulin A, Benedetti L, Rizza M, Rupnik PJ, Gosar A et al. The Dinaric fault system: Large-scale structure, rates of slip, and Plio-Pleistocene evolution of the transpressivnortheastern boundary of the Adria microplate, Tectonics (2016) 35: DOI : 10.1002/2016TC004188

[22] Argnani A, Barbacini G, Bernini M, Camurri F, Ghielmi M, Papani G, Rizzini F., Rogledi S, Torelli L. Gravity tectonics driven by quaternary uplift in the northern Apennines: Insights from the La Spezia-Reggio Emilia geo-transect. Quaternary Int (2003) 101-102:13-26.

[23] Boccaletti M, Corti G, Martelli L. Recent and active tectonics of the external zone of the northern Apennines (Italy). Int J Earth Sci (Geologische Rundschau) (2010): doi: 10.1007/s00531-010-0545-y

[24] Cerrina Feroni A, Martelli L, Martinelli P, Ottria G, Sarti G. The Romagna Apennines, Italy: An eroded duplex. Geological Journal (2001) 36: 39-54.

[25] Pizzi A. Plio-quaternary uplift rates in the outer zone of the central Apennines fold-and-thrust belt, Italy. Quaternary Int (2003) 101-102:229-237.

[26] Elter FM, Elter P, Eva C, Eva E, Kraus RK, Padovano M, Solarino S. An alternative model for the recent evolution of northern-central Apennines (Italy). Journal of Geodynamics (2012) 54: 55-63.

[27] Pizzi A, Galadini F. Pre-existing cross-structures and active fault segmentation in the northern-central Apennines (Italy). Tectonophysics (2009) 476: doi:10.1016/j.tecto.2009.03.018

[28] Amadori C, Toscani G, Di Giulio A, Maesano FE, D'Ambrogio C, Ghielmi M, Fantoni R. From cylindrical to non-cylindrical foreland basin: Pliocene–Pleistocene evolution of the Po plain–northern Adriatic basin (Italy). Basin Research (2019) 31: DOI: 10.1111/bre.12369

[29] Ascione A, Cinque A, Miccadei E, Villani F. The Plio-quaternary uplift of the Apennines chain: New data from the analysis of topography and river valleys in Central Italy. Geomorphology (2008) 102: https://doi.org/10.1016/j.geomorph.2007.07.022

[30] Mazzoli S, Pierantoni PP, Borraccini F, Paltrinieri W, Deiana G. Geometry, segmentation pattern and displacement variations along a major Apennine thrust zone, Central Italy. Journal of Structural Geology (2005) 27: https://doi.org/10.1016/j.jsg.2005.06.002

[31] Catalano S, Monaco C, Tortorici L. Neogene-quaternary tectonic evolution of the southern Apennines. Tectonics (2004) 23: doi:10.1029/2003TC001512

[32] Ferranti L, Santoro E, Mazzella ME, Monaco C, Morelli D. Active transpression in the northern Calabria Apennines, southern Italy. Tectonophysics (2009) 476: https://doi.org/10.106/j.tecto.2008.11.010

[33] Ferranti L, Burrato P, Pepe F, Santoro E, Mazzella ME, Morelli D, Passaro S, Vannucci G. An active oblique-contractional belt at the transition between the Southern Apennines and Calabrian Arc: The Amendolara Ridge, Ionian Sea, Italy. Tectonics (2014), 33:doi:10.1002/2014TC003624

[34] Maschio L, Ferranti L, Burrato P. Active extension in Val d'Agri area, southern Apennines, Italy: Implications for the geometry of the seismogenic belt. Geophysical Journal International (2005) 162: https://doi.org/10.1111/j.1365-246X.2005.02597.x

[35] Ascione A, Caiazzo C, Cinque A. Recent faulting in southern Apennines (Italy): Geomorphic evidence, spatial distribution and implications for rates of activity. Boll Soc Geol It (2007) 126:293-305.

[36] Brozzetti F. The Campania–Lucania extensional fault system, southern Italy: A suggestion for a uniform model of active extension in the Italian Apennines. Tectonics (2011) 30: TC5009, https//doi.org/10.1029/2010TC002794

[37] Amoruso A, Crescentini L, Scarpa R. Inversion of source parameters from near and far field observations: An application to the 1915 Fucino tarthquake, central Apennines, Italy. Journal of Geophysical Research (1998) 103: 29989-29999.

[38] Galadini F, Messina P. Early-middle Pleistocene eastward migration of the Abruzzi Apennine (Central Italy) extensional domain. Journal of Geodynamics (2004) 37:57-81.

[39] Piccardi L, Gaudemer Y, Tapponnier P, Boccaletti M. Active oblique extension in the central Apennines (Italy): Evidence from the Fucino region. Geophysical Journal International (1999) 139:499-530.

[40] Piccardi L, Tondi G, Cello G. "Geo-structural evidence for active oblique extension in south-Central Italy". In: Pinter N, Grenerczy G, Weber J, Stein S, Medak D. editors. The Adria Microplate: GPS Geodesy, Tectonics and Hazard. NATO Science Series IV-Earth and Environmental Sciences, Springer (2006). p. 95-108.

[41] Brozzetti F, Boncio P, Lavecchia G, Pace B. Present activity and Seismogenetic potential of a low-angle Normal fault system (Città Di Castello, Italy): Constraints from surface geology, seismic reflection data and seismicity. Tectonophysics (2009) 463:31-46.

[42] Mantovani E, Viti M, Babbucci D, Tamburelli C, Cenni N, Baglione M, D'Intinosante V. Seismotectonics and present seismic hazard in the Tuscany-Romagna-Marche-Umbria Apennines (Italy). Journal of Geodynamics (2015a) 89: http://dx.doi.org/10.1016/j.jog.2015.05.001

[43] Martini IP Sagri M. Tectono-sedimentary characteristics of late Miocene-quaternary extensional basins of the northern Apennines, Italy. Earth Science Reviews (1993) 34: http://dx.doi.org/10.1016/0012-8252(93)90034-5

[44] Costa M. The buried, Apenninic arcs of the Po plain and northern Adriatic Sea (Italy): A new model. Boll Soc Geol It (2003) 122:3-23.

[45] Mantovani E, Viti M, Babbucci D, Tamburelli C, Cenni N. Possible location of the next major earthquakes in the northern Apennines: Present key role of the Romagna-Marche-Umbria wedge. IJG (2017) 8: https://doi.org/10.4236/ijg.2017.811075.

[46] Sorgi C, Deffontaines B, Hippolyte JC, Cadet JP. An integrated analysis of transverse structures in the northern Apennines, Italy. Geomorphology (1998) 25:193-206.

[47] Viti M, Mantovani E, Babbucci D, Tamburelli C, Cenni N. Seismotectonics of the Padanian region and surrounding belts: Which driving mechanism? International Journal of Geosciences (2016) 7: http://dx.doi.org/10.4236/ijg.2016.712100

[48] Mantovani E, Viti M, Cenni N, Babbucci D, Tamburelli C. Present velocity field in the Italian region by GPS data: Geodynamic/tectonic implications. In J Geosciences (2015b) 6: http://dx.doi.org/10.4236/ijg.2015.612103.

[49] Viti M, Mantovani E, Babbucci D, Tamburelli C, Cenni N, Baglione M, D'Intinosante V. Belt-parallel shortening in the northern Apennines and seismotectonic implications. International Journal of Geosciences (2015a) 6: http://dx.doi.org/10.4236/ijg.2015.68075.

[50] Cenni N, Mantovani E, Baldi P, Viti M. Present Kinematics of central and northern Italy from continuous GPS measurements. Journal of Geodynamics (2012) 58: doi:10.1016/j.jog.2012.02.004

[51] Cenni N, Viti M, Baldi P, Mantovani E, Bacchetti M, Vannucchi A. Present vertical movements in central and northern Italy from GPS data: Possible role of natural and anthropogenic causes. Journal of Geodynamics (2013) 71: http://dx.doi.org/10.1016/j.jog.2013.07.004

[52] Viti M. Strain diffusion from the 30 October 2016 Norcia (Central Italy) earthquake. Tectonics (2019): https://doi.org/10.1029/2018TC005464

[53] Viti M, Mantovani E, Cenni N, Vannucchi A. Post-seismic relaxation: An example of earthquake triggering in the Apennine belt (1915-1920). J Geodynamics (2012) 61: http://dx.doi.org/10.1016/j.jog.2012.07.002

[54] Viti M, Mantovani E, Cenni N, VannucchI A. Interaction of seismic sources in the Apennine belt. Journal Physics Chemistry Earth (2013) 63: http://dx.doi.org/10.1016/j.pce.2013.03.005

[55] Wells D, Coppersmith KJ. New empirical relationships among magnitude, rupture length, rupture width, rupture area and surface displacement. Bull Seismol Soc Am (1994) 84:974-1002.

[56] Mantovani E and Albarello D. Medium-term precursors of strong earthquakes in southern Italy. Physics of the Earth and Planetary Interiors (1997) 101(1-2):49-60.

[57] Mantovani E, Viti M, Babbucci D, Albarello D, Cenni N., Vannucchi A. Long-term earthquake triggering in the southern and northern Apennines. J. Seismology (2010) 14: 53-65.

[58] Mantovani E, Viti M, Babbucci D, Cenni N, Vannucchi A. Middle-term prediction of earthquakes in Italy: Some remarks on empirical and deterministic approaches. Boll Geof Teor Appl (2012) 53:89-111.

[59] Mantovani E., Viti M., Babbucci D., Tamburelli C., Cenni N., Baglione M., D'Intinosante V. Recognition of

periAdriatic seismic zones most prone to next major earthquakes: Insights from a deterministic approach. In: D'Amico S. (Ed.), Earthquakes and their Impact on Society. Springer Natural Hazard Series, Springer (2016) 43-80. DOI 10.1007/978-3-319-21753-6_2

[60] Viti M, D'Onza F, Mantovani E, Albarello D, Cenni N. Post-seismic relaxation and earthquake triggering in the southern Adriatic region. Geophysical Journal International (2003) 153:645-657.

[61] Viti M, Mantovani E, Babbucci D, Cenni N, Tamburelli C. Where the next strong earthquake in Italy: Possible insights by a deterministic approach. Boll Geof Teor Appl (2015b) 56: doi:10.4430/bgta0137

[62] Moretti I. and Royden L. Deflection, gravity anomalies and tectonics of doubly subducted continental lithosphere: Adriatic and Ionian seas. Tectonics (1988) 7:875-893.

[63] Benetatos C and Kiratzi A. Finite-fault slip models for the 15 April 1979 (mw 7.1) Montenegro earthquake and its strongest aftershock of 24 may 1979 (mw 6.2). Tectonophysics (2006) 421:129-143.

[64] Ascione A, Cinque A, Improta L, Villani F. Late quaternary faulting within the southern Apennines seismic belt: New data from Mt. Marzano area (southern Italy). Quaternary International (2003) 101-102: https://doi.org/10.1016/S1040-6182(02)00127-1

[65] Freed AM. Earthquake triggering by static, dynamic, and postseismic stress transfer. Annual Reviews of Earth and Planetary Sciences (2005) 33(1): https://doi.org/10.1146/annurev.earth.33.092203.122505

[66] Luo G, Liu M. Stress evolution and fault interactions before and after the 2008 great Wenchuan earthquake. Tectonophysics (2010) 491(1-4): https://doi.org/10.1016/j.tecto.2009.12.019

[67] Pollitz FF, Burgmann R, Romanowicz B. Viscosity of oceanic asthenosphere inferred from remote triggering of earthquakes. Science (1998) 280:1245-1249.

[68] Pollitz FF, Stein RS, Sevilgen V, Burgmann R. The 11 April 2012 East Indian Ocean earthquake triggered large aftershocks worldwide. Nature (2012) 490(7419): https://doi.org/10.1038/nature11504

[69] Rydelek PA, Sacks IS. Asthenospheric viscosity and stress diffusion: A mechanism to explain correlated earthquakes and surface deformation in NE Japan. Geophysical Journal International (1990) 100(1): https://doi.org/10.1111/j.1365-246X.1990.tb04566.x

[70] Le Pichon X., Biju-Duval B. Les fonds de la Mediterranee. Hachette-Guides bleus (1990) Paris, sud offset-Rungis

[71] Barka AA. Slip distribution along the north Anatolian fault associated with the large earthquakes of the period 1939 to 1967. Bull Seism Soc Am (1996) 86:1238-1254.

[72] Ambraseys NN and Jackson JA. Faulting associated with historical and recent earthquakes in the eastern Mediterranean region. Geophysical Journal International (1998) 133:390-406.

[73] Finetti I, Del Ben A. Crustal tectono-stratigraphy of the Ionian Sea from new integrated CROP seismic data. In: Finetti IR, editor. CROP PROJECT: Deep Seismic Exploration of the Central Mediterranean and Italy. Elsevier Sciences (2005) Chapter 19:447-470.

[74] Cenni N, D'Onza F, Viti M, Mantovani E, Albarello D, Babbucci D. Post seismic relaxation processes in the Aegean-Anatolian system: Insights from space geodetic data (GPS) and

geological/geophysical evidence. Boll Geof Teor Appl (2002) 43:23-36.

[75] Mantovani E, Viti M, Cenni N, Albarello D, Babbucci D. Short and long-term deformation patterns in the Aegean-Anatolian systems: Insights from space geodetic data (GPS). Geophysical Research Letters (2001) 28:2325-2328.

Chapter 3

Seismological Data Acquisition and Analysis within the Scope of Citizen Science

Ewald Brückl, Peter Carniel, Stefan Mertl and Rita Meurers

Abstract

From 2017 till 2020 a low cost seismic sensor network was built in the southern Vienna Basin, Lower Austria, as a part of ongoing educational and citizen science projects. The purpose of the project is to inform society about the seismic activity in this area and to include authorities and interested citizens into data acquisition and exploitation. Near real time (NRT) seismic data are made accessible online. Seismic events are detected and archived automatically. The visualization of these events online facilitates instantaneously estimates of the extent of the shaking area and potential damage. Peak ground velocities (PGV) are related to macroseismic intensities (EMS-98) derived from reports about ground motion felt in the vicinity of the network stations. Observed amplitudes and travel times are modeled by simple, but effective relations. Traditional and innovative localization methods based on travel times and amplitudes are applied and analyzed with respect to data quality and localization accuracy. All results are accessible online and the computer code is open and applicable, e.g. for educational purposes.

Keywords: public seismic network, NRT ground motion watching, peak ground velocity, macroseismic intensity, earthquake localization

1. Introduction

Instrumental seismology started worldwide at the beginning of the 20th century. Data acquired by the continuously improved seismometers built and still build the basis for our present-day knowledge about seismic waves, the structure of the Earth's interior, the origin of earthquakes and their impact on infrastructures and humans. However, earthquake phenomena have been fascinating and even threatening mankind from time immemorial. Systematic documentation and classification was based on observations and reports of educated persons, officers, chroniclers, clerics, and presumably only a small fraction of scientists and specialists. To summarize, seismological research before 1900 was only possible with the contribution of the public.

Nowadays, the evaluation of reports about felt ground motion and damage caused by earthquakes is treated by a seismological subdiscipline. Historic macroseismic intensity scales (e.g., Rossi - Forel, Mercalli, Cancani, Medvedev Sponheuer – Karnik) have been refined (e.g. EMS-98) and correlations with instrumentally recorded ground motions have been established. Near real-time (NRT) preparation of so-called instrumental intensity maps is a scientific task to support mitigation in case of an earthquake. However, the reports of citizens on their perceptions of ground motions during earthquakes and damage is still an essential scientific input. We concentrate on the following tasks to promote the interest of Citizen Science in seismological data acquisition and analysis:

Citizens are frequently prepared to report their perceptions about ground motion. However, they also want to immediately know, if a ground motion was caused by an earthquake, a blast in a nearby quarry or only by a very local source such as traffic or construction works. The public is interested in whether damage to buildings occurred or health and safety were at stake. We intend to answer these questions in NRT and intuitively interpretable information via the internet, based on the data provided by a public low-cost seismic network.

The stations of this low-cost sensor network are installed in private homes and industrial buildings, schools and offices. These locations are representative of places where people observe ground motion and report it. At best we get reports from citizens about felt ground motion directly from station locations. We take advantage of these circumstances to establish a very close correlation of instrumental data and intensity classifications.

Students of polytechnics were and are still involved in the production of low-cost sensors, coding of digitizers, and developing special tools for data visualization. We intend to maintain these cooperations, but also to demonstrate that seismological data analysis must not be a black box for students of polytechnics or grammar schools, alumni, and interested citizen. We will show that accurate hypocenter localization is possible with data from the low-cost sensor network. We try to achieve this goal with easily understandable algorithms.

2. Public seismic sensor network

2.1 Area

We chose the southern Vienna Basin and its surroundings for the installation of a low-cost seismic sensor network. This area belongs to the zone of relative high seismic hazard in Austria and is densely populated and industrialized. Therefore we reasonably presume that ground motion caused by earthquakes or other sources is an interesting issue for officials and citizens.

The Vienna basin is a representative example of a pull apart basin well-explored and documented in geological literature (e.g. [1]). The basin was created by lateral extrusion of the most eastern part of the Eastern Alps from the compressional zone in the west to the extensional Pannonian Basin in the east during Miocene [2]. The basin reaches a maximum depth of 6 km. It is surrounded by Austroalpine Crystalline, the Northern Calcareous Alps, and Flysch. Shallow quaternary sediments not outlined in the schematic geological map (**Figure 1**) may significantly influence the seismic response. The Vienna basin transfer fault (VBTF) corresponds to the southern strike-slip boundary of the pull apart basin. It is still active and constitutes the main tectonic process responsible for the seismicity in this area.

Since 1200 AD about 460 earthquakes have been documented as felt and classified according to the European macroseismic scale EMS-98 in or near the southern Vienna Basin [3]. The highest epicentral intensities have been evaluated for the Schwadorf (8th October 1927, I0 = VIII) and the Seebenstein (16th April 1972, I0 = VII-VIII) earthquakes.

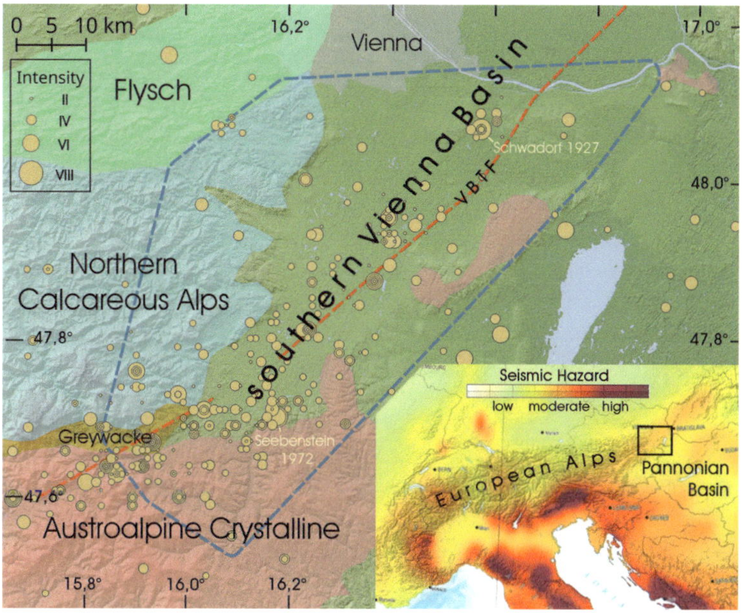

Figure 1.
Topography, geology (VBTF … Vienna Basin transfer fault) and felt earthquakes since 1200 (solid circles); the dashed polygon delimits the low-cost seismic sensor network. Insert: European seismic Hazard map, rectangle marks the extent of the map.

2.2 Sensor and network

The development of our low-cost seismic sensor started within the scope of the national educational project „Schools & Quakes"in 2015 [4]. One goal was the design and assembly of seismic low-cost sensors from scratch until final operation by students of polytechnic schools. This activity was inspired by the Quake-Catcher Network [5], where low-cost MEMS accelerometers either integrated into computers or in external units are used to form a world wide seismic network. However, we could not reach the desired sensitivity on the basis of low-cost MEMS (Micro-Electrical–Mechanical-Sensor) accelerometers. Following the Raspberry Shake seismograph [6] we changed to classical geophones to transform ground motions into electrical signals. Our low-cost sensor is dedicated to collecting quantitative ground motion data of felt local earthquakes. Therefore, we call it "MacroSeismic Sensor" or MSS in order to emphasize its purpose. The term MSS will be used for our sensor throughout this chapter.

The essential MSS components are two orthogonally oriented, horizontal geophones, two 16bit Analogue to Digital Converters (ADC), and a single board computer (SBC), specifically a Raspberry Pi (**Figure 2a**). The 4.5 Hz natural frequency and the 0.7 damping coefficient of the geophones and first order 12.5 Hz RC low-pass filters determine the frequency response of the MSS (**Figure 2b**). The whole assembly is protected by a robust casing. The SBC controls signal processing and provides internet connectivity. Depending on the programmable pre-amplification gain of the ADC the sensitivity ranges from 0.28 μm/s/count to 2.24 μm/s/count. Accurate time information is provided by Network Time Protocol (NTP). Seismic data is formatted to MSEED (100 Hz sample rate) and sent every 10 s to the MSS-Server.

The MSS deployment started in 2017. Up until October 2020 a total of 48 MSS were installed in the southern Vienna Basin and the surrounding area in the province Lower Austria. The selection and deployment of the MSS stations received much support from the federal warning center, local authorities, schools, one

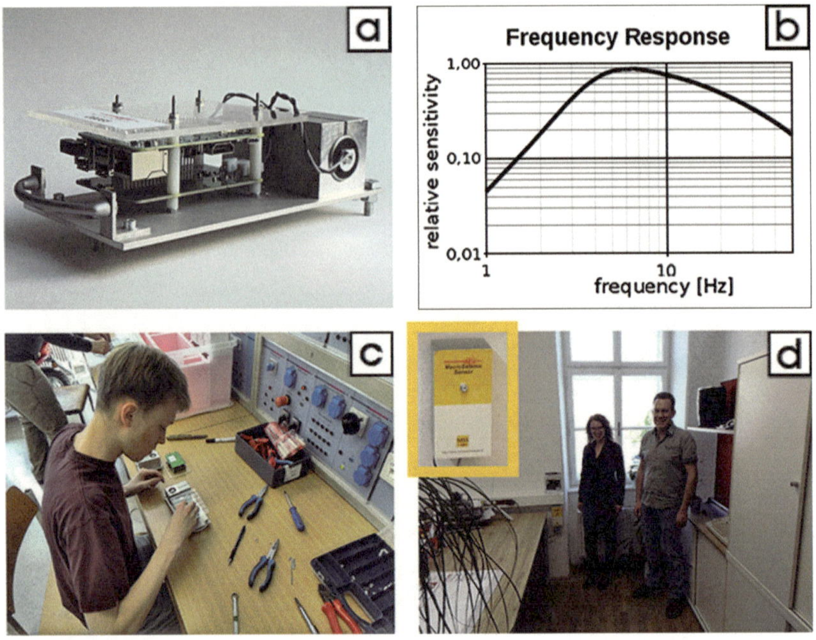

Figure 2.
MSS – MacroSeismic sensor; (a) geophones, ADC and SBC mounted on base plate; (b) frequency response between f = 1 Hz and $f_{Nyquist}$ = 50 Hz, (c) student at polytechnic Wiener Neustadt assembling MSS; (d) MSS mounted in office of district exchange Bruck an der Leitha together with contact persons.

quarry operator, and private citizens. The MSS were mounted by a single plug to a vertical, preferably a retaining wall in solidly constructed buildings, mainly at basement, ground floor, or first floor level.

2.3 Near real-time ground motion watching

A third essential component of the MSS network, beside the MSS stations and the MSS-Server, is the MSS-homepage that provides data visualizations and access to numeric data (https://www.macroseismicsensor.at/). The MSS-network is meant to inform communities, governmental administration, civil protection organizations and last but not least citizens about the felt or presumed seismic activity in the southern Vienna Basin. Ground motions take 10–30 seconds to travel over the whole area of the MSS-net from the epicenter. Immediately people are curious to know the source of the vibration. The authorities contacted should be able to answer, at least preliminarily, these questions on the basis of the information and visualization provided on the internet. In the case of a stronger earthquake (intensities > = V), the staff of the civil protection organization should know if panic could arise or if there was damage to buildings and sensitive infrastructure. Therefore, visualization of the essential seismic data should be swiftly available and understandable.

We mainly try to meet these demands by using a map of the MSS data. We determine Peak Ground Velocity (PGV) as the maximum resultant horizontal ground velocity over a time interval of 1 sec and visualize it at each MSS station with symbols. The map is updated every 10 seconds. **Figure 3** shows seismic recordings and the essential components of the PGV-map for a 60 s time window covering the M_L = 2.5 earthquake on 14th June 2019.

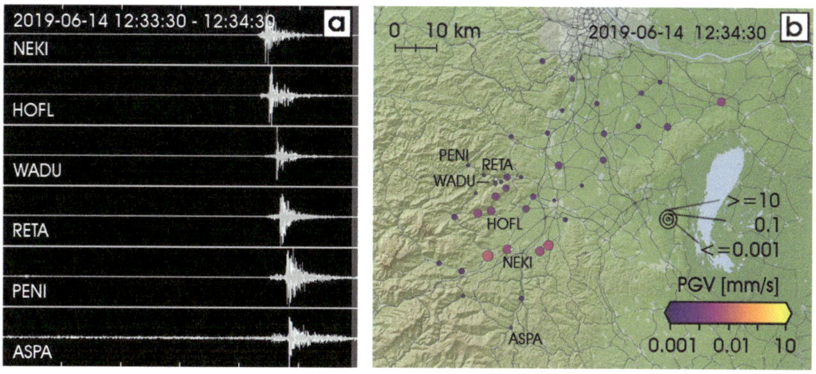

Figure 3.
(a) Recordings (stacked horizontal components) of the time interval 2019-06-14 12: 33:30–12:34:30; (b) corresponding NRT PGV-map; opaque symbols show PGV of last second (gray background in a), transparent symbols PGV of last minute.

3. Event detection, visualization, and archiving

The PGV map as shown in **Figure 3** is transient. Significant seismic events should be detected and saved in order to keep this information and to make it available for more detailed analysis. The definition of a seismic event and proper trigger criteria should take the data quality into consideration. MSS stations are intentionally mounted in buildings where people potentially experience ground motions and report their observations. These places are frequently noisy and even high PGVs may be recorded due to nearby activities (e.g. traffic, construction work, washing machine, etc.). The main objective of a detection algorithm is to distinguish high amplitude noise at individual stations and regional events like earthquakes or quarry blasts. We perform a "Delaunay" triangulation [7] of the MSS station network and examine the triples of PGV values belonging to the different triangles. Once the minimum PGV value within one triple exceeds a preselected threshold the recorded PGV at all MSS-stations are classified as a seismic event. The duration of the seismic event is expanded by the triggering of other Delaunay triangles and prolongated by a listening time window. This time window takes care of the propagation of the maximum amplitude seismic waves over the network area. **Figure 4** shows the temporal sequence of the trigger status for an entire seismic event.

As soon as a seismic event ends, the seismic data of the respective time window is archived. We offer two options for the visualization of the whole seismic event:

- Coloring the Voronoi regions [8] of each MSS station according to the event PGV (**Figure 5a**),

- contouring the PGVs at the MSS-stations by the Kriging method (**Figure 5b**).

Both visualizations are available in NRT after the seismic event.

The PGV values observed during an earthquake are strongly affected by specific geological and technical peculiarities at the individual MSS stations. In the Section 5.1 we introduce station amplification factors "SA" to improves the fit of PGV to a power law amplitude - distance relation. The application of SA significantly improve the spatial correlation of PGV. Contours become much smoother and better delineate the areas of felt ground motions and maximum shaking. Therefore, we also offer the aforementioned data display alternatively based on PGV/SA instead of PGV (**Figure 5c, d**).

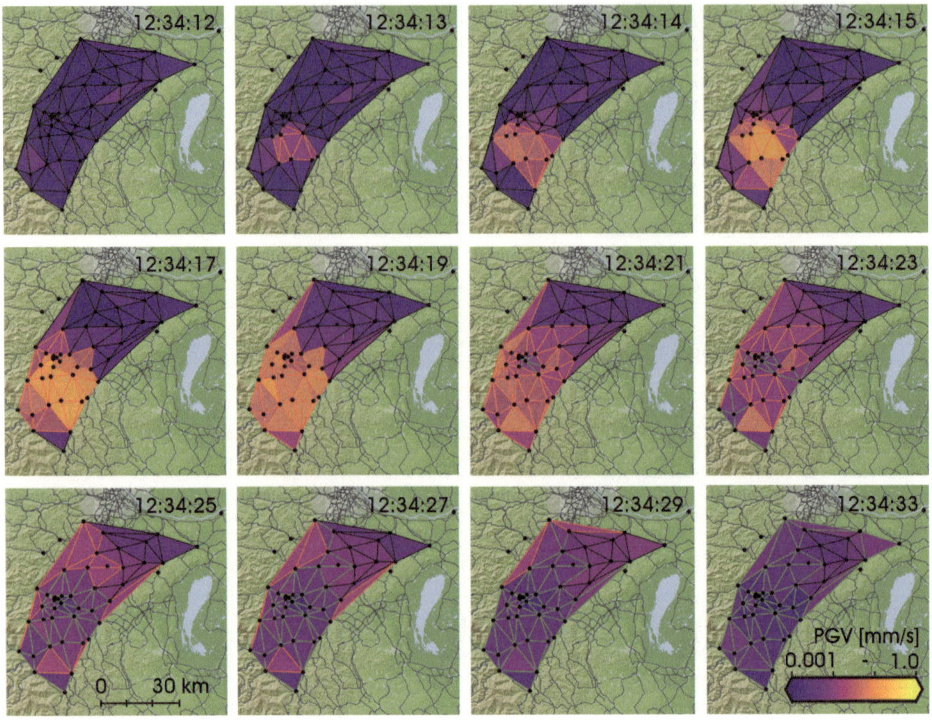

Figure 4.
Sequence of visualizations of the trigger status of Delaunay triangles during the M_L = 2.5 earthquake near the center of the MSS-network, 14th June 2019.

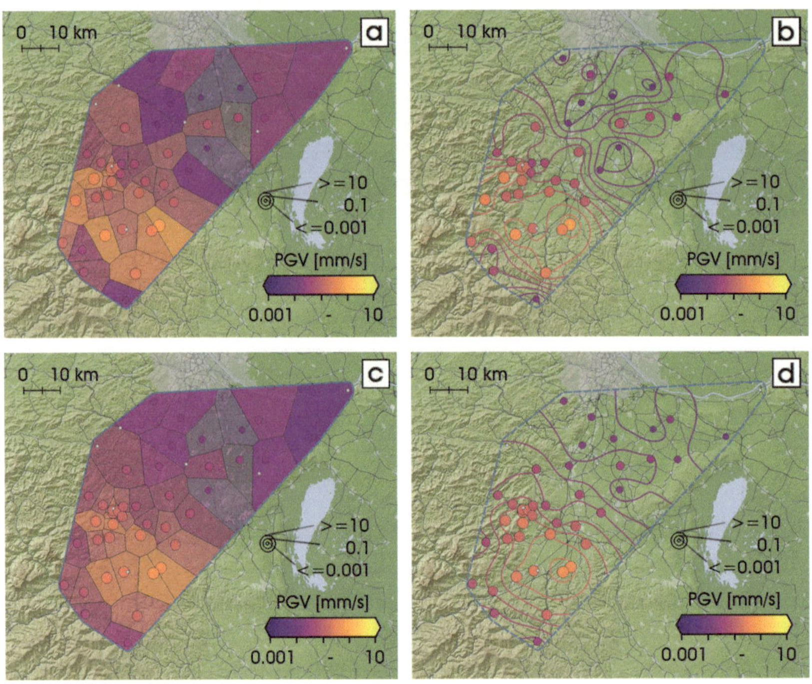

Figure 5.
Visualization of the ML = 2.5 earthquake near the center of the MSS-network, 14th June 2019; (a) coloring the Voronoi regions according PGV, (b) contouring PGV by Kriging, (c) coloring the Voronoi regions according PGV/SA, (d) contouring PGV/SA by Kriging.

Beside the maximum PGV values of each event and the visualization options, the archive also provides quick and easy access to the PGV-time series and the original seismic traces (MSEED, 100 Hz). Interactive data analysis by seismologists and motivated citizen scientists is supported by this portal.

4. PGV and intensity

Since the beginning of MSS recording, the macroseismic intensities of 16 earthquakes in the area of the MSS-network were evaluated according to EMS-98 by the Seismological Service of ZAMG. These intensities were assigned to macroseismic data points corresponding to municipalities. We relate each macroseismic data point to PGV values recorded within a circumference of 5 km. In total 120, PGV intensity pairs were found by this procedure.

The maximum epicentral intensity V was assigned for a magnitude M_L = 3.7 earthquake with PGV up to 8.44 mm/s. The minimum PGV associated with intensity ≥ II amounted PGV = 0.05 mm/s. On the other hand, we observed PGV values up to 0.42 mm/s related to earthquakes with no reports about felt ground shaking. The number of PGV values, binned to PGV classes, is opposed for 'felt' and 'not felt' in **Figure 6a**. The number for 'felt' overtakes 'not felt' from the class 0.03–0.1 mm/s to the class 0.1–0.3 mm/s. As a first estimate, we set PGV = 0.1 mm/s as the lower valueof felt earthquakes in exceptional instances (higher floors, night time, at rest, etc.), or with intensity II.

Figure 6b shows the cross plot of intensity over PGV. The scatter of PGV within intensity classes is considerable and exceeds, in part one decade. Preliminarily we assume a non-linear relation between the logarithm of PGV and intensity. According to the macroseismic detection threshold derived before, the relation is fixed to PGV = 0.1 mm/s at intensity II. The PGV values corresponding to the isoseismals III, IV and V are 0.3 mm/s, 1.0 mm/s, and 10 mm/s. This correlation allows for the interpretation of the corresponding contours in our PGV maps (**Figure 5d**) as isoseismals.

The estimate of macroseismic intensity from instrumental data and vice versa is an important issue for the preparation of shake maps (e.g., https://earthquake.usgs.gov/data/shakemap/). These maps provide just in time information about the area and magnitude of ground shaking of an earthquake and its effect on human perception as well as the intactness of infrastructure derived from instrumental observatory data. In principle, we attempt the same procedure with our PGV contour maps. However, the correlations implemented worldwide differ significantly from our relation (e.g., [9]). The attenuation relation used by the Swiss Seismological

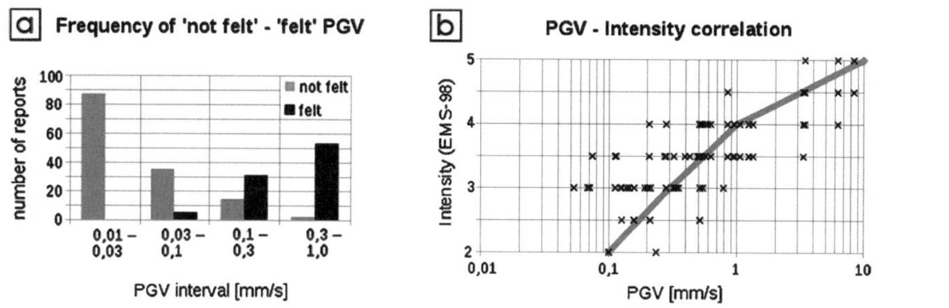

Figure 6.
(a) Frequency of PGV values of 'not felt' versus 'felt' earthquakes recorded at MSS stations between October 2017 and October 2020; (b) intensity (EMS-98) over PGV for the same data set; gray line shows a preliminary PGV-intensity relation.

Service for the shake maps (http:\\shakemapa.ethz.ch; visited on 23th October 2020) mimics ours. PGV = 0.8 mm/s, 3 mm/s and 9 mm/s corresponds to intensities II – III, IV, and V. PGV < 0.2 mm/s are classified as 'not felt'.

5. Modeling of PGV and travel-time data

5.1 PGV - distance decay

We use PGV recorded by MSS in our network area and hypocentre coordinates from the ZAMG bulletin. We model PGV caused by earthquake "e" and observed at station "i" by a power law and station amplification factors. The power law considers geometric spreading and damping of the maximum amplitude seismic waves, the station amplification factors the local geological and technical conditions.

$$PGV_{e,i} = A0_e * r_{e,i}^{n} * SA_i \tag{1}$$

$A0_e$... source strength of earthquake "e".
$r_{e,i}$.... hypocentral distance of earthquake "e" to station "i".
n .....constant.
SA_i ... ..amplification factor of station "i".

We calculate the station amplification factors SA according to [10, 11]. The unknown source strength is eliminated by the ratio of SA at stations "i" and "j". We calculate the mean ratio derived from the different earthquakes:

$$SA_i/SA_j = MEAN\left(PGV_{e,i}/PGV_{e,j} * \left(r_{e,j}/r_{e,i}\right)^{n}\right) \tag{2}$$

The logarithm of Eq. 2 forms a linear equation system for log(SA), which is solved using least squares and the additional condition the geometrical mean of all SA is unity. We vary the exponent "n" to minimize the standard deviation of ($PGV_{observed} - PGV_{calculated}$). So far, the optimum exponent is n = −2.2 based on the data available.

Next, we consider a single earthquake and omit the index "e". Given the exponent "n", the hypocentral distance "r_i" and the station amplification factor SA_i, the logarithm of the source strength related to station i can be estimated by Eq. 3. The average of log10($A0_i$) over all stations defines the magnitude MSS_M of the earthquake (Eq. 4).

$$\log 10(A0_i) = \log 10(PGV_i) - \log 10(SA_i) - n * \log 10(r_i) \tag{3}$$

$$MSS\_M = MEAN(\log 10(A0_i)) \tag{4}$$

The units are PGV [nm/s] and r [deg] according to the definition of M_L (local magnitude).

Figure 7 shows a) log10(PGV) and log10(PGV/SA) reduced to MSS_M = 3 versus log10(r), and b) the histograms of log10(PGV) and log10(PGV/SA) reduced to MSS_M = 3 and r = 10 km. The standard deviation (STD) derived from the histogram data are $\sigma 0 = 0.332$ and $\sigma 1 = 0.221$, respectively. The STD $\sigma 1$ quantifies the accuracy of log10 (PGV/SA) calculated by Eq. 1, given A0, the exponent "n", and the hypocentral distance "r". We calculate log10(r) from Eq. 3, given "A0", "n", and PGV/SA, the STD $\sigma 1/n = 0.1$ quantifies the uncertainty. The accuracy of the hypocentral distance "r" calculation is therefore about 25% or +/− 5 km at $r \sim 20$ km.

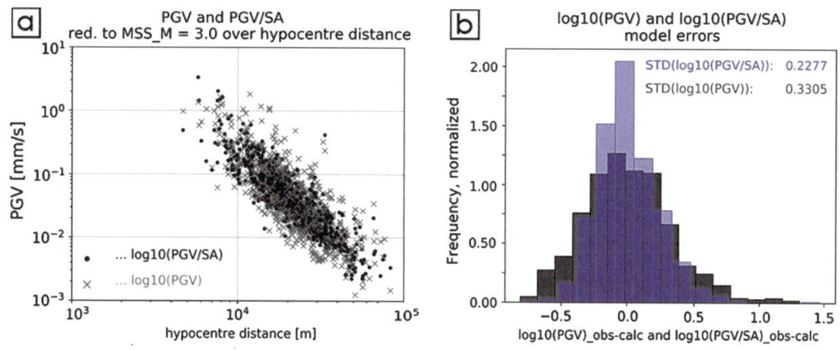

Figure 7.
(a) PGV and PGV/SA reduced to MSS_M = 3 over log10(r); (b) histograms of PGV and PGV/SA reduced to MSS_M = 3 and r = 10 km.

5.2 P-wave travel time

The acquisition, processing and analysis of ground motion amplitudes is the main task of the MSS network. However, the seismic traces show clear first P-wave arrivals and frequently very distinct S-wave phases (see **Figure 3a**). Therefore, it is worthwhile to analyze and evaluate these data.

We model the P-wave arrival times Tp_i using the following linear relation:

$$Tp_{e,i} = T0_e + r_{e,i}/Vp + SDp_i \tag{5}$$

$T0_e$ ... focal time of earthquake “e”
Vp ... constant P-wave velocity
SDp_i ... P-wave station delay at station “i”

To calculate the station delays SDp, we eliminate T0 by considering differences of Eq. 5.

$$SDp_i - SDp_j = MEAN\left(Tp_{e,i} - Tp_{e,j} - (r_{e,i} - r_{e,j})/Vp\right) \tag{6}$$

Eq. 6 has the same form as the logarithm of Eq. 2. Therefore we proceed analogue to the calculation of the station amplification factors SA and solve the equation system by least squares and the condition SUM(SDp_i) = 0. We apply SDp to Tp and find the minimum standard deviation of ($Tp_{observed} - Tp_{calculated}$) with Vp = 5700 m/s. With Vp and the station delays SDp given we calculate the focal time of the earthquake e by equation Eq. 7:

$$T0_e = MEAN\left(Tp_{e,i} - r_{e,i}/Vp - SDp_i\right) \tag{7}$$

Figure 8a and **b** show Tp and (Tp -SDp) reduced to T0 = 0 versus log10(r) and the histograms of Tp and (Tp - SDp) reduced to T0 = 0 and r = 10 km. The STDs derived from the histogram data are $\sigma0$ = 0.235 s for Tp and $\sigma1$ = 0.161 s for (Tp - SDp). In case we want to calculate the hypocentral distances “r” from Tp observations at specific stations, we expect an error of about $Vp^*\sigma1 \sim 0.9$ km.

5.3 P-wave to S-wave arrival time difference

Now, we focus on the P-wave to S-wave arrival time difference Tps. This value is insensitive to the time drift of the MSS, which could occasionally occur if the

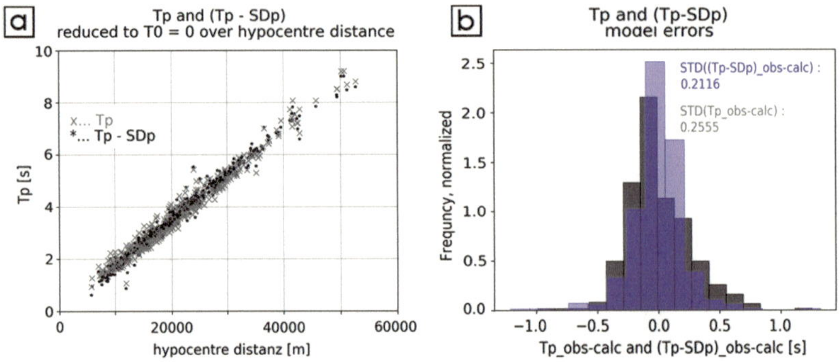

Figure 8.
(a) Tp and Tp - SD reduced to To = 0 over log10(r); (b) histograms of T and Tp - SD reduced to To = 0 and r = 10 km.

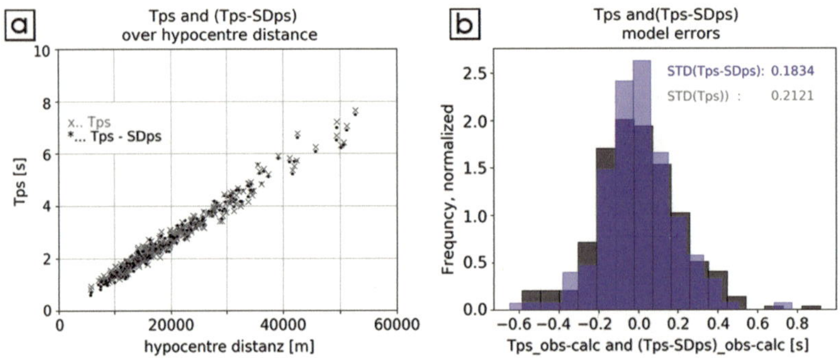

Figure 9.
(a) Tsp and Tsp - SDsp over hypocenter distance); (b) histograms of Tsp and Tsp - SDsp.

connection to NTP servers is interrupted. We assume that P- and S-waves are generated simultaneously and the focal time difference Tps(r = 0) is zero. We model Tps as follows:

$$Tps_{e,i} = r_{e,i}/Vps + SDps_i \tag{8}$$

Vps = $(1/Vs - 1/Vp)^{-1}$ … ..difference velocity

$SDps_i$ .....P- wave to S-wave station delay at station “i”

The assumption Tps(r = 0) makes the calculation of the station delays SDps very simple:

$$SDps_i = \text{MEAN}\left(Tps_{e,i} - r_{e,i}/Vps\right) \tag{9}$$

As before with Tp, we apply the station delays SDps to Tps and find the minimum standard deviation of ($Tps_{observed}$ - $Tps_{calculated}$) with Vps = 7300 m/s. **Figure 9** shows a) Tps and (Tps -SDps) versus log10(r) and b) the histograms of Tps and (Tps - SDps) reduced to r = 10 km. The STDs derived from the histogram data are $\sigma0$ = 0.212 s for Tps and $\sigma1$ = 0.183 s for (Tps – SDps). Consequently the error of hypocenter distances r derived from Tps observations is about Vps*$\sigma1 \sim$ 1.3 km.

6. Locating seismic events

In the following we consider amplitude and travel time data (PGV, Tp, Tsp) of one particular seismic event (earthquake) after the application of station corrections (SA, SDp, SDsp). We search for the hypocenter within a 3D grid, centered at the maximum PGV or minimum travel time station. In our special case the grid extends from −20 km to +20 km in W-E and S-N with a grid spacing of 0.5 km around this center. We consider 17 hypocenter depth levels from −16 km to 0 km. The indices of the grid in the east, north and upward directions are (k, l, m).

6.1 Cost function methods

These methods share the definition of a proper cost function followed by the search of its minimum within the 3D volume of a grid with the indexes (k, l, m) in X (east), Y (north), and Z (upward) directions. The computational complexity comprises four nested loops, three over the grid dimensions (k, l, m) and one fourth over the number of stations that recorded particular data. We present three cost function methods based on each of the three data sets (PGV, Tp, Tps) described before. We take the liberty to name the various methods after scientists who defined the principles of the relevant cost functions.

6.1.1 Kanamori

Hiroo Kanamori [12] introduced earthquake locating based on amplitudes with application to real-time seismology. He identified the optimum epi- or hypocenter with the location of the minimum standard deviation of the magnitudes calculated from the amplitudes recorded at the different seismic stations. He implemented an empirical 1D model for the amplitude (acceleration) decay with distance. Thus, we take PGV/SA and the amplitude – distance power law according to Eq. 1 and choose the following cost function according to Eq. 3 and equivalent to Kanamori's principle:

$$\text{cost function} = \text{STD}(\log 10(\text{A0}_\text{i})). \tag{10}$$

6.1.2 Geiger

Ludwig Geiger [13] was the first to present a method to locate an earthquake through minimization of the differences between observed and calculated arrival times. The corresponding cost function for P-wave arrival times according our notation Eq. 5 is:

$$\text{cost function} = \text{STD}(\text{T0}_\text{i}) = \text{STD}\left(\text{Tp}_\text{i} - \text{SDp}_\text{i} - \text{r}_\text{i}/\text{Vp}\right) \tag{11}$$

6.1.3 Hopkins

In 1848, William Hopkins [14] identified two types of seismic waves traveling with different velocities. He pointed out the relation between the corresponding arrival time difference and the distance from the point of observation to the origin of seismic waves. At that time the great advantage of using time differences derived from single recordings was the insensitivity to clock reading errors. The definition of the time difference Tps Eq. 8 and the station delays SDps Eq. 9 implies that (Tps - SDps_i) is zero at the origin. Hence, we define the following cost function:

$$\text{cost function} = \text{MEAN}\left(\text{ABS}\left(\text{Tps}_\text{i} - \text{SDps}_\text{i} - \text{r}_\text{i}/\text{Vps}\right)\right) \tag{12}$$

6.2 Cell-hit methods

The cell-hit methods apply to the various graphical location methods. PGV or arrival times observed at single or couples of seismic stations restrict the possible hypocenter locations to surfaces within the 3D volume. Each surface may hit a distinct assembly of cells of the 3D grid. The number of hits are added up for each cell and the variety of 3D surfaces given by seismic data. The cell with the maximum hit count is taken as the hypocenter location.

Graphical location methods are based on the simple PGV or travel-time distance relations we derived in the previous section. In case the observed data are exactly equal to data calculated by these relations, all surfaces would have one common point and the hit count of the corresponding cell would be the number of surfaces. In fact, the simple relations cannot exactly predict the observed data and these data may also include observational errors. We consider these uncertainties by weighting cell-hits according to their proximity to the surfaces containing the hypocenter. Instead of rating cell-hits as either 1 or 0 we define the following Gaussian weighting schema.

$$\text{hitweight} = \exp.\left(-\text{CD}^2/\left(2 * \text{sigma}^2\right)\right) \tag{13}$$

Both, CD and sigma have the dimension of a length. CD quantifies the proximity of the center of the grid cell [k, l, m] to the surface and will be specified for each cell-hit method. The parameter sigma considers the data uncertainty.

6.2.1 Apollonius-circle

Given PGV-values at two stations, these values determine the ratio of the distances from these two stations to the hypocenter according to Eq. 1. The geometrical loci of the hypocenters that fulfill this condition are Apollonius circles in 2D or spheres in 3D [10]. The examination of the accuracy of the Apollonius circle method suggests combining only stations with relative high PGV values with stations of similar or lower PGV values. The combination of low PGV stations does not contribute to an accurate localization.

The proximity length CD of the Apollonius circle method is:

$$\text{CD} = \text{R}_\text{p} - \text{D}_{\text{p,klm}} \tag{14}$$

R_p ... radius of index “p” Apollonius circle

$D_{p,klm}$ ... distance of the center of grid cell [k, l, m] to the Apollonius circle center C_p

The formulae for how to calculate the Apollonius radii R_p and the coordinates of the circle centers C_p are given in the Appendix.

6.2.2 Hyperbola

Next, we consider the application of the hyperbola method to P-wave travel times Tp. This method dates back to the work of Andrija Mohorovičić ([15]) during his analysis of the Kupa earthquake und the detection of the crust–mantle boundary, the Moho. The travel time differences (Tp_i -Tp_j) between the stations i and j define L_{ij}, the difference in length of the ray paths from these stations to the hypocenter. Assuming, that Vp = constant, the geometrical loci of the hypocenters are hyperbolas in 2D, or hyperboloids in 3D. The proximity length CD for the hyperbola method is:

$$CD = L_{ij} - D_{ij,klm} \quad (15)$$

$L_{ij} = (Tp_i - Tp_j) * Vp$...difference of the travel path lengths of stations i and j to hypocenter

$D_{ij,klm}$ ...difference of the distances from the center of grid cell [k,l,m] to stations i and j.

6.2.3 *PS-circle*

The PS-circle method is the graphical complement to the Hopkins cost function method. The distance R_i of station i to the hypocenter follows from Eq. 8. The radius R_i defines a circle, in 3D a sphere, which describes the geometrical loci of possible hypocenters. The corresponding proximity length CD is:

$$CD = R_i - D_{i,klm} \quad (16)$$

$R_i = (Tps_i - SDps_i) * Vps$...radius of the PS-circle

$D_{i,klm}$.....distance of station i to the center of the grid cell [k,l,m].

6.2.4 *Additional comment to cell-hit methods*

In 3D Apollonius and PS-circles expand to spheres and hyperbolas to hyperboloids. Spheres with radii smaller than the hypocenter depth do not reach that depth level, and hyperboloids with high eccentricity may not intersect within the grid at greater depth levels. We normalize the cell-hit counts by the sum of cell-hits at each depth level to account for this characteristic.

6.3 Visual check of the location quality

The location methods presented in the previous subsection work without interaction of the user. Numeric output comprises the hypocenter coordinates (longitude, latitude, focal depth) and in case of amplitude based methods (Kanamori, Apollonius) the magnitude MSS_M. Focal time, which could be an output of the Geiger and Hyperbola methods, is not documented because we use the trigger time for event identification. However, it would be desirable to get information about the uniqueness and accuracy of the solution. Professionally used location methods like 'HYPOELLIPSE' [16] or 'NonLinLoc' [17] derive confidence ellipsoids from the analysis of the linearization applied to solve the non-linear location problem or by using the probability density function. Here, we content ourselves with visual checks of the location quality.

The question is, how to visualize the 3D volume of the cost-function or the cell-hit values in order to identify the global extremum and to value it against competing local extremes. We chose presentation of the grid values in the different search grid depth levels. First, we show these graphics with synthetic data for an earthquake in the center of the MSS network and at focal depth of 8 km. We take the Kanamori method as an example for the cost function methods (**Figure 10a**) and the Hyperbola method as a representative of the cell-hit method (**Figure 10b**). The cost function low and the maximum cell-hit count areas are clearly und uniquely confined. The visual identification of the optimum hypocenter depth is possible with an accuracy of +/− 1 km. Of course, presuming synthetic PGV, Tp, Tps data and correct program codes presumed, all six methods find the correct hypocentre.

Next, we test all six methods with PGV, Tp, and Tps data of the M_L = 2.5 earthquake on 14th June 2020 in the center of the MSS-network. **Figure 11** shows

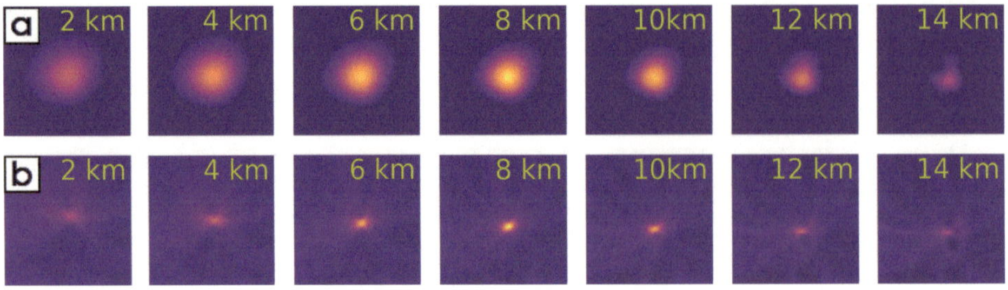

Figure 10.
Depth slices through the search grid visualizing (a), the cost function calculated by the Kanamori method and (b), the cell-hit count calculated by the hyperbola method; bright colors mark the cost function low and the cell hit height.

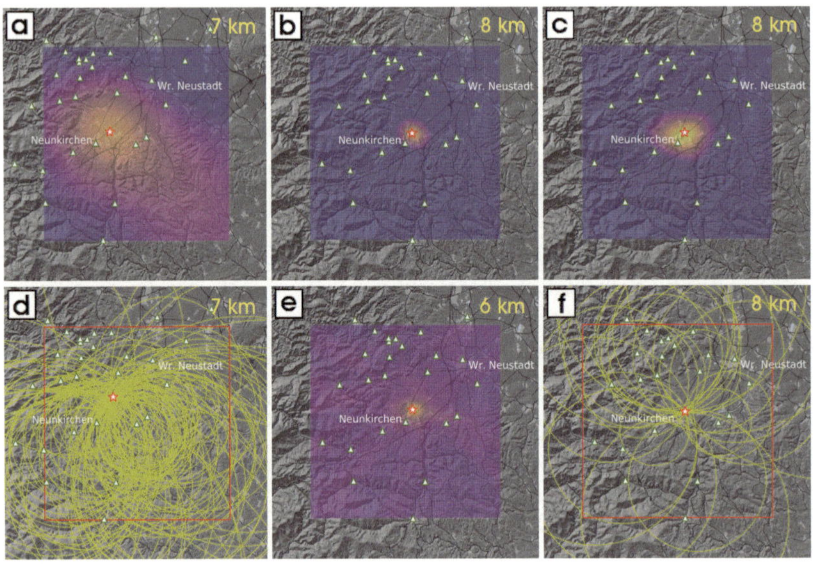

Figure 11.
Location of the M_L = 2.5 earthquake, 14th June 2019 by the (a) Kanamori, (b) Geiger, (c) Hopkins, (d) Apollonius, (e) hyperbola, and (f) PS-circle methods; depth slices through the search grids at the optimum focal depth levels are shown; data points (MSS-stations) and the extent of the search grid are marked; the cost function is shown for (a), (b), and (c), the cell-hit count for (e); Apollonius circles (d) and PS-circles (f) visualize the corresponding focal solutions.

depth slices of the search grid at each optimum depth level for all methods. The visualization of the density of cell hits by Apollonius circles and PS-circles was applied for the corresponding methods. The hypocenter solutions differ not only between the methods based on the different data type, but also between the cost-function and cell-hit methods using the same data. The latter discrepancy is due to the different weighting of data by the cost function and cell-hit methods. Therefore the variance of the hypocenter solutions obtained by minimum-cost and cell-hit methods using the same data type may be an indicator of the accuracy or significance of the focal solution. The epicenter localizations of the sample earthquake by the six methods scatter within a circle with a radius of 1.6 km. The focal depth varies between 6 km and 8 km.

6.4 Evaluation of the different location methods

At first we consider the epicenter solutions. **Figure 12** shows a representative sample of epicenter localizations by the six different methods (Kanamori,

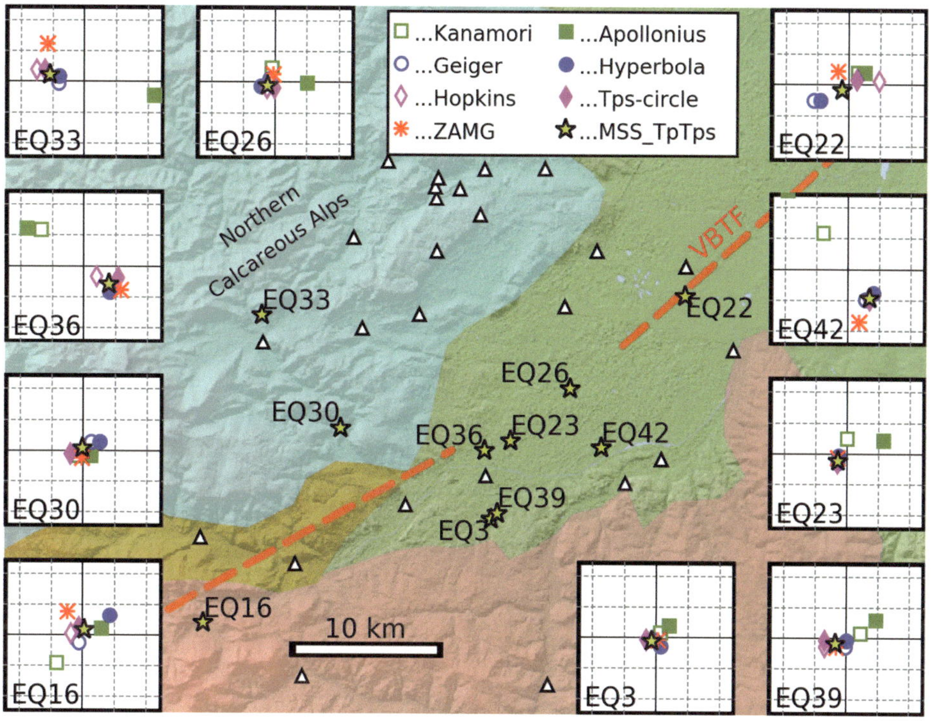

Figure 12.
*Epicenter solutions of 10 earthquakes in the southern half of the MSS-network: EQ23 is the acronym of the M_L = 2.5 earthquake, 14th June 2019, addressed in **Figures 3, 4, 5,** and **11**; triangles mark MSS-stations, gridline spacing of insert plots is 2 km.*

Apollonius, Geiger, Hyperbola, Hopkins, PS-circle) as described before. Subplots centered up the average of these epicenters show the particular solutions. Furthermore, the bulletin epicenters published by ZAMG (epi_ZAMG) are included in the subplots. Generally, the four epicenters based on the travel-time data Tp and Tps (Geiger, Hyperbola, Hopkins, PS-circle) cluster together well. We calculate an average of these solutions (epi_TpTps) and plot it on the map and the subplots. We also calculate the mean of the two epicenter solutions based on PGV data (Kanamori, Apollonius) and term it epi_PGV.

The epicenter data compiled in **Figure 12** allow for a preliminary assessment of the accuracy of the solutions presented. We take epi_TpTps as reference und calculate the lateral distances to the four travel-time based epicenter solutions (Geiger, Hyperbola, Hopkins, PS-circle), to epi_PGV and to epi_ZAMG. Statistical data about these differences are compiled in **Table 1**.

Disregarding outliers, the statistics compiled in **Table 1** indicates that the accuracy of epi_TpTps (mean of epi_Geiger, epi_Hyperbola, epi_Hopkins, epi_PS-circle) mimics the spacing of the search grid spacing (~0.5 km). The accuracy of the bulletin solution (epi_ZAMG) corresponds to the limitation to two decimals of longitude and latitude [0.01°] in the report.

Next, we consider the focal depth solutions for ten selected events. **Figure 13** shows the individual solutions gained by the six methods (Kanamori, Apollonius, Geiger, Hyperbola, Hopkins, PS-circle), the mean value of the travel-time based methods MSS_TpTps, and the bulletin focal depth values from ZAMG. The bulletin solution fits to MSS_TpTps for seven earthquakes in the Vienna Basin near the VBTF (Vienna BasinTransfer Fault) within the 1 km vertical spacing of the search grid. Foci at depths more than 3 km deeper than MSS_TpTps are indicated by the

Epicenter determination method	Median	Mean	Maximum
Geiger, Hyperbola, Hopkins, and PS-circle	0.6 km	0.7 km	2.4 km
epi_ZAMG	0.8 km	1.0 km	2.0 km
epi_PGV (mean of epicenter solutions gained by Kanamori and Apollonius method)	1.9 km	3.0 km	6.9 km

Table 1.
Statistics of distances from epi_TpTps to different epicenter solutions.

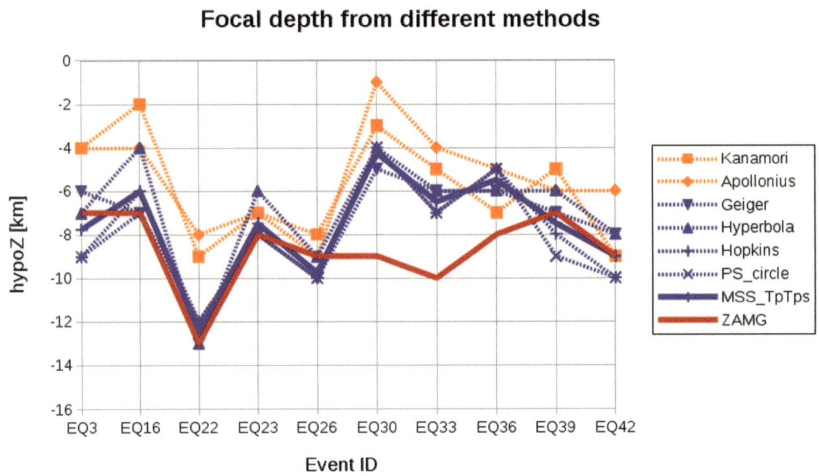

Figure 13.
Focal depth solutions for 10 events EQ3 – EQ42; dotted lines are the individual solutions determined by the six methods, solid blue line (MSS_TpTps) is the mean of the travel-time based solution, solid red line focal depth solutions from the ZAMG-bulletin.

bulletin for two earthquakes within the Northern Calcareous Alps. The focal depths resolved by the amplitude based methods (Kanamori and Apollonius) follow the trend of MSS_TpTps but show systematically lower focal depths.

The locations by the Kanamori- and Apollonius methods include the determination of the magnitudes Eqs. 3, 4. We term the mean of both MSS_M. This magnitude correlates well (correlation coefficient 0.96) with M_L (bulletin magnitude, ZAMG). We derived the following relation:

$$M_L = 0.97 * MSS\_M - 0.36 \quad (17)$$

The difference between M_L and MSS_M could be explained by the constant $C = -0.30$ used by ZAMG in the local magnitude formula and not added to our magnitude calculation. The remainder may be caused by the difference in ground coupling between the MSS in buildings and the seismometers at observatories. The factor 0.97 instead of 1.00 may be due to the MSS frequency response limited by the 4.5 Hz geophones.

7. Conclusion

A low-cost seismic sensor network has been installed in the southern Vienna basin, an area of moderate seismic hazard on a global scale, but high in Austria. Students of polytechnics in Wiener Neustadt (Lower Austria) and Vienna have been integrated into the development and production of the sensors, called 'MacroSeismic Sensors' or MSS. The federal warning center of Lower Austria, local authorities, one quarry operator, and private people supported the selection and deployment of the MSS. We intentionally selected locations where citizens live or work and would be able report about felt ground motion. Up until October 2020 a total of 48 MSS were installed in our study area. All data collected by the MSS network are principally accessible for the public.

Citizens willing to report on their perceptions of ground motion also want immediately to know about the source of their perceptions. Civil defense authorities need NRT information about the intensity and range of ground shaking for an instananeous organization of possibly necessary mitigation measures. Other authorities contacted are confronted with the problem of informing the public, within a few minutes of the event, about the impact of the seismic waves on people and infrastructure. We attempt to supply this information through the visualization of the MSS-network data on the internet in an intuitively ascertainable format. NRT observation of peak ground motion (PGV) at each MSS station is made possible by the visualization in a map (**Figure 3**). In case specific robust trigger criteria are met seismic data are defined as seismic events and archived. Some few seconds after the maximum amplitude seismic waves spread over the network area visualizations of the event PGV are available on internet. Interested parties and potential respondents (such as officials and civil protection personnel) are able to immediately assess the significance of the seismic event using the graphic facilities we offer (https://www.macroseismicsensor.at/).

The correlation of instrumental data with intensity values based on reports about felt ground motions and their effect on infrastructure or nature is a general seismological issue and an essential task of our project. The conversion of PGV (or peak ground acceleration) into intensity or vice versa is fundamental for the just-in-time preparation of shake maps. So far, we have been able to correlate 120 macroseismic data points from 16 earthquakes with PGV data from the MSS network in the intensity range II – V. The correlation used by the Swiss Seismological Service for shake maps fits well with our relation at intensity V, but indicates higher PGV for the intensity range II – IV. We interpret this discrepancy as a commitment to extend our database in order to get a better knowledge of the correlation of PGV with intensity. Of course, we need the contribution of citizens, who are ready to report their perceptions of ground motion.

The integration of seismology in the curriculum at schools and in general is the third main goal of our project. So far, classes at polytechnic schools have produced several MSSs. They programmed the CNC-machine for the manufacturing of the mechanical parts and assembled the ADC board keeping electronic industry standard. Finally they assembled and tested the complete sensor. Another polytechnical class is still involved in programming of special add-on's for the visualization of the MSS data. During these courses, however, students focused more on general technical or technological skills (as the curriculum demanded) than on a deeper understanding of seismological phenomena. In order to compensate for this deficit we presented in this chapter elementary methods for seismic data analysis. These methods can be understood once principles in physics and mathematics at high school level are acquired. Despite the easy theoretical background of these methods, the simple amplitude-, or travel-time-distance relations, and computer codes, we determined locations and magnitudes of earthquakes in the area of the MSS network at an accuracy level comparable to the bulletin data of the ZAMG. We present our solutions on the homepage and citizens involved in the maintenance should be satisfied that their contributions to the installation and maintenance of the MSS network lead to results of scientific value. A further step could be a regional initiative to supply volunteers with equipment to install and maintain a MSS-station, to perform their own data analysis aided by our computer programs, and to share "their" results with the community. We gladly support such initiatives.

Acknowledgements

The development of the MacroSeismic Sensor (MSS) in cooperation with teachers and students of the polytechnic TGM Wien was funded by the Federal Ministry of Education, Science and Research, Austria, the Austrian Academy of Sciences, and the Natural History Museum Vienna. Funds of the Federal State Government of Lower Austria allowed for the buildup of the MSS network in cooperation with teachers and students of the polytechnic HTBLuVA Wiener Neustadt. Additional sponsoring granted the Baumit GmbH (operator of a large quarry). Funding of the new MSS data visualizations is granted by netidee (https://www.netidee.at/) for MSS Outreach. Map images in this publication have been created using data from OpenStreetMap (www.openstreetmap.org, © OpenStreetMap-Contributors), OE3D (www.oe3d.at, © Rechenraum e.U.) and BEV (www.bev.gv.at, © BEV 2020). We express our thanks to mayors, officials, and private persons who supported the installation of the MSS stations in their communities. Furthermore we thank Wolfgang Lenhardt for critically reading and commenting the manuscript.

Appendix

We present the necessary formula to calculate the Apollonius circle centers und radii. Please read [10] for a detailed description.

HI, LO...position vectors of the higher and lower PGV station

PGV_HI, PGV_LO...PGV/SA at the HI and LO stations

P...position vector of unknown source locations

From Eq. 1 follows:

$$\text{ratio} = |\text{HI–P}|/|\text{LO–P}| = (\text{PGV\_HI}/\text{PGV\_LO})^{1/n}$$

Please note that n is negative.

$$\text{P1} = (\text{HI} + \text{ratio} * \text{LO})/(1 + \text{ratio})$$
$$\text{P2} = (\text{HI} - \text{ratio} * \text{LO})/(1 - \text{ratio})$$

P1, P2...position vectors of possible source locations at the straight line connecting the HI and LO stations

R = |P2 – P1| / 2...radius of the Apollonius circle or sphere

C = (P1 + P2) / 2...center of the Apollonius circle or sphere

Author details

Ewald Brückl[1]*, Peter Carniel[2], Stefan Mertl[3] and Rita Meurers[4]

1 Technische Universität Wien, Vienna, Austria

2 Private, Vienna, Austria

3 Mertl Research GmbH, Vienna, Austria

4 Zentralanstalt für Meteorologie und Geodynamik, Vienna, Austria

*Address all correspondence to: ewald.brueckl@geo.tuwien.ac.at

References

[1] Allen PA, Allen JR. Basin Analysis Principles and Applications, Oxford: Blackwell Science; 1990. 451 p; ISBN 0-632-02422

[2] Ratschbacher L, Frisch W, Linzer H-G, Merle O. Lateral extrusion in the Eastern Alps, Part 2: Structural analysis. Tectonics. 1991; 10:2, 257–271

[3] Austrian Earthquake Catalogue, computer file, Austrian Seismological Service, Zentralanstalt für Meteorologie und Geodynamik, Vienna

[4] Brückl E, Köberl C, Lenhardt W, Mertl S, Rafeiner-Magor W, Stark A, Stickler G, Weber R. Scientific goals of SCHOOLS & QUAKES. Geophysical Research Abstracts. 2015; 17, EGU General Assembly 2015, EGU2015–3401-4

[5] Cochran ES, Lawrence J, Christensen C, Jakka RS. The Quake-Catcher Network: Citizen science expanding seismic horizons. Seismological Research Letters. 2009; 80:1, 26–30, DOI: 10.1785/gssrl.80.1.26

[6] Anthony RE, Ringler AT, Wilson DC, Wolin E. Do low-cost seismographs perform well enough for your network? An overview of laboratory tests and field observations of the OSOP Raspberry Shake 4D. Seismological Research Letters. 2018; 90:1, 219–228, DOI:10.1785/02200220180251

[7] Delaunay B. Sur la sphère vide. Bulletin de l'Académie des Sciences de de URSS, Classe des Sciences Mathématiques et Naturelles. 1934; 6, 793–800

[8] Voronoï G. Nouvelles applications des paramètres continus à la théorie des formes quadratiques. Premier mémoire. Sur quelques propriétés des formes quadratiques positives parfaites. Journal für die Reine und Angewandte Mathematik. 1908; 133, 97–178. DOI: 10.1515/crll.1908.133.97

[9] Wald D, Quitoriano V, Heaton T. Kanamori H. Relationships between Peak Ground Acceleration, Peak Ground Velocity, and Modified Mercalli Intensity. Earthquake Spectra. 1999; 15: 3, 557–564. ISSN 8755–2930

[10] Brückl E. Graphical Location of Seismic Sources Based on Amplitude Ratios. Seismological Research Letters. 2019; 90:22, 784–789. DOI: 10.1785/0220180216

[11] Brückl E, Filz K, Hochwartner R, Mertl S, Stickler G, Zöhling S. Monitoring von Sprengerschütterungen mit seismischen „low-cost "Sensoren. BHM Berg- und Hüttenmännische Monatashefte. 2019; 164:10, 431–437. DOI: 10.1007/s00501-019-00892-z

[12] Kanamori H. Locating earthquakes with amplitude: Application to real-time seismology. Bull. Seismol. Soc. Am. 1993; 83, 264–268

[13] Geiger L. Probability method for the determination of earthquake epicenters from the arrival time only. Bull. St. Louis Univ. 1912; 8, 60–71

[14] Hopkins W. Report on the geological theories of elevation and earthquakes. Report of the British Association for the Advancement 0f Science. 1848; 21, 33–92

[15] Mohorovičić A. Die Bestimmung des Epizentrums eines Nahbebens. Gerlands Beitr. zur Geophysik. 1916; 14, 199–205

[16] Lahr JC. HYPOELLIPSE - Version 2.0: A computer program for determining local earthquake hypocentral parameters, magnitude and first motion pattern. U.S. Geological

Survey Open-File Report. 1989; 99. 119 p

[17] Lomax A. Probabilistic, non-linear, global-search earthquake location in 3D media. 2017. Available from http://alomax.free.fr/nlloc/ [Accessed: 2020-10-31]

Chapter 4

Seismicity at Newdigate, Surrey, during 2018–2019: A Candidate Mechanism Indicating Causation by Nearby Oil Production

Rob Westaway

Abstract

During 2018–2019, oil was intermittently produced from the Late Jurassic Upper Portland Sandstone in the Weald Basin, southeast England, via the Horse Hill-1 and Brockham-X2Y wells. Concurrently, a sequence of earthquakes of magnitude $\leq$3.25 occurred near Newdigate, $\sim$3 km and $\sim$8 km from these wells. The pattern, with earthquakes concentrated during production from this Portland reservoir, suggests a cause and effect connection. It is proposed that this seismicity occurred on a patch of fault transecting permeable Dinantian limestone, beneath the Jurassic succession of the Weald Basin, hydraulically connected to this reservoir via this permeable fault and the permeable calcite 'beef' fabric within the Portland sandstone; oil production depressurizes this reservoir and draws groundwater from the limestone, compacting it and 'unclamping' the fault, reaching the Mohr-Coulomb failure criterion and causing seismicity. In principle this model is fully testable, but required data, notably the history of pressure variations in the wells, are not currently in the public domain. Quantitative estimates are, nonetheless, made of the magnitudes of the variations, arising from production from each well, in the state of stress on the seismogenic Newdigate fault. The general principles of this model, including the incorporation of poroelastic effects and effects of fault asperities into Mohr-Coulomb failure calculations, may inform understanding of anthropogenic seismicity in other settings.

Keywords: anthropogenic seismicity, geomechanics, calcite 'beef', Weald Basin, Jurassic, surrey

1. Introduction

Highlights.
Earthquakes at Newdigate in 2018–2019 correlate with oil production from Portland sst.
A cause-and-effect relation via a high-permeability hydraulic connection is proposed.
The model seismogenic fault incorporates internal structure with asperities.
Testing by Mohr-Coulomb failure analysis incorporates effects of poroelasticity.

A 'swarm' of earthquakes with magnitudes up to ~3, starting on 1 April 2018, has affected the Newdigate area of Surrey, in the Weald Basin of southeast England (**Figures 1–3**). As is detailed in the online supplement, a potential connection with

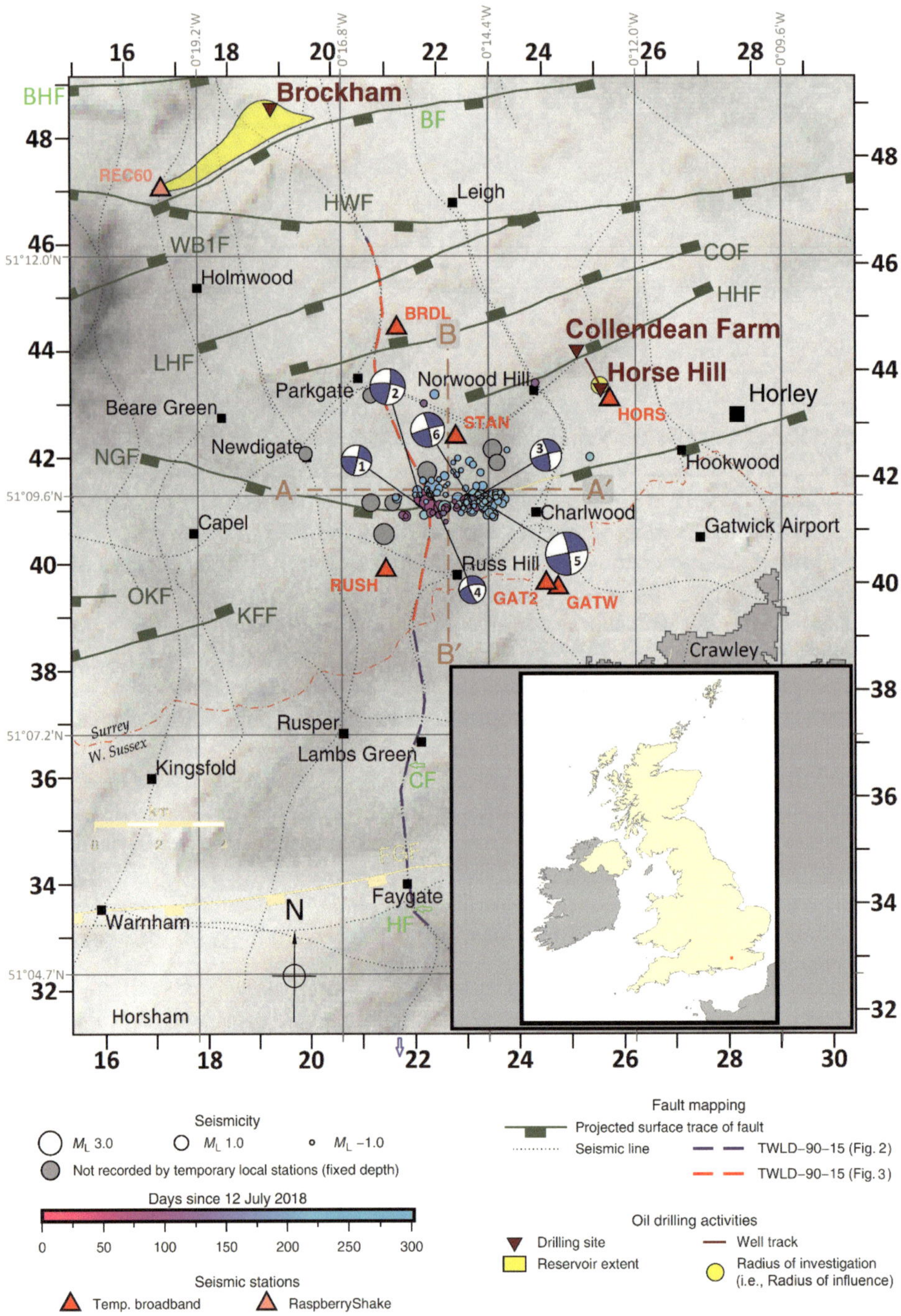

Figure 1.
Map of the study area, modified from Figure 2(a) of Hicks et al. [1]. The original geographical (latitude-longitude) co-ordinate system has been retained, but 'greyed out', with a new co-ordinate system added, indexed to the British National Grid (BNG). Inset shows location. As is discussed in the main text, the original scale bar by Hicks et al. [1] is much too small and has also been 'greyed out'. Faults are identified thus: BF, Brockham fault;

local oilfield activities, in the nearby Brockham-X2Y (BRX2Y) and Horse Hill-1 (HH1) wells, was immediately suspected, but dismissed by petroleum developers (e.g., [2, 3]). Concerns about the possibility that activities in these oilfields were indeed causing these earthquakes were raised through correspondence in The Times newspaper in August 2018 [4]. A workshop, convened by the Oil & Gas Authority (OGA), followed on 3 October 2018, the OGA being a UK government body with responsibilities that include the licencing of exploration and development of onshore oil and gas resources in England, including managing the risk of seismicity from such operations. A summary of the proceedings of this workshop was reported [5], including the statement that 'the workshop participants concluded that, based on the evidence presented, there was no causal link between the seismic events and oil and gas activity although one participant was less certain and felt that this could only be concluded on "the balance of probabilities" and would have liked to see more detailed data on recent oil and gas surface and subsurface activity' ([5], p. 1). It has subsequently been argued that there is indeed no such cause and effect connection (e.g., [1, 6]), developers having repeatedly issued strong public statements to this effect (e.g., [3, 7, 8]). However, a major issue, not noted in any of the above-mentioned works, is the clear temporal pattern of earthquake occurrence (**Figure 4**), with earthquakes strongly concentrated at times when oil is being produced from the Upper Portland Sandstone via the HH1 and/or BRX2Y wells. Production will reduce the fluid pressure in the reservoir being pumped. Fluid pressure changes within faults are well known as a cause of anthropogenic seismicity (e.g., [9, 10]); however, rather than a decrease, the causative change is usually an increase in fluid pressure as, for example, for the Preese Hall earthquake sequence in 2011, caused by injection of water under pressure during 'fracking' for shale gas (e.g., [11]).

Given the geology of the Weald Basin [12, 13], a conceptual model can be envisaged whereby pressure reduction the Portland reservoir might bring nearby faults to the Mohr-Coulomb condition for slip, as illustrated in **Figure 5**. As summarised in the figure caption, this model also provides a natural explanation for why production from the deeper Kimmeridge reservoir does not have an equivalent effect. Nonetheless, testing this model is difficult, for several reasons. The map and cross-section reported by Hicks et al. [1] provide the most detailed documentation of the Newdigate seismicity available (**Figures 1** and **2**), and thus serve as a basis for further discussion. However, a first reason why model testing is difficult is that use of these outputs is problematic because of mistakes in their preparation; before they can be used their geolocation has to be improved (as discussed in the present online supplement, also below). A second reason is uncertainty in the hydraulic properties of elements of the proposed model; this includes the distribution of the permeable 'calcite "beef"' fabric within clay-dominated lithologies that are otherwise impermeable. Each of these aspects will be investigated in this study. A third reason why

---

*BHF, Box Hill fault; CF, Crawley fault; COF, 'Collendean fault'; FGF, 'Faygate fault'; HF, Holmbush fault; HHF, Horse Hill fault; HKF, Hookwood fault; HWF, Holmwood fault; KFF, Kingsfold fault; LHF, Leigh fault; NGF, Newdigate fault; OKF, Ockley fault; WB1F, and Whiteberry-1 fault. Most of these structures are depicted as shown by Hicks et al. [1], although some are misplaced or misidentified, as discussed in the text. The Crawley and Holmwood faults, not recognised by Hicks et al. [1], are shown schematically (from [15]) where they cross seismic line TWLD-90-15, the southward continuation of which (beyond the excerpt shown in **Figure 2**) is also shown schematically. The 'Faygate fault' is a mistaken concept by Hicks et al. [1], so is shown 'greyed out' (see text and online supplement). Horse Hill 1 well track is from https://ukogl.org.uk/map/php/well_deviation_survey.php?wellId=3041. Positions of seismic lines, including line TWLD-90-15, are from the schematic location map provided by the UK onshore geophysical library (https://ukogl.org.uk/), which is indexed to the BNG, and was transformed to geographical co-ordinates by Hicks et al. [1]. Seismograph station GATW ceased operation on 17 May 2019 due to equipment theft. It was replaced by station GAT2, ~230 m northwest, from 6 June.*

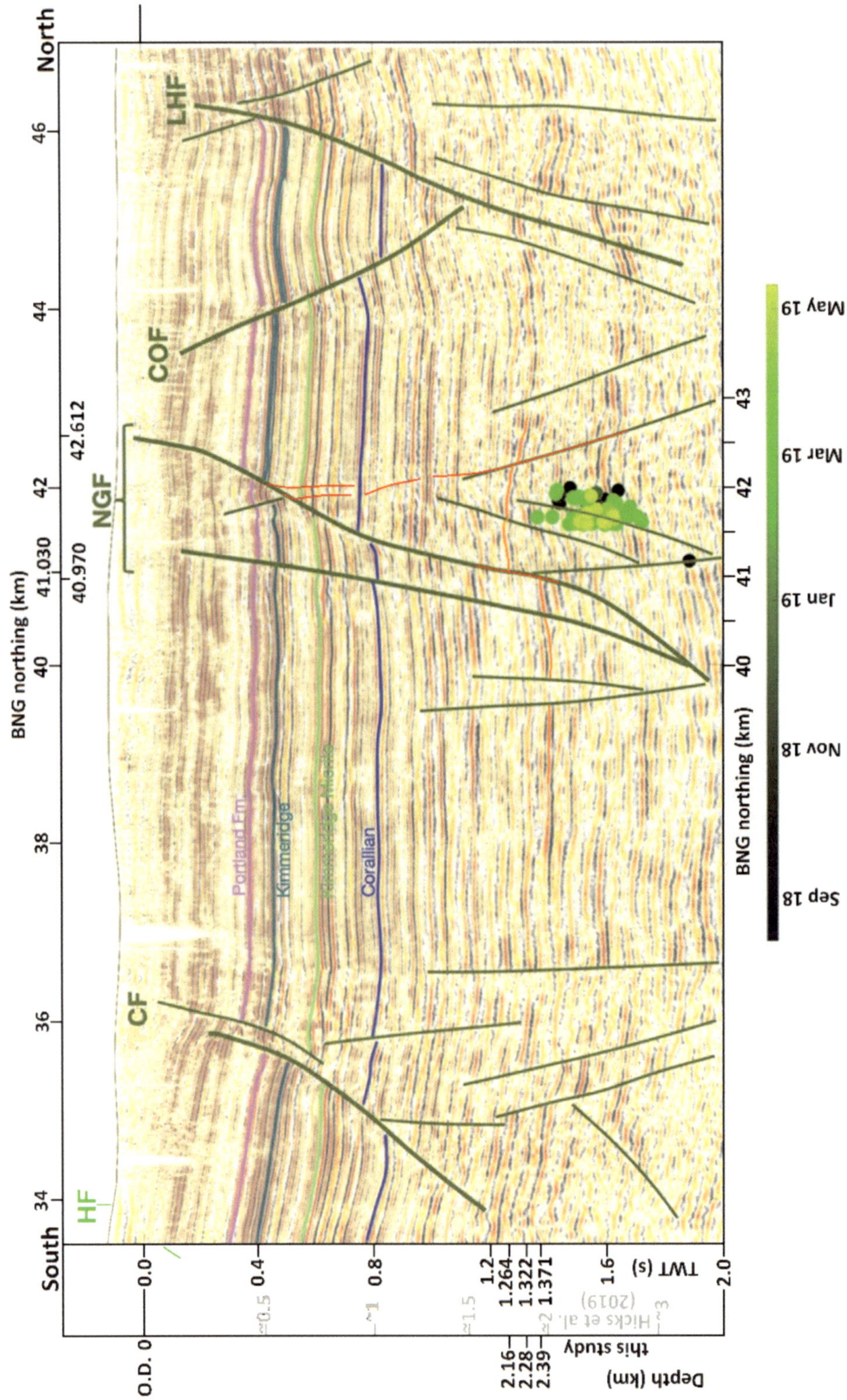

Figure 2.
*2-D seismic section along seismic line TWLD-90-15, modified from **Figure 6** of Hicks et al. [1]. The original was indexed by Hicks et al. [1] to the British National Grid (BNG), rather than the geographical co-ordinates used for **Figure 1**; it has been geolocated for this study using documentation provided by the UKOGL.*

testing the proposed model is difficult is that key operational data, such as pressure variations in oil wells and logs of wellsite activities that might affect reservoir conditions, have been found to be unavailable. Indeed, preparation of this manuscript was delayed pending an attempt to obtain such data from the OGA under UK law using a Freedom of Information (FOI) request; this request was unsuccessful on the basis that the OGA did not hold such data, notwithstanding the scope of their statutory duties. In the absence of pressure data, testing the proposed model will be limited to investigating the magnitudes of pressure variations (and associated changes to the state of stress) that can be anticipated in the model fault and in each well, as a result of operations in the other well, and the time delays for their propagation, subject to the assumed hydraulic properties for each lithology, which are informed by the limited available data.

2. Geological structure and stratigraphy

The study area is in southeast England, near the boundary between the counties of Surrey and West Sussex, ~40 km WSW of central London, on the northern flank of the Weald Basin (**Figures 1** and **6**). The outcrop geology and shallow subsurface structure of this area are documented by Dines and Edmunds [14] and Gallois and Worssam [15]; Trueman [16], DECC [17], and others have discussed the history of petroleum exploration. Many authors have discussed the origin and structure of the Weald Basin, or Weald sub-basin of the wider Wessex Basin (e.g., [12, 13, 18–25]). As these and many other works demonstrate, this basin has developed near the northern margin of the Variscan orogenic belt, Variscan reverse faults having been reactivated as normal faults during the Mesozoic. Chadwick [19] resolved two phases of Mesozoic extension, during the Early Jurassic (Hettangian to Toarcian; extension factor, β, 1.12) and Late Jurassic and earliest Cretaceous (late Oxfordian to Valangian; β = 1.10). The Jurassic and Cretaceous sedimentary formations that accompanied and followed this extension are documented in many works and summarised in the British Geological Survey (BGS) stratigraphic lexicon (https://www.bgs.ac.uk/lexicon/); **Table 1** summarises the local stratigraphy, based on the HH1 well record. This basin experienced Cenozoic inversion, when some of the Mesozoic normal faults were reactivated as reverse faults (e.g., [23]; **Figure 6**). As a result of this history, some faults have normal offsets within the syn-rift succession but show reverse slip in younger sediments, as illustrated in **Figure 2**.

Most oil reservoirs in the study area are in the Upper Portland Sandstone, a shallow marine sandstone of Upper Jurassic age, sealed above by the overlying impermeable Purbeck Anhydrite, deposition of which reflects isolation of the Weald Basin interior from the sea (e.g., [21]; **Table 1**). The oil–water contact for the Horse Hill reservoir has been inferred as ~580 m TVDSS [26], thus roughly at the mid-point of the Upper Portland Sandstone. The modelled extent of this ~3 km square reservoir is illustrated in **Figure 7**; to the north and south it is sealed by

The labelled horizons were not explained by Hicks et al. [1], but appear to be the top Portland group, top Kimmeridge clay formation, and top coralline Oolite formation (cf. ***Table 4****). Faults designated by Hicks et al. [1] are identified thus: COF, Collendean fault; LHF, Leigh fault; and NGF, Newdigate fault. CF denotes the Crawley fault. The depth scale from Hicks et al. [1], which appears to be based on their velocity model for earthquake location (****Table 2****), is 'greyed out', being now considered inaccurate. The new depth scale, using the seismic interval velocities from well HH1 (****Table 1****) and now considered more accurate, is emphasised. Additional interpretation has also been added, including the interpreted top Penarth group / base Lias group reflector and its offset by the main strand of the Newdigate fault, and some of the additional lesser fault strands forming the multi-stranded Newdigate fault zones, other strands being evident in* ***Figure 3*** *and in the uninterpreted version of this seismic section provided by Hicks et al. [1] in their online supplement.*

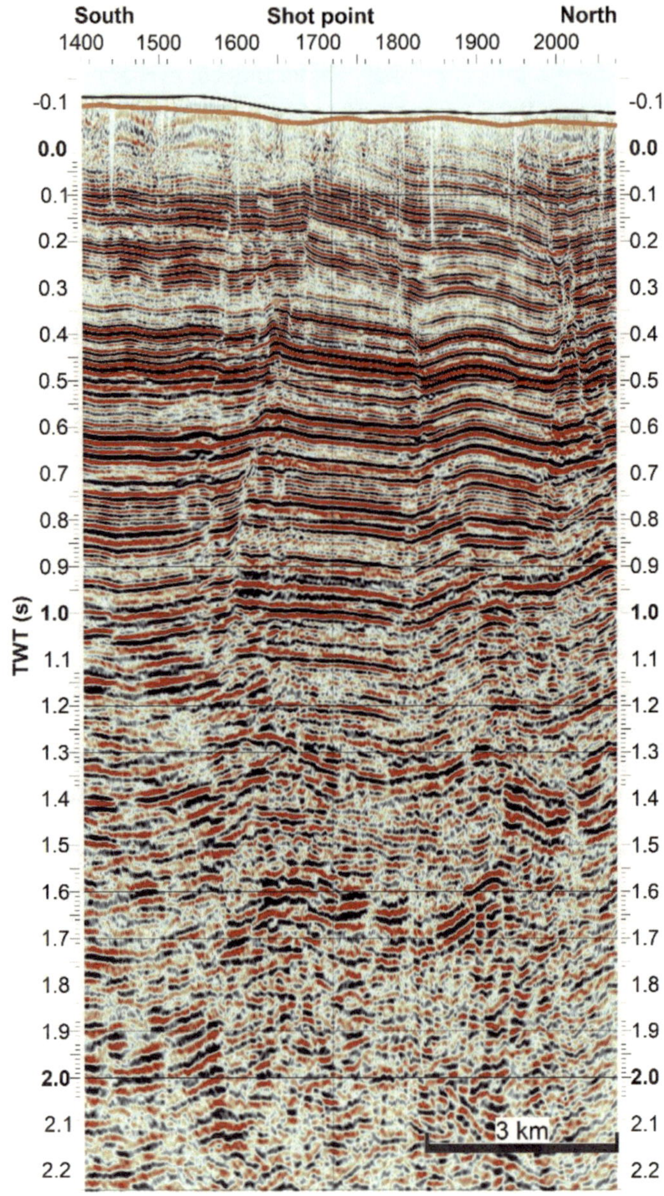

Figure 3.
Excerpt from the record section for seismic line TWLD-90-15 between shot points 1400 and 2070, as provided by UKOGL, illustrating the Newdigate fault. As geo-located using documentation provided by UKOGL, this excerpt extends between BNG references TQ 21618 38952 and TQ 20381 46852, a distance of ~8 km. The location of this seismic line adjoining the Newdigate fault is illustrated in ***Figures 1*** *and* ***7****.*

normal faults (the Horse Hill Fault and Hookwood Fault) across which the Purbeck Anhydrite is downthrown, whereas in other directions it is sealed by the dip of the impermeable cap rock. This reservoir, hydraulically connected to the HH1 and HH2Z wells, from which production began in mid 2018 and late 2019, respectively (see supplement), is thus much larger than the hydraulic 'radius of influence' depicted by Hicks et al. [1] in **Figure 1**. The neighbouring Collendean Farm reservoir is to the north, separated by the Collendean Farm Fault, was reached by the Collendean Farm-1 (CF1) borehole (drilled in 1964; BGS ID TQ24SW1; at TQ 2480 4429; **Figure 7**). Oil has also been produced by the HH1 well from a deeper reservoir in the Kimmeridge Limestone (e.g., [26]) (the 'Kimmeridge micrites' in **Table 1**). However, the production from this reservoir shows no correlation with seismicity (**Figure 4**), indicating that this deeper reservoir is not hydraulically connected to the seismogenic Newdigate Fault.

The smaller (~2 km × ~1 km) Brockham reservoir (**Figure 8**) has a permeability of up to ~200 mD (e.g., [27]), its base at 582 m TVDSS [27]. As **Figure 8** shows, it is sealed to the southeast by the Brockham Fault and in other directions by the dip of the Purbeck Anhydrite. However, although the Brockham Fault has ~40 m of throw and acts as an impermeable reservoir seal, ~1 km farther SW, where the top Portland Sandstone is deeper than the reservoir base, this fault has low throw and the permeable sandstone is juxtaposed on both sides. This aspect of the fault geometry (discussed in more detail in the supplement) is key to the proposed conceptual model (**Figure 5**), as it makes possible a permeable hydraulic connection between the Brockham reservoir and the seismogenic Newdigate Fault to the southeast. In contrast, Hicks et al. [1] depicted the Brockham Fault as sealing the reservoir to the SE for ~3 km distance (**Figure 1**). Together with the different geometry of other faults in the area, this portrayal would preclude the possibility of any permeable hydraulic connection between this reservoir and the Newdigate Fault. The Angus [27] fault reconstruction is favoured in the present study, being assessed as more soundly based than either the Hicks et al. [1] version or the earlier interpretation by Europa [28] (see supplement). As detailed in the supplement, during prolonged production from Brockham well BRX2Y from 1998 to 2016, amounting to 36,900 m^3 [11], the reservoir pressure had decreased from ~900 to ~500 psi. Angus [29] reported that at the end of production in 2016 the reservoir pressure was ~490 psi or ~ 3.4 MPa, indicating a decline by ~2.7 MPa below the expected hydrostatic pressure of ~6.1 MPa at its ~622 m depth.

The geometry of the Newdigate Fault also differs radically between the Xodus [26] and Hicks et al. [1] interpretations (**Figures 1** and 7). Hicks et al. [1] depict it as a continuous structure with total E-W length ~ 13 km, extending from the Newdigate area to the vicinity of Horley, its eastern part thus forming the southern seal to the Horse Hill reservoir. In contrast, Xodus [26] depicted it as dying out eastwards ~3 km SW of the HH1 well, at the eastern end of the Newdigate seismicity (**Figure 7**). In this interpretation the fault that forms the southern seal of the Horse Hill reservoir (here designated the Hookwood Fault, HKF; **Figure 7**) is separated from the Newdigate Fault (NGF) by a ~ 1 km 'gap' around Charlwood. To investigate this difference, the records from seismic lines C79–36 and TWLD90–21, which run N-S through this area, were obtained from the UK Onshore Geophysical Library (OGL; https://ukogl.org.uk/) archive. Neither seismic line shows any stratigraphic offset transecting this area, favouring the Xodus [26] interpretation, thus also casting doubt on other aspects of the Hicks et al. [1] analysis where they differ from other interpretations (such as for the aforementioned Brockham Fault), and initiating thorough checking of their article, which identified other problematic aspects (see **Figure 1** caption, also below).

The base of the Jurassic sequence lies at ~2100–2200 m depth in the study area (e.g., [12, 13]; **Figure 6**). This sequence is locally underlain by thin Triassic deposits overlying pre-Variscan (Palaeozoic) 'basement' at depths of > ~ 2200 m [30]. The uppermost 'basement' in much of this area is known from borehole records to be Dinantian (Early Carboniferous) limestone [13, 30]. Thus, the HH1 well log (**Table 1**) indicates that the Jurassic Lias Group is underlain by ~60 m of latest Triassic Penarth Group ('Rhaetic') rocks, then ~50 m of the Triassic Mercia Mudstone, then ~10 m of dolomitic conglomerate of uncertain age, then ~70 m of Dinantian limestone, above a mudstone-dominated Upper Devonian succession.

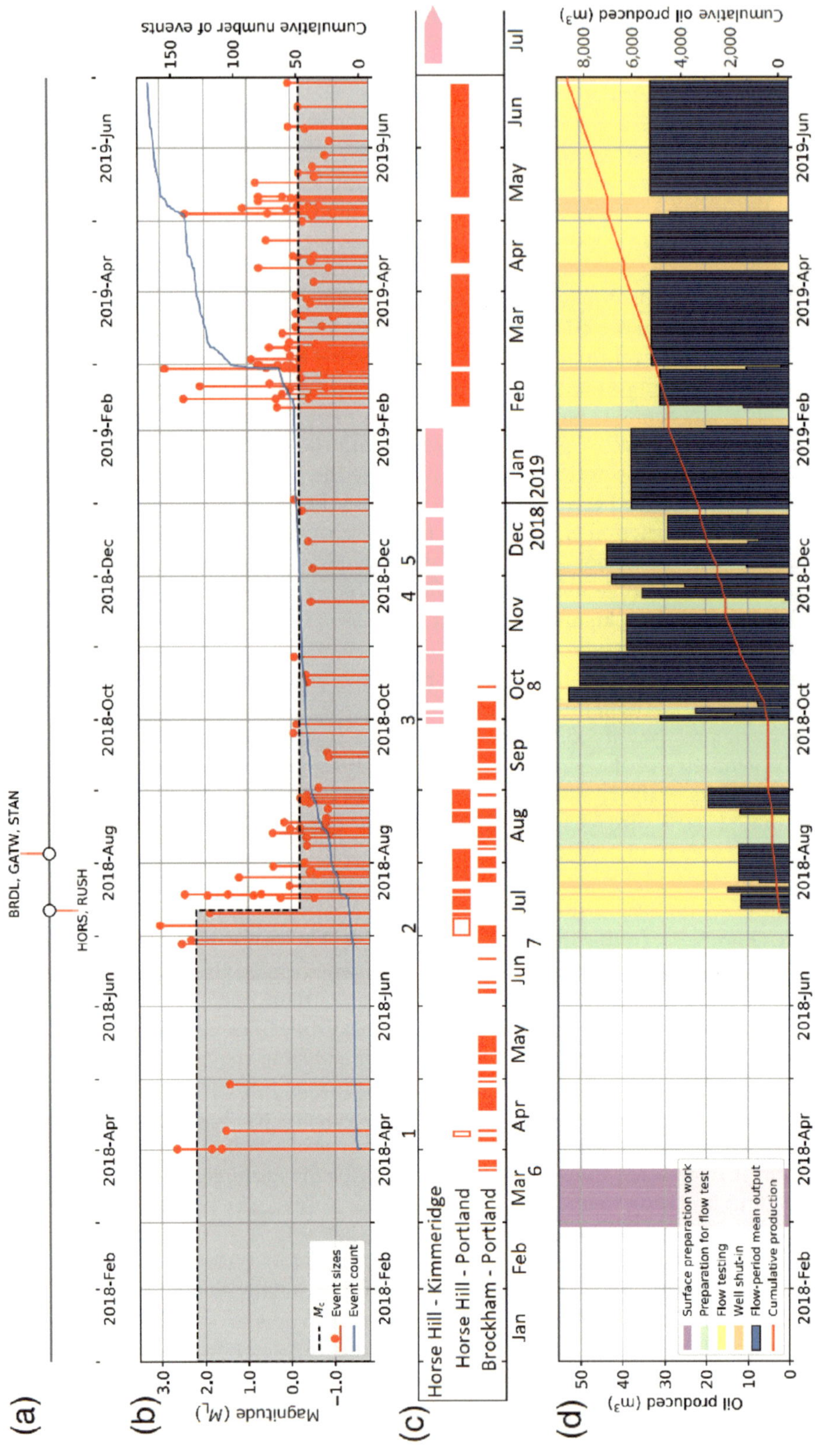

Figure 4.
Time series of Newdigate earthquakes and other activities affecting the Horse Hill 1 and Brockham X2 wells, based on Figure 3 of Hicks et al. [1]. Note that their original Fig. 6(d) has been omitted as it depicts an incorrect

Busby and Smith [30] estimated using gravity modelling that these Devonian rocks are typically ~1–2 km thick, their base at a typical depth of ~3.5–4 km, being underlain in the central Weald Basin by many kilometres of Lower Palaeozoic metamorphic basement. Around the northern margin of the basin and the southern margin of the adjoining London Platform the Dinantian limestone and underlying Devonian rocks are well imaged seismically at <1 s two way time (TWT), indicating depth < ~1500 m (e.g., [18]). Both these subdivisions are locally several hundred metres thick, the limestone being relatively unreflective and the Devonian succession highly reflective. Moving southward, as the overlying Mesozoic succession thickens, the Dinantian limestone gradually becomes thinner and its boundaries become more difficult to interpret seismically (e.g., [18]). Busby and Smith [30] noted reports of this limestone in many boreholes beneath the Weald Basin; in their view it persists southward beneath most of the basin, almost to the English Channel coastline.

More recently, Pullan and Butler [13] have presented a new map (their Fig. 21) showing the pre-Variscan subcrop beneath much of the Weald Basin (including most of the area of **Figure 1**) as Devonian, the Dinantian limestone being inferred to be absent. As interpreted by these workers, this limestone dies out ~2.5 km SW of the HH1 well, indicating that it is absent in the vicinity of the Newdigate fault and the associated seismicity. However, seismic lines in this area (e.g., that in **Figure 2**, which is part of line TWLD90–15), as well as lines C79–36 and TWLD90–21 that have also been scrutinised as part of this study, do not clearly resolve whether this limestone is present or not; as Pullan and Butler [13] showed, there is no well control for ~20 km distance SW of the HH1 well, so no direct evidence either way. Pullan and Butler [13] noted dipmeter evidence that in the HH1 well this limestone dips northward at ~20–30°; their inference that it dies out not far away seems based on structural projection given its thickness (**Table 1**) and assuming continued northward dip. However, it is clear from other seismic sections (e.g., [18]) that in other parts of the basin the Dinantian limestone is folded. At this stage it is unclear whether this lithology is present across the study area or not. The proposed conceptual model (**Figure 5**) requires a highly permeable lithology, such as this, beneath the Mesozoic sediment in this area.

The two key issues already noted will now be addressed. The distribution and properties of calcite 'beef' will first be discussed. Second, the geolocation of features, mislocated by Hicks et al. [1], will be considered.

---

timeline for production at Brockham up to 2016: The correct timeline is shown in Fig. 11 of Angus [27].
***(a)** Installation dates of stations forming the local temporary seismic monitoring network. From Fig. 3(a) of Hicks et al. [1]. **(b)** Detected earthquakes, cumulative number of events, and inferred variations in the completeness threshold magnitude M_C. From Fig. 3(b) of Hicks et al. [1]. Note that some of the magnitudes M_L depicted here differ slightly from those listed in supplementary Fig. S2 of Hicks et al. [1], which feature in discussion in the text. **(c)** Summary timeline for activities at the Horse Hill 1 well, indicating (based on the information sources discussed in the text and details in part (d)) the phases of production from each reservoir. Notes refer to details discussed in the text, thus, regarding the HH1 well: 1, first known intervention affecting the well, 5 April 2018; 2, removal of bridge plug that had isolated the Portland reservoir from the surface, 4 July 2018; 3, production from KL3; 4, production from KL4; and 5, 'co-mingled' production from both KL3 and KL4. Regarding the BRX2Y well, based on the timeline reported by Hicks et al. [1]: 6 denotes the restart of production on 22 march 2018; 7 denotes a later resumption, with injection of water starting on 25 June and (net) production restarting on 28 June (but with both injection and production occurring on 27 June); and 8 denotes the end of production on 15 October 2018. **(d)** More detailed operations timeline for activities at the Horse Hill 1 well, with flow-period averaged production and cumulative production over time. From Fig. 3(c) of Hicks et al. [1], with further details, including dates, provided in their supplementary Table S4. The information provided by Hicks et al. [1] is much more detailed than that which has been otherwise released into the public domain, and must have been obtained from the developer. However, their reporting of the information does not identify the hydrocarbon reservoirs to which the activities relate (see part (c)), which is essential to reveal the pattern of correlation between seismicity and activities affecting the Portland reservoir (see also the online supplement).*

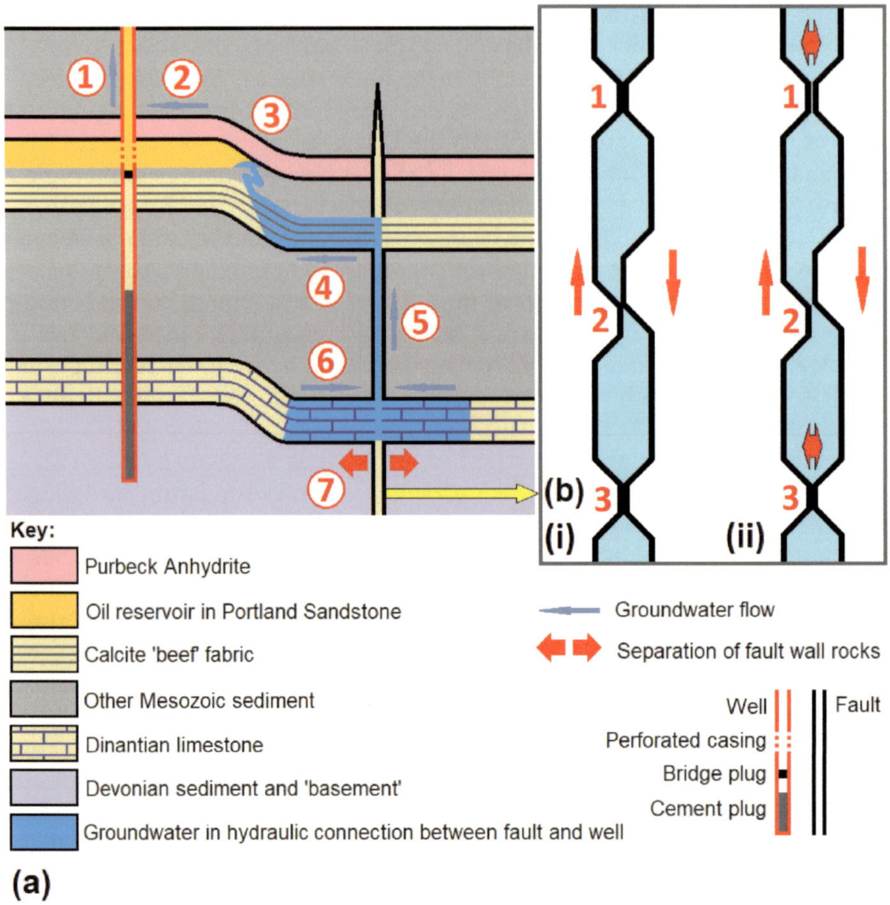

Figure 5.
Cartoons summarising the proposed conceptual model linking the Newdigate seismicity to reductions in fluid pressure in oil wells. ***(a)*** *Cross-section showing large-scale processes. Production of oil (1) will reduce the pressure within the Portland reservoir near the production well. This will cause flow of oil towards the well from more distal parts of the reservoir (2). This will be accompanied by flow of groundwater into the volume vacated by the oil, which is inferred to be hydraulically connected (presumably by fractures) to the 'hyper-permeable' calcite 'beef' fabric in the underlying lower Portland sandstone a short distance below (3), which will cause flow within this fabric (4). This 'beef' fabric is inferred to be continuous as far as the Newdigate fault zone. The flow within it will therefore draw groundwater from greater depths (5), up one or more strands of this fault zone, which is assumed permeable, reducing the pressure in the section where this fault transects the Dinantian limestone. This pressure reduction will act to draw groundwater from the permeable Dinantian limestone into the fault (6). The associated compaction of the Dinantian limestone will cause separation of its two surfaces across the seismogenic fault (7). Surface interventions affecting the wells, such as bleeding pressure following shut-in, will reduce the pressure inside the well and have a similar overall effect. In contrast, the Kimmeridge reservoir is below the 'beef' layer, so no corresponding downward hydraulic connection from this reservoir exists, therefore production from this reservoir causes no equivalent effect on the hydrology and state of stress within the Dinantian limestone.* ***(b)*** *Plan view of a patch of the seismogenic fault showing processes on a micro-structural scale on the seismogenic fault, where separation of the fault surfaces by a small distance Δx, from configuration (i) to configuration (ii), affects three model asperities (1, 2 and 3). After this change, at asperity 1 the fault surfaces are no longer in contact, at asperity 2 what was an interference fit between the fault surfaces has become a clearance fit, and the rocks forming at asperity 3 have decompressed elastically, so they remain in contact but with a reduced normal stress and thus a reduced limiting shear stress that can maintain fault stability.*

2.1 Calcite 'beef' and its significance

Calcite 'beef', first reported in 1826 by Webster [31], consists of bedding-parallel veins of diagenetic calcite (e.g., [32, 33]). In 1835, Buckland and De la Beche [34] adopted this nomenclature for veins of fibrous calcite within claystone beds in what is now known as the Purbeck Group in Dorset, the term 'beef' having originally

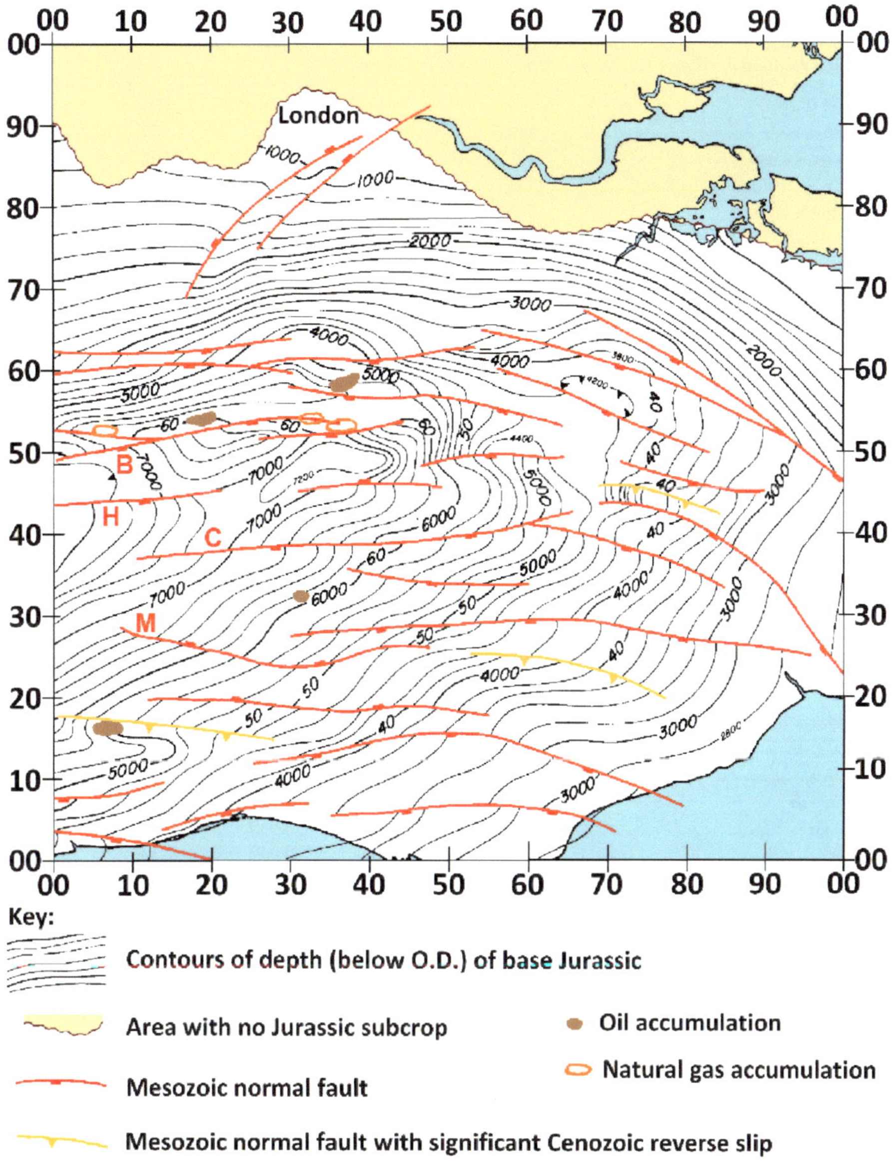

Figure 6.
Map of the structure of British National Grid 100 km × 100 km quadrangle TQ, showing the depth of base Jurassic (in feet below O.D., with contours at 200 ft. or ~ 60 m intervals) and locations where the base Jurassic is offset by faults. Faults in the vicinity of the present study area are named: C and M correspond to the Crawley and Maplehurst faults [15]; B and H appear to denote the Box Hill and Holmwood faults (cf. ***Figure 1****), although the latter is misplaced. Modified from part of Fig. 4(a) of Butler and Pullan [12].*

been used by quarry workers on account of similarity to the fibrous structure of meat. This fabric (illustrated by many authors, including [35–37]), is now recognised in mudstone formations worldwide (e.g., [35]). Following the above-mentioned early reporting its mode of origin was widely debated; the view has become accepted relatively recently that 'beef' develops by natural hydraulic fracturing associated with overpressure during hydrocarbon maturation and migration (e.g., [32, 33, 35, 36, 38–41]; cf. [36, 42]). This fabric is indeed sometimes designated as 'hydrocarbon-expulsion fractures' (e.g., [43]). The conditions for calcite 'beef' development include palaeo-temperature ~ 70–120 °C [35]. In the central Weald Basin, such conditions are expected throughout the Jurassic succession,

Subdivision	MD (m)	TVDSS (m)	TWT (s)	V_I (m s^{-1})	Notes
Younger subdivisions (Early Cretaceous; Berriasian to Barremian)					
Weald Clay	7.6	-66.9	NR	ND	
Hastings Beds	158.5	84.0	NR	ND	
Grinstead Clay	211.8	137.3	NR	ND	
Lower Tunbridge Wells Sands	234.7	160.2	NR	ND	
Wadhurst Clay	245.4	170.8	NR	ND	
Ashdown Beds	298.1	223.6	NR	ND	
Purbeck Group (latest Jurassic and earliest Cretaceous; Tithonian and Berriasian)					
Purbeck Durlston Beds	396.8	322.3	NR	ND	1
Purbeck Carbonates	464.8	390.3	NR	ND	
Purbeck Main Anhydrite	604.7	530.2	0.370	2500	
Portland Group (Late Jurassic; Tithonian)					
Upper Portland Sandstone	622.4	547.7	0.384	2531	
Lower Portland Sandstone	708.4	632.5	0.451	5011	
Ancholme Group (latest Middle Jurassic and Late Jurassic; Callovian to Tithonian)					
Kimmeridge Clay	755.9	677.6	0.469	2787	
Kimmeridgian Micrite 1	851.3	765.4	0.532	2861	
Kimmeridgian Micrite 2	939.7	835.5	0.581	2961	
Top Corallian	1359.1	1139.0	0.786	3289	
Corallian Limestone	1523.7	1272.2	0.867	3743	
Oxford Clay	1539.2	1285.3	0.874	3540	
Kellaways Beds	1666.0	1403.9	0.941	3725	
Great Oolite Group (Middle Jurassic; Bathonian and Callovian)					
Cornbrash	1681.6	1418.8	0.949	2600	2
Main Great Oolite	1682.8	1420.1	0.950	5095	
Fuller's Earth	1732.8	1468.5	0.969	4886	
Inferior Oolite Group (Middle Jurassic; Aalenian and Bajocian)					
Inferior Oolite	1767.5	1502.7	0.983	5584	
Lias Group (Early Jurassic; Hettangian to Toarcian)					
Upper Lias	1941.6	1675.8	1.045	4244	
Middle Lias	2048.0	1781.9	1.095	4796	
Lower Lias	2158.3	1892.2	1.141	4301	
Older subdivisions (Triassic and older)					
Rhaetic	2470.1	2204.0	1.286	5318	
Mercia Mudstone	2528.6	2262.5	1.308	4434	3
Dolomitic Conglomerate	2581.7	2315.6	NR	ND	
Carboniferous Limestone	2593.2	2326.8	1.337	ND	
Upper Devonian	2659.4	2393.3	NR	ND	4
TD	2686.8	2420.7	NR	ND	5

Data for tops of stratigraphic subdivisions (as used by UKOGL; not all expressed using modern formal stratigraphic nomenclature, which is available from https://www.bgs.ac.uk/lexicon) are from the online well log (https://ukogl.org.uk/map/php/pdf.php?subfolder=wells\tops&filename=3041.pdf), supplemented by values from Pullan and Butler [13], NR indicating 'not reported'. Measured Depth (MD) is measured below a datum at 66.9 m O.D., below the local ground level of 74.5 m O.D. at the wellhead, at TQ 25254 43600. TVDSS is True Vertical Depth below O.D.; TWT is echo time. Values of interval velocity, V_I, are determined in this study, ND indicating 'not determined'. Notes: 1. The Durlston Beds (or Durlston Formation) are nowadays regarded as earliest Cretaceous; the rest of the Purbeck Group is Late Jurassic. 2. The Cornbrash Formation is too thin here for its interval velocity to be reliably determined. 3. Interval velocities for the Mercia Mudstone and the Dolomitic Conglomerate combined. 4. TVDSS for the top Devonian estimated given the vertical orientation of the deepest part of the well. 5. The well bottoms (at TD) in Upper Devonian mudstone.

Table 1.
Stratigraphy of the Horse Hill 1 borehole.

given the estimated ≥2 km of burial during the Cretaceous, before the Cenozoic denudation (e.g., [18]). The idea that the properties of calcite 'beef' enable the Newdigate Fault and neighbouring oil reservoirs to be hydraulically connected was suggested by Geosierra [44], but the present study proposes a different physical mechanism as the cause of the hydraulic connection (**Figure 5**).

In southern England, calcite 'beef' is best known in the Early Jurassic Shales-With-Beef Member (https://www.bgs.ac.uk/lexicon/lexicon.cfm?pub=SHWB) of the Charmouth Mudstone Formation, part of the Lias Group, which crops out around Lyme Regis in Dorset (e.g., [36, 45–47]). Calcite 'beef' is also well known in the Late Jurassic of the Weald Basin from both outcrop and borehole sections (e.g., [48]). In the Howett [48] stratigraphy, this fabric occurs within the 'Shales with Beef and Clay-ironstone' unit, which occurs at the top of the Middle Purbeck succession and is typically ~20 m thick.

This fabric (reported as 'calcite veining') is also known from older Late Jurassic deposits, for example in core recovered between 701 and 710 m depth (below ground level 80.3 m O.D., so at 621–630 m TVDSS) in the Collendean Farm borehole near the Horse Hill site (**Figure 1**), in glauconitic sandstone forming the lower part of the Portland Group. Gallois and Worssam [15] placed this stratigraphic level in what they regarded as the sandy upper part of the underlying Kimmeridge Clay Formation. Nonetheless, in recent petroleum exploration reports (e.g., [26]), as in **Table 1**, this glauconitic sandstone with calcite 'beef' is reinstated within the Lower Portland Sandstone. Its inclusion within the Portland Group explains why this group is portrayed as much thicker in the recent petroleum-oriented literature (e.g., ~130 m thick in **Table 1**) than by Gallois and Worssam [15], who stated its thickness as only 54 m at Collendean Farm. As these latter authors noted, the Portland Group in the Weald Basin is not well correlated with the 'type' Portlandian of the Portland area of Dorset, which is in the Portland – South Wight Basin (e.g., [21]). The 'type' Portlandian includes the Portland Limestone (now known as the Portland Stone Formation), an important building stone, the sediments of this age not being sandstone-dominated as in the Weald Basin.

The significance of all the above for the present study is as follows. It has previously been noted that the processes responsible for 'beef' formation will create permeability anisotropy, permeability being far greater parallel to the fabric and bedding than in the perpendicular direction (e.g., [39, 49]). Various workers have estimated the permeability of such bedding-parallel fractures, the highest estimate identified during the present work, ~900 mD (~9×10^{-13} m^2), being by Carey et al. [50] for the Ordovician Utica Shale of eastern North America. This is many orders-of-magnitude higher than the expected nanodarcy permeability of shale perpendicular to bedding, and is quite a high value for rocks in general.

2.2 Geolocation

The study area has been illustrated using the map (**Figure 1**), and seismic cross-section (**Figure 2**) from Hicks et al. [1]. However, the original versions of both these figures have required significant amendment regarding accuracy issues. This map was originally geolocated using geographical co-ordinates; to make it easier to use British National Grid (BNG) co-ordinates have been added. This map also shows seismic lines and faults. The information source for seismic lines, including line TWLD90–15 that is illustrated in **Figures 2** and **3**, was not reported by Hicks et al. [1]; it is evident that they are from the UKOGL location map. Hicks et al. [1] also explained that (in lieu of using the existing literature) they identified faults in the study area through their own interpretation of seismic lines. As already

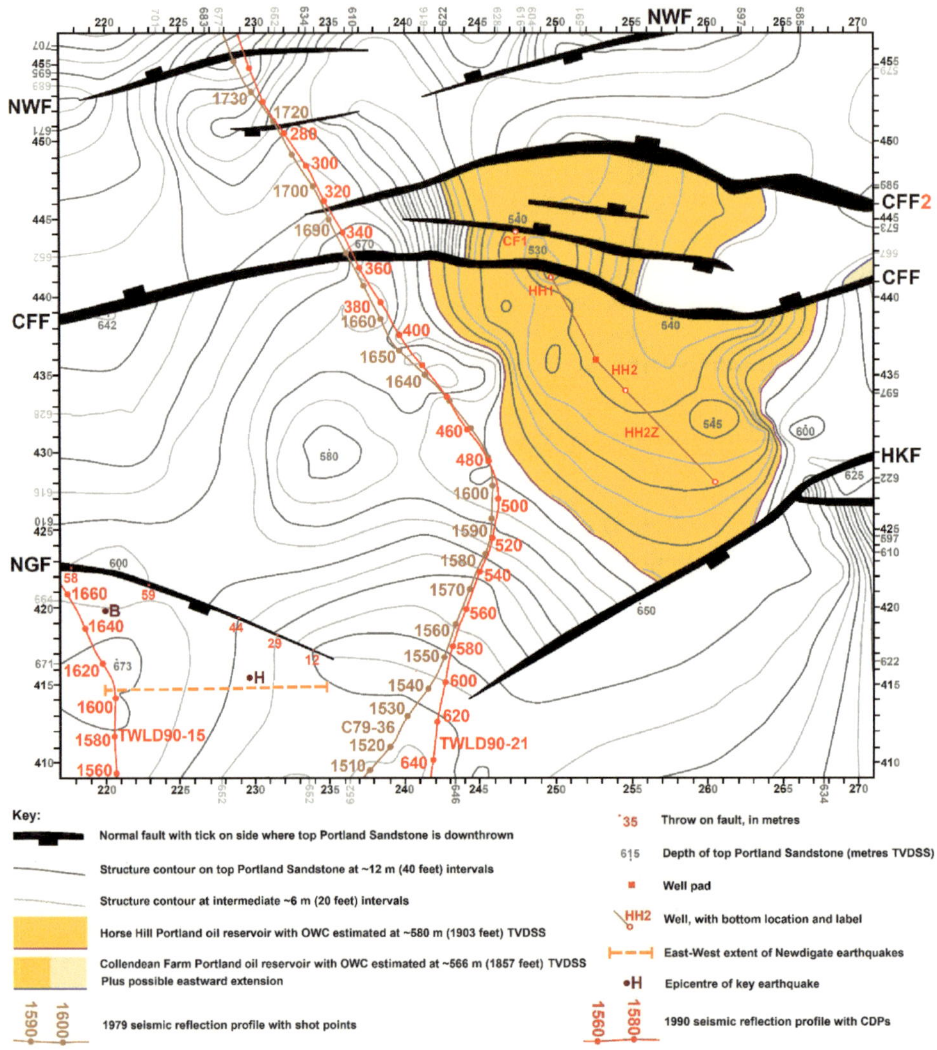

Figure 7.

Structure contour map of the top Portland sandstone in the vicinity of the Horse Hill oilfield. Redrawn after part of Fig. 6.6 of Xodus [26]. The contours within the range of reservoir depths are at 530.4, 536.4, 542.5, 548.6, 554.7, 560.8, 566.9 m, 573.0, and 579.1 m TVDSS. Faults are labelled thus: CFF, Collendean farm fault; CFF2, northern strand of Collendean farm fault; HKF, Hookwood fault; NGF, Newdigate fault; and NWF, Nalderswood fault. The key earthquakes depicted are those that initiated the first and third 'bursts' of seismicity, for which analysis of subsurface pressure changes is undertaken: For the first at 11:10 on 1 April 2018 (B; TQ 21992 41976; depth 3.08 km; M_L 2.66; M_W 2.76); and for the third at 07:43 on 14 February 2019 (H; TQ 21959 41543; depth 2.05 km; M_L 2.47; M_W 2.52). The event that initiated the second 'burst' of seismicity occurred at 12:28 on 27 June 2018 (TQ 23230 42421; depth 2.39 km; M_L 2.52). Earthquake details are after Hicks et al. [1]. Locations of seismic lines are from UKOGL.

noted, regarding key aspects of the structure the existing interpretations are favoured over these revisions by Hicks et al. [1].

The seismic section in **Figure 2** clearly has higher resolution than older ones, including those which informed earlier fault maps such as that by Butler and Pullan [12] (**Figure 6**). Some of the faults depicted in **Figure 2** are, thus, recognised for the first time. However, additional faults are also evident in the uninterpreted version provided by UKOGL (**Figure 3**). It is thus evident that in the lower part of the Jurassic sediment and upper part of the underlying Palaeozoic basement, the Newdigate Fault consists of multiple fault strands distributed across a zone with

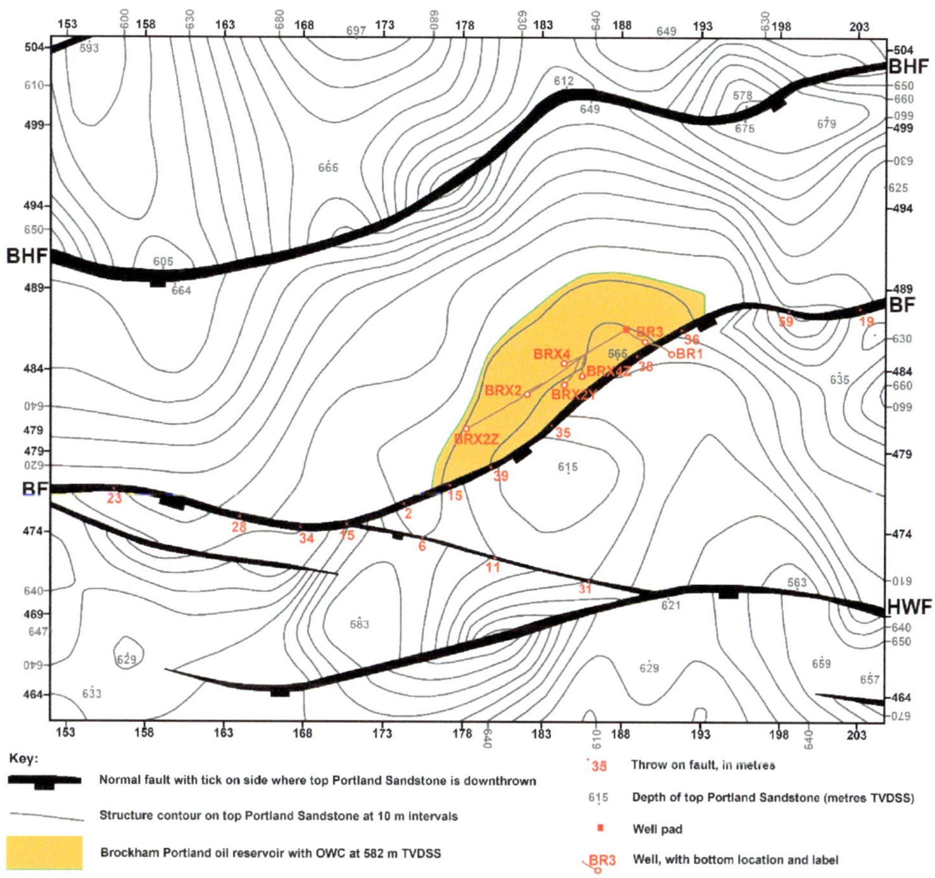

Figure 8.
Structure contour map of the top Portland sandstone in the vicinity of the Brockham oilfield. Redrawn after part of Fig. 4 of Angus [27]. Faults are labelled thus: BF, Brockham fault; BHF, Box Hill fault; and HWF, Holmwood fault.

H (km)	V_P (km s^{-1})	V_S (km s^{-1})
0.0	2.2	1.2
0.2	2.4	1.4
0.4	2.6	1.5
0.7	2.7	1.5
1.2	3.1	1.8
1.5	3.6	2.0
1.8	4.7	2.7
2.1	5.0	2.8
2.4	5.5	3.1
7.6	6.4	3.7
18.9	7.0	4.1
34.2	8.0	4.6

This velocity model was used by Hicks et al. [1] for earthquake relocation and moment tensor inversion. H denotes the depth to the top of each layer; V_P and V_S denote the P-wave and S-wave velocities. Note that this velocity model is significantly slower that that in ***Table 1****; it results in a two-way time to depth 2326.8 m, corresponding to the top of the Carboniferous Limestone at Horse Hill, of 1.466 s rather than the actual 1.337 s.*

Table 2.
Velocity Model from Hicks et al. [1]

N-S width approaching ~2 km. Careful inspection of supplementary Figure S13 of Hicks et al. [1] and **Figure 3** indicates broken and offset seismic reflectors which delineate these subsidiary strands of the Newdigate fault zone, some evidently near the limit of seismic resolution (cf. [51]), which merge upwards by ~0.5 s two way time (TWT).

A significant issue to have emerged from checking the Hicks et al. [1] analysis concerns their velocity model used for earthquake location (**Table 2**). As **Figure 9 (a)** shows, this is significantly slower than is expected from the sonic logs for wells BR1, CF1 and HH1, and from the recent depth-conversion analysis by Pullan and Butler [13]. The Hicks et al. [1] velocity model is also significantly slower that that obtained from moveout analysis during the processing of seismic line TWLD90–21

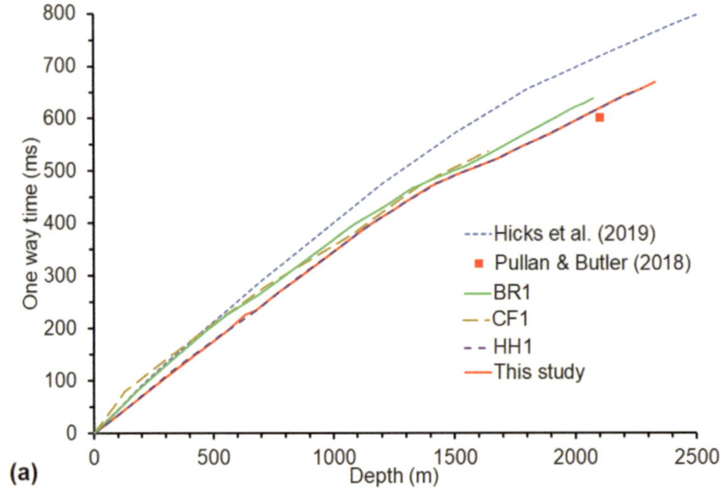

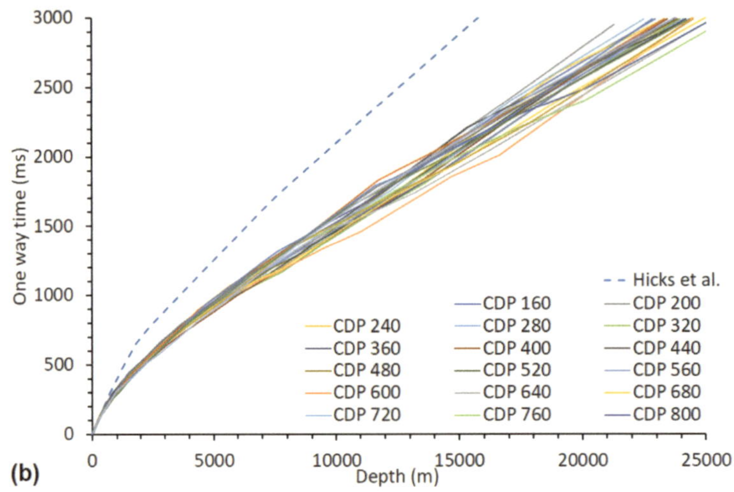

Figure 9.
Comparisons of one-way time versus depth for different seismic velocity models. ***(a)*** *Comparison of the Hicks et al. [1] velocity model with the sonic logs from wells CF1, HH1, and BR1 (from [24], and UKOGL), the velocity model (based on well HH1) adopted for this study, and a representative point from the analysis by Pullan and Butler [13] where a ~ 600 ms one-way time corresponds to a depth of ~2100 m.* ***(b)*** *Comparison of the Hicks et al. [1] velocity model with velocity models derived from interval velocities obtained from the moveout analysys of seismic line TWLD90–21 (**Fig. 7**), for a suite of representative CDPs. These interval velocities were calculated in this study using Dix's formula (e.g., [52, 53]) from root mean square velocities provided by UKOGL. UKOGL did not have such data for the other seismic lines analysed for this study.*

No.	Date	Time (UTC)	Epicentre (BNG reference)	Depth (m)	ΔN (m)	ΔE (m)	Δz (m)	$\Delta\alpha$ (°)	N_P	N_S	Δt_o (s)	z_{DD} (m)	M_L	M_W 1	M_W 2	z_C (m)	Strike (°)	Dip (°)	Rake (°)	f_C (Hz)	r_o (m)	$\Delta\sigma$ (MPa)
1	18 Jul 2018	03:59:56	TQ 22005 41393	1990	1397	803	1011	151	15	13	0.04	ND	2.01	2.20	2.03	2.00	282	74	178	6.4	48	0.25
2	18 Jul 2018	13:33:18	TQ 21920 41474	1860	1463	737	1014	145	15	15	0.02	ND	2.54	2.56	2.45	2.20	276	75	169	4.3	25	0.27
3	14 Feb 2019	07:43:33	TQ 22959 41543	2220	297	330	379	98	9	7	0.09	2050	2.47	2.52	2.27	2.80	255	86	173	7.7	142	1.05
4	19 Feb 2019	17:03:57	TQ 22872 41538	2050	220	429	393	106	5	4	0.04	2040	1.98	1.95	1.77	2.20	256	61	-163	11.2	86	0.56
5	27 Feb 2019	03:42:21	TQ 22622 41517	2110	286	352	316	98	14	11	0.13	2300	3.18	3.25	2.87	3.60	260	78	178	5.8	169	2.62
6	04 May 2019	00:19:19	TQ 22796 41516	2190	143	165	294	94	13	10	0.11	2440	2.35	2.31	2.17	2.40	255	85	167	16.9	64	8.06

Data from supplementary Table S2 of Hicks et al. [1]. The events are numbered to match ***Figure 1****. For events 1 and 2 only conventional hypocentral locations were determined, which yielded the epicentral co-ordinates and focal depths. Double difference focal depths (z_{DD}) were not determined (ND). For the other events, the epicentral co-ordinates and z_{DD} are determined from the double difference location procedure and the 'Depth' by conventional location. ΔN, ΔE and Δz are the uncertainties in northing, easting, and depth, based for events 3–6 on the double difference solutions. $\Delta\alpha$ is the maximum gap between ray path azimuths to seismograph stations that recorded each event, N_P and N_S being the numbers of P- and S-wave records. Δt_o is the rms residual in origin time. M_L and M_W are local magnitude and moment magnitude. M_W 1 and the centroid depth z_C are determined from the moment tensor; M_W 2 is from P-wave spectra. Strike, Dip and Rake are for the nodal plane of the focal mechanism that is regarded as the fault plane, being subparallel to the Newdigate Fault (****Figure 2****). Mean corner frequency f_C, source radius r_O, and stress drop $\Delta\sigma$ are determined from seismogram spectra.*

Table 3.
Source parameters for Newdigate earthquakes with focal mechanisms.

(**Figure 9(b)**). Hicks et al. [1] explained that they based their velocity model on existing regional models, but 'then improved [it] using sonic logs from the Brockham and Horse Hill wells'. However, as **Figure 9(a)** shows, despite this workflow, their velocity model is inconsistent with the records from these wells; they evidently somehow made a mistake over this aspect. Being too slow, their velocity model causes hypocentres to be mislocated too shallow. **Figure 2** includes a depth scale using the preferred velocity model in **Table 1**, based on the HH1 log. At the position of the earthquakes, this depth scale is ~400 m deeper than that provided by Hicks et al. [1]. Furthermore, as is discussed in the supplement, Hicks et al. [1] appear to have depth-converted this seismic section using the same velocity model (**Table 2**) that they used for earthquake location. As is also discussed in the supplement, location of the earthquakes using the velocity model in **Table 1** would adjust their hypocentres downward by an estimated ~400 m. The hypocentres are thus positioned more-or-less correctly relative to the detail in the seismic section, but they and this detail are now placed ~400 m deeper than Hicks et al. [1] envisaged. This adjustment moves the earthquake population from within the Jurassic sedimentary section to within the Palaeozoic 'basement'. These earthquakes presumably occurred on one or more of the steeply north-dipping subsidiary strands of the Newdigate fault zone, given the steeply north-dipping nodal planes, identified as the fault planes, of the focal mechanisms (**Figure 1** and **Table 3**), rather than on the main Newdigate Fault that dips south.

3. Seismicity and its correlation with well activities

As already noted, multiple publications have already documented the 2018–2019 Newdigate 'earthquake swarm', notably those by Baptie and Luckett [54], Verdon et al. [6], and Hicks et al. [1]. Baptie and Luckett [54] presented a preliminary analysis of 14 earthquakes between 1 April and 18 August 2018; their results informed the OGA workshop. The more extensive analysis by Hicks et al. [1] will now be appraised. These latter authors determined hypocentres and other source parameters for 168 earthquakes between 1 April 2018 and 28 June 2019, some with local magnitude $M_L < -1$, their location patterns and timeline being depicted in **Figures 1, 2** and **4** and summarised in **Table 3**, with **Table 4** listing events that post-date their study. The first nine earthquakes up to 10 July 2018 (including one of the largest, with M_L 3.02, on 5 July) were located before any local seismograph stations were operational, using only data from regional stations. Hicks et al. [1] explained that due to the limited available data these events were located by assigning each a fixed focal depth. The resulting reported depths vary between 2.33 and 3.08 km (see Table S2 of [1]), it being unclear on what basis different depths were assumed for different events. The next sixteen events, until 11 July, were located conventionally but including data from local stations. For the rest of the events, both 'double difference' relocations (after [55]) and conventional locations were determined, using the velocity structure in **Table 2**. As detailed in **Figures 1** and **2**, most of the earthquakes in a zone ~1.5 km long (E-W) by ~300 m wide (N-S). The compact width of this zone suggests that many patches of a single strand of the Newdigate fault zone were reactivated.

Focal mechanisms were determined by Hicks et al. [1] for six events, including the largest, of M_L 3.18 and moment magnitude M_W 3.25, on 27 February 2019, as illustrated in **Figure 1** and listed in **Table 3**. All six events have a nodal plane striking roughly east–west and dipping steeply north. As already noted, this plane is inferred to be the fault plane, indicating predominant right-lateral slip. Available data regarding the state of stress in the Weald Basin are extremely limited; Kingdon et al. [56] and Fellgett et al. [57] provided syntheses of in situ stress data across much of Britain. However, these authors wrote little about the Weald Basin, Fellgett et al. [57] noting that many hydrocarbon wells in this area have yielded stress data but these data had not yet been placed in the public domain. The stress dataset available for the Weald Basin thus remains that presented by Evans and Brereton [58]. As is detailed in the supplement, this limited dataset indicates a NW-SE maximum principal stress and a NE–SW minimum principal stress. The Newdigate focal mechanisms (**Figure 1**) are consistent with this stress field orientation, given the standard requirement for the maximum principal stress to lie within dilatational quadrants [59].

3.1 Temporal clustering

As detailed by Hicks et al. [1], the Newdigate seismicity between April 2018 and June 2019 involved four 'bursts' of activity (**Figure 4**). The first began at 11:10 on 1 April (M_L 2.66), followed by two events later on the same day, another on 9 April, and a final event on 28 April. The smallest of these events (on 9 April)

Date	Time (UTC)	Latitude (°N)	Longitude (°W)	BNG	Depth (km)	M_L	Note
9 June 2019	02:43:18.2	51.159	0.237	TQ 23382 41449	2.7	-0.5	
9 June 2019	23:00:15.0	51.133	0.295	TQ 19393 38462	3.3	-0.1	
6 July 2019	01:03:20.4	51.161	0.242	TQ 23027 41663	2.5	-0.7	
6 July 2019	01:03:23.7	51.161	0.242	TQ 23027 41663	2.5	-0.7	
6 July 2019	01:03:30.1	51.159	0.241	TQ 23102 41442	2.1	-0.8	
6 July 2019	01:03:40.2	51.159	0.241	TQ 23102 41442	2.2	-0.7	
6 July 2019	03:57:15.3	51.160	0.239	TQ 23239 41557	2.5	0.1	
20 July 2019	22:02:26.0	51.158	0.251	TQ 22405 41315	2.1	-0.6	
29 July 2019	03:35:25.5	51.160	0.242	TQ 23029 41552	2.2	-0.1	
6 Aug 2019	02:32:00.9	51.157	0.239	TQ 23247 41223	2.2	-0.5	
12 Aug 2019	00:46:46.6	51.160	0.241	TQ 23099 41554	2.1	-0.7	
12 Aug 2019	00:46:49.2	51.160	0.241	TQ 23099 41554	2.1	-1.4	
2 Sep 2019	05:13:04.9	51.160	0.237	TQ 23379 41560	2.0	1.1	1
3 Sep 2019	20:19:13.2	51.161	0.237	TQ 23376 41672	2.0	0.2	
6 Sep 2019	07:09:45.5	51.161	0.237	TQ 23376 41672	2.0	1.0	
21 Sep 2019	14:43:45.2	51.160	0.237	TQ 23379 41560	2.2	0.6	
31 Oct 2019	19:25:16.4	51.160	0.238	TQ 23309 41558	2.0	0.8	
29 Apr 2020	00:11:25.6	51.172	0.256	TQ 22019 42863	3.1	0.3	
24 May 2020	15:16:56.9	51.157	0.250	TQ 22478 41205	2.4	0.6	

Cataloguing here is complete to 27 August 2020. Data are from https://earthquakes.bgs.ac.uk; these earthquakes have been located using standard BGS procedures, as reported by the International Seismological Centre (http://www.isc.ac.uk). Co-ordinate transformations to the British National Grid, as part of this study, use https://www.bgs.ac.uk/data/webservices/convertForm.cfm
Note: 1. Felt in Newdigate; maximum EMS intensity 2.

Table 4.
Newdigate seismicity since the start of June 2019.

had M_L 1.28. No local seismograph stations were then in operation; Hicks et al. [1] estimated that the completeness threshold for earthquake detection was circa M_L 2, so many smaller events were undoubtedly missed.

The second 'burst' (**Figure 4**) began at 12:28 on 27 June (M_L 2.52), and included four other events above M_L 2.0 (on 29 June and 5 July, and two on 18 July) including the second largest event overall (M_L 3.02), at 10:53 on 5 July. The installation of local seismograph stations in mid June and early July lowered the completeness threshold for earthquake detection to below M_L 0 [1], resulting in many small events being thereafter detected and enabling use of the aforementioned relative location procedure. After these initial relatively large events this 'burst' of earthquakes began to tail off, in terms of both magnitude and frequency of occurrence. The last event with $M_L > 0$ occurred at 03:21 on 18 August (M_L 0.30), with infrequent smaller events persisting into early 2019.

The third 'burst' (**Figure 4**) began on 14 February 2019 with a relatively large event at 07:43 (M_L 2.47), followed by two other events of $M_L \geq \sim 2$, at 17:03 on 19 February (M_L 1.98) and at 03:42 on 27 February (M_L 3.18), this being the largest event of the overall sequence. After these initial relatively large events this 'burst' also tailed off, although two events with $M_L > 0$ occurred during April 2019 (on 11 and 22 April; M_L 0.73 and 0.56).

The fourth 'burst' (**Figure 4**) began with a relatively large event (M_L 2.35) at 00:19 on 4 May 2019. As for the preceding 'bursts', this seismicity thereafter began to tail off, although events with $M_L \sim 0$ persisted until the end of June 2019. Locations by BGS confirm the tailing-off trend through July and August 2019 (**Table 4**), with a M_L 1.1 event on 2 September, three smaller events later that month, one during October, and none more before the end of 2019 (**Table 3**).

Overall, this pattern of seismicity, consisting of 'bursts' of events, each involving activity tailing off after a peak, with the largest event increasing during successive 'bursts', bears a striking resemblance to other earthquake swarms that are inferred to be caused by fluid pressure changes in a fault (e.g., [60]). However, the Newdigate earthquake population is insufficient to permit statistical testing of the patterns expected for this mechanism.

3.2 Correlation of seismicity with well activities

Figure 4(c) indicates how these four 'bursts' of earthquakes correlate with activities affecting the Portland reservoir in the HH1 or BRX2Y wells. Production from well BRX2Y resumed in late March 2018: from Hicks et al. [1] $\sim$4.0 m^3 ($\sim$25 barrels) of oil were produced on 23 March followed by $\sim$1.1, $\sim$0.9 and $\sim$1.0 m^3 ($\sim$7, $\sim$6 and $\sim$6 barrels) on 25–27 March. Reservoir pressure during this and subsequent production has not been reported, but from standard theory (e.g., [61, 62]) one expects it to have again decreased. This start of production occurred nine days before the first Newdigate earthquake on 1 April 2018. Furthermore, as is detailed in **Figure 4** and in the online supplement, other brief 'pulses' of production occurred from well BRX2Y in June, respectively 20, 19, 16 and 6 days before the start of the second 'burst' of seismicity on 27 June.

Although the activities that were planned in the HH1 well in 2018 have been disclosed [63], most of the actual activities that took place, and any associated pressure variations in the well, have not been, other than in the very general terms reported by Hicks et al. [1]. An attempt is made in the supplement to piece together the sequence of events, based on fragments of information available. It is thus evident that before 4 July 2018, the Portland reservoir was reported as isolated from the surface by a removable bridge plug. Claims have been made that the reservoir might have been influenced before this date by surface activities and by activities in the shallow part of the well [2]; if so, this would imply that the bridge plug had failed.

It is evident from **Figure 4** that production ceased from well BRX2Y in October 2018; production at HH1 switched from the Portland reservoir to the Kimmeridgian reservoirs around the same time. Around this time, seismicity at Newdigate tailed off significantly. The conceptual model in **Figure 5** provides a natural explanation for such a variation.

The seismicity then re-intensified as the third 'burst', recognised by Hicks et al. [1], starting on 14 February 2019, which followed the resumption on 11 February 2019 of production from well HH1, now at rates of up to 220 bopd, from the Portland reservoir. As is detailed in the online supplement, production from this reservoir continued until late June 2019, after which it switched back to the Kimmeridgian reservoir, then during December 2019 to the newly-completed horizontal lateral, off well Horse Hill-2 (designated HH2Z), in the Portland reservoir. Seismicity at Newdigate remained significant during this phase of production from the Portland reservoir at HH1. However, production was not continuous; Hicks et al. [1] reported shutdowns during 9–12 April and 4–10 May, the latter corresponding to the start of the fourth 'burst' of seismicity as recognised by these authors. Seismicity subsequently tailed off following the end of production at HH1 from the Portland reservoir in late June 2019 and the switch to production from the Kimmeridgian reservoir in early July (**Figure 4** and **Table 4**). Furthermore, seismicity did not resume during the initial flow testing of well HH2Z in December 2019, even though the production rates from the Portland reservoir were much greater, up to 1087 barrels of fluid per day, than they had been from well HH1 (see supplement).

Overall, the correlation between phases of production from the Portland reservoir, from well HH1 or well BRX2Y or both, and 'bursts' of seismicity has been compelling (**Figure 4**). Hicks et al. [1] did not recognise this pattern, apparently because they did not differentiate between the Portland and Kimmeridgian reservoirs as sources of production from well HH1, as is now done (based on details in the supplement). There are particularly clear patterns for the first and third 'bursts': the first began 9 days after the resumption of production from well BRX2Y in March 2018; the third began 3 days after the resumption of production from well HH1 in February 2019. However, the patterns are less clear for the other two 'bursts' of seismicity, nor has the flow testing of well HH2Z, starting in December 2019, been associated with any significant seismicity.

4. Conceptual geomechanical model

The conceptual geomechanical model already summarised (**Figure 5**), which might account for seismicity beneath Newdigate, caused by pressure decreases in the Portland reservoir resulting from production (or other activities) from the HH1 or BRX2Y wells, will now be developed quantitatively. The basis of this model (**Figure 5**) is as follows. The Upper Portland Sandstone reservoir adjoining these wells is assumed to make a subhorizontal hydraulic connection with the seismogenic Newdigate fault zone via a permeable fabric formed in calcite 'beef' in the stratigraphically adjacent Lower Portland Sandstone. The seismogenic fault is assumed highly permeable and to provide a downward hydraulic connection to the rocks beneath the Jurassic succession. These rocks are assumed to include the dolomitic conglomerate and Dinantian limestone, as in the HH1 well (**Table 1**), which are themselves permeable. It is further assumed that the Newdigate seismicity has occurred at locations where these permeable lithologies are in contact across this fault. Pressure reduction in the Portland reservoir will thus be communicated via the 'beef' fabric, reducing the fluid pressure in this fault, which will cause flow from within the adjoining permeable lithologies into the fault. The associated reduction in fluid volume within these lithologies will cause them to compact. This will result in surfaces that were previously in contact across this fault to separate slightly, reducing the normal stress across the fault. This will 'unclamp' the fault (as in **Figure 5(b)**), moving it closer to the Mohr-Coulomb failure condition. The fault is itself assumed to be 'critically stressed', already near this failure condition, potentially enabling relatively small changes in the state of stress to cause coseismic slip (cf. [64, 65]).

Regarding the assumptions thus made, the presence of calcite 'beef' within the Portland Group sediments and its permeability have already been discussed. The permeability of faults is a major issue in Earth science (e.g., [66–70]). There is no information regarding the permeability of any strand of the Newdigate fault zone; however, although counterexamples exist (e.g., faults made impermeable by cemented fault gouge [71]), the view that faults are generally permeable, especially when critically stressed (e.g., [72]) is widely accepted, as is the precautionary principle that faults are assumed permeable, in the absence of contrary evidence, when assessing the possibility of subsurface fluid migration (e.g., [73, 74]). In the present study area, faults with offsets of tens of metres, where the permeable Portland Sandstone is juxtaposed against the impermeable Purbeck Anhydrite, act as seals for oil reservoirs (e.g., [26, 27]) and are obviously impermeable. However, low-offset faults with the permeable Portland Sandstone juxtaposed on both sides can be expected to be permeable. The question of the continuity of the Dinantian limestone from the HH1 area to the vicinity of the seismogenic strand of the Newdigate fault zone has already been discussed. The uncertainty regarding the state of stress in the Weald Basin is considered in the online supplement. As will become clear below, if the differential stress here is anything like as high as it is the Preese Hall area (after [11]), then any fault with the orientation of that which slipped will be very close to the Mohr-Coulomb failure condition.

The model thus includes three elements: pressure drawdown caused by production in wells, and its communication via the 'beef' fabric; compaction of the permeable rocks alongside the Newdigate fault caused by fluid withdrawal accompanying depressurization; and the associated Mohr-Coulomb failure analysis.

4.1 Pressure drawdown accompanying production

In general variations in fluid pressure, P, in a porous medium, are solutions to the diffusion equation

$$\frac{\partial P}{\partial t} = D\,\nabla^2 P, \tag{1}$$

where t is time, ∇^2 is the Laplacian operator, and D is the hydraulic diffusivity of the medium (e.g., [75]). The value of D depends on properties of the medium and fluid and on solution details, such as boundary conditions for pressure and strain (e.g., [76]). For pressure diffusion,

$$D \equiv \frac{kM}{\eta} \tag{2}$$

(e.g., [77]), where k and η are the permeability of the medium and viscosity of the fluid and M is the Biot modulus of the fluid-rock combination. M can be expressed as

$$\frac{1}{M} \equiv \frac{1}{B_R} - \frac{B_E}{{B_R}^2} + \phi\left(\frac{1}{B_F} - \frac{1}{B_R}\right), \tag{3}$$

where B_R and ϕ are the bulk modulus and porosity of the rock, B_F is the bulk modulus of the fluid, and B_E is the 'effective' bulk modulus of the combination, defined as

$$\frac{1}{B_E} \equiv \frac{1-\phi}{B_R} + \frac{\phi}{B_W}. \tag{4}$$

Costain [78] expressed D as

$$D \equiv \frac{k\,B_E}{\eta}, \tag{5}$$

which can be compared with Eq. (2); in practice, the difference in choice of elastic modulus (B_E or M) makes little practical difference (see below); from Eqs. (3) and (4),

$$\frac{1}{M} = \frac{1}{B_E} - \frac{B_E}{{B_R}^2}. \tag{6}$$

The pressure variations in the calcite 'beef' layer hydraulically connected to, and surrounding, the Brockham and Horse Hill Portland reservoirs can be assumed to have circular symmetry; analysis in terms of cylindrical polar co-ordinates is thus required. Thus, if the flow does not vary azimuthally or across the vertical extent h of the 'beef', Eq. (1) reduces to

$$\frac{\partial P}{\partial t} = D\left(\frac{\partial^2 P}{\partial r^2} + \frac{1}{r}\frac{\partial P}{\partial r}\right). \tag{7}$$

As others (e.g., [79, 80]) have noted, if production at rate Q starts at time t = 0, the variation in P, ΔP, after t = 0 takes the form

$$\Delta P = \frac{Q\eta}{4\pi k_B h_B} E_1\left(\frac{r^2}{4 D_B t}\right), \tag{8}$$

where the subscripts B denote values of D, h and k appropriate for the layer of 'beef'. Here, E_1 is the Exponential Integral Function, defined as

$$E_1(x) \equiv \int_x^\infty \frac{\exp(-s)}{s} ds. \tag{9}$$

E_1 is not supported directly in Microsoft Excel, but using its relationship to other functions (discussed, e.g., by [81]) it can be evaluated indirectly. This is possible because

$$E_1(x) \equiv \lim_{\psi \to 0} \Gamma(\psi, x) \tag{10}$$

where $\Gamma(\psi, x)$ is the Upper Incomplete Gamma Function. As Schurman [82] has noted, this means that an Excel formula providing a good approximation to $E_1(x)$ can be written as

$$EXP(GAMMALN(B1)) * (1 - GAMMA.DIST(A1, B1, 1, TRUE) \tag{11}$$

where cell A1 contains x and cell B1 contains a small positive number (say, 10^{-8}), representing ψ. Values of E_1 thus calculated were checked against the tables by Harris [83] and Stegun and Zucker [84] and against the power series approximations for $E_1(x)$ for the limit of $x \ll 1$,

$$E_1(x) \approx -\gamma - \ln(x) + x - \frac{x^2}{4} + \cdots \tag{12}$$

(e.g., [79]), where $\gamma = 0.5772156649...$ is Euler's Constant, and for the limit $x \gg 1$,

$$E_1(x) \approx \frac{\exp(-x)}{x}\left(1 - \frac{1}{x} + \cdots\right) \tag{13}$$

(e.g., [85]), and were found to be accurate to four significant figures or better. This method for evaluating $E_1(x)$ for all values of x is, thus, more accurate in general than the overall approximation formula proposed by Barry et al. [85].

In the near-wellbore volume the pressure variation will be in the Portland sandstone, not in the 'beef'. Thus, from Eq. (8), at time t the variation at the well rim, of radius r_W, is ΔP_W where

$$\Delta P_W = \frac{Q\eta}{4\pi k_P h_P} E_1\left(\frac{r_W{}^2}{4D_P t}\right), \tag{14}$$

where the subscripts P denote values of D, h and k appropriate for the Portland sandstone. Given the high value of D_P for the Portland sandstone, for most of the durations of the production pulses at BRX2Y and HH1, $r_W{}^2/(4\,D_P\,t) \ll 1$. One may thus approximate $E_1(x)$ using Eq. (12). Indeed, x will be so small that only the $-\ln(x)$ term need be considered. The resulting logarithmic dependence of ΔP_W on t means that ΔP_W remains roughly constant, as observed during the HH1 well test (see supplement).

Using the same general approach, a brief episode of production at rate Q starting at time $t = 0$ and ending at time Δt causes a pressure variation given by

$$\Delta P = \frac{Q\eta}{4\pi k_B h}\left(E_1\left(\frac{r^2}{4D_B t}\right) - E_1\left(\frac{r^2}{4D_B(t - \Delta t)}\right)\right). \tag{15}$$

Using the definition of E_1 in Eq. (9), for $t \gg \Delta t$, Eq. (8) can be approximated as

$$\Delta P \approx \frac{Q\eta\Delta t}{4\pi k_B h t}\exp\left(\frac{-r^2}{4D_B t}\right). \tag{16}$$

This equation can be differentiated with respect to t; by setting the resulting partial derivative to zero one may solve for the time delay t_D for the maximum pressure variation, thus:

$$t_D = \frac{r^2}{4D_B}. \tag{17}$$

The maximum pressure variation ΔP_M at distance r and time t_D is thus

$$\Delta P_M(r) = \frac{Q\eta D_B \Delta t}{\pi k_B h e r^2}, \tag{18}$$

indicating that ΔP_M varies inversely with r^2.

4.2 Compaction alongside the Newdigate fault

The Newdigate fault is envisaged as extremely permeable, such that a pressure variation ΔP applied to any point of it by via the layer of 'beef' is transmitted downward, with no significant time delay, to depths where it transects the permeable Dinantian limestone, the presumed seismogenic layer. This model fault is vertical, the permeable seismogenic layer being assigned thickness H_D, hydraulic diffusivity D_D, permeability k_D, and porosity ϕ_D. Depressurization at the point where the 'beef' layer intersects this fault is thus inferred to cause a reduction in groundwater pressure ΔP_O at each point below on the fault within the permeable seismogenic layer. The resulting groundwater pressure variation in the Dinantian limestone will be governed by Eq. (1). However, if this variation is independent of vertical position and position parallel to the fault, the one-dimensional variant

$$\frac{\partial P}{\partial t} = D\frac{\partial^2 P}{\partial x^2}, \tag{19}$$

will require solution, where t is time and x is distance from the fault.

As a solution to Eq. (19), the drawdown δP in groundwater pressure within such a permeable layer is given (e.g., [78]) by

$$\delta P = \Delta P_O \, erfc\left(\frac{x}{2\sqrt{(Dt)}}\right) \tag{20}$$

erfc() denoting the Complementary Gaussian Error Function. For z > 0, erfc(z) decreases as z increases, reaching ~0.0047 when z = 2. As Detournay and Cheng [77] noted, this condition indicates an effective outer limit to significant pressure variations, at distance x_M from the model fault, where

$$x_M = 4\sqrt{(Dt)}. \tag{21}$$

As time progresses, as a result of continuing production from a well that is hydraulically connected to the model fault, an ever-widening volume of rock, in the x direction perpendicular to the model fault, will thus become depressurized, water previously stored within this volume being released into the fault. **Figure 10** illustrates this effect for D = 1 $m^2 \, s^{-1}$.

In general, poroelastic strain responses to changes in fluid pressure can be highly complex (e.g., [77, 86–89]). Segall [87] noted that in the limit where $\Delta\sigma_{kk}$ = 0, a reduction in fluid pressure by δP will cause a contractional strain $\varepsilon = \alpha\, \delta P / B_E$ where α is Biot's coefficient, defined as

$$\alpha \equiv 1 - \frac{B_E}{B_R}. \tag{22}$$

Contraction will occur in the vertical z direction as well as in the x direction. Partitioning the contractional strain as $\Delta\varepsilon_{xx} = \gamma\, \varepsilon$ and $\Delta\varepsilon_{zz} = (1 - \gamma)\, \varepsilon$,

$$\Delta\varepsilon_{xx} = \frac{\gamma\, \alpha\, \delta P}{B_E}. \tag{23}$$

Many studies of poroelastic deformation have treated petroleum or geothermal reservoirs as inclusions embedded in surrounding rocks and have treated faults as idealised planes within the inclusion or its surroundings (e.g., [86, 88, 90, 91]). Because the present study aims to explore the effect of fault asperities, the fault cannot be treated in this idealised manner. The fault is instead envisaged as a vertical 'cut' in the poroelastic medium, which has (distant) boundaries in both directions perpendicular to the fault plane. In this configuration, with the outer ends of the blocks on either side of the fault fixed or 'pinned', depressurization of pore fluid will cause their inner ends, facing each other across the fault, to separate slightly, by distance Δx, as depicted schematically in **Figure 5(b)**. In contrast, if the

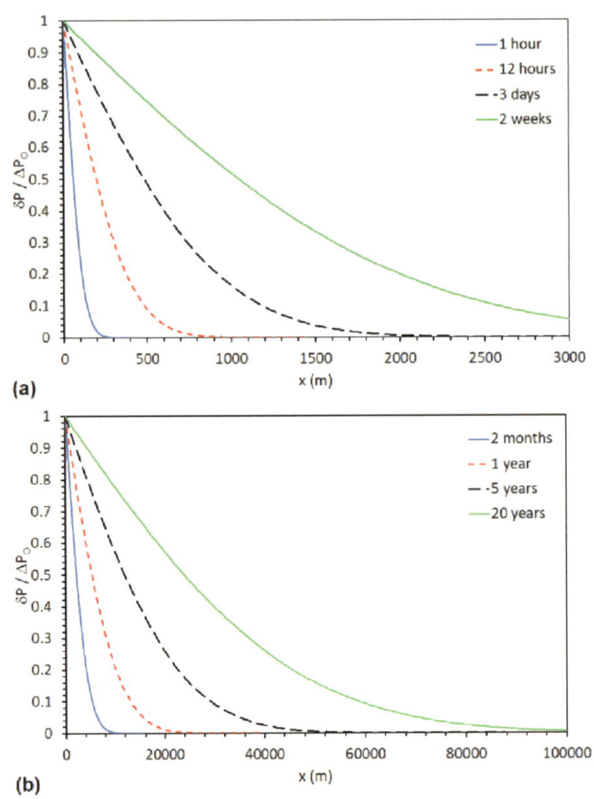

Figure 10.
Graphs of the predicted variation in pressure $\delta P / \Delta P_O$ with distance x from the model seismogenic fault, calculated using Eq. (20) for hydraulic diffusivity $D = 1\ m^2\ s^{-1}$, representing Dinantian limestone. ***(a)*** *after times t of 1 hour, 12 hours, 3 days and 2 weeks.* ***(b)*** *after times t of 2 months, 1 year, 5 years, and 20 years.*

rocks on either side of a vertical fault at the mid-point of a continuum model (such as that by [88]) were depressurised, these rocks would move towards the fault, the opposite sense of motion to what is required to provide the ability to make an assessment of the effect of asperities on the fault. It is for this essential reason that new theory is derived here for the pressure and stress response in the Dinantian limestone, rather than using existing published theory.

Subject to the above model definition, Δx can be estimated as

$$\Delta x = 2\int_{x=0}^{x\to\infty} \Delta\varepsilon_{xx} dx, \tag{24}$$

the factor of 2 taking into account that the rocks on both sides of the fault will move away from it. Evaluation of Δx requires the integral of erfc() (cf. Eq. (20)). From Abramowicz and Stegun [92], p. 299 and Weisstein [93],

$$E(x) \equiv \int_{z=0}^{z=x} erfc(z)dz = x\,erfc(x) + \frac{(1-\exp(-x^2))}{\sqrt{\pi}} \tag{25}$$

so

$$E(\infty) \equiv \int_{z=0}^{z\to\infty} erfc(z)dz = \frac{1}{\sqrt{\pi}} \tag{26}$$

Using Eqs. (20), (23) and (26), one obtains.

$$\Delta x = \frac{4\gamma\alpha\Delta P_O}{B_E}\sqrt{\left(\frac{DT}{\pi}\right)}, \tag{27}$$

this quantity being positive for a reduction in pressure by ΔP_O.

Costain [78] also showed that if, rather than persisting indefinitely, the pressure change ΔP_O imposed at x = 0 persists for duration δt, the resulting pressure variation δP is given by

$$\delta P(x,t) = \Delta P_O\left(erfc\left(\frac{x}{2\sqrt{(Dt)}}\right) - erfc\left(\frac{x}{2\sqrt{(D(t-\delta t))}}\right)\right). \tag{28}$$

At times t ≫ Δt, the pressure variation δP is given to a good approximation by

$$\delta P(x,t) = \frac{\Delta P_O D x\,\delta t\exp(-x^2/(4Dt))}{2\sqrt{\pi(Dt)^3}}. \tag{29}$$

The maximum pressure variation at distance x occurs after a time delay t_D given by

$$t_D = \frac{x^2}{6D}. \tag{30}$$

It follows that the maximum pressure variation at distance x and time t_D is given by δP_M where

$$\delta P_M(x,t) = \frac{3\sqrt{6}\,\Delta P_O\, D\,\delta t}{\sqrt{(\pi e^3)}x^2}. \tag{31}$$

These results, in Eqs. (30) and (31), can be compared with those for the radially symmetric case in Eqs. (17) and (18). In both cases, t_D is proportional to the square of distance and inversely proportional to D, but differs by a numerical factor. Alternative empirical analysis by Hettema et al. [94], based on 'rules of thumb' rather than derivation from first principles, predicts a value for t_D that likewise differs by a numerical factor, but does not predict pressure variations. The pressure variation varies inversely with the square of distance for both the radially symmetric case and for the one-dimensional case (cf. Eqs. (18) and (31)).

Δx can thus be calculated using the exact formula for δP (Eq. (28)) as

$$\Delta x = \frac{4\gamma\,\alpha\,\Delta P_O}{B_E}\sqrt{\left(\left(\frac{Dt}{\pi}\right) - \left(\frac{D(t-\delta t)}{\pi}\right)\right)}, \tag{32}$$

and using the approximate formula (Eq. (29)) as

$$\Delta x = \frac{\gamma\alpha\,\Delta P_O\,\delta t}{B_E}\sqrt{\left(\frac{D}{\pi t}\right)}. \tag{33}$$

Segall [87] reported that, for the $\Delta\sigma_{kk} = 0$ boundary condition applicable for this analysis,

$$D \equiv \frac{k}{\eta}\frac{2\mu(1-\nu)}{(1-2\nu)}\frac{B(1+\nu)}{3\alpha(1-\nu) - 2B\alpha^2(1-2\nu)}, \tag{34}$$

where μ is the shear modulus of the rock and B is its Skempton coefficient.

One may likewise quantify the loss of volume ΔV as a result of the vertical compaction Δz of the Dinantian limestone. By analogy with Eq. (24), an upper bound to Δz can be estimated as

$$\Delta z = \int_{x=0}^{x=H} \frac{(1-\gamma)\alpha\,\Delta P_O}{B_E}dz, \tag{35}$$

where H is the thickness of the Dinantian limestone. If the volume of limestone thus affected has dimensions L_y parallel to and L_x perpendicular to the model fault, so $\Delta V = L_x \times L_y \times \Delta z$ or

$$\Delta V = \frac{(1-\gamma)\,\alpha\,\Delta P_O\, H\, L_x L_y}{B_E}. \tag{36}$$

4.3 Mohr-Coulomb failure analysis

The tendency for coseismic slip on the seismogenic fault is analysed using the standard Mohr-Coulomb approach. The Mohr-Coulomb failure parameter Φ:

$$\Phi = \tau - c(\sigma_N - P), \tag{37}$$

will thus be evaluated where σ_N, τ and c are the resolved normal stress, shear stress and coefficient of friction on the fault plane, and P is the fluid pressure in the fault. $\Phi = 0$ marks this condition, with $\Phi < 0$ indicating frictional stability. This analysis can also be visualised using the standard Mohr circle construction, as a graph of τ against effective normal stress σ_N', defined as $\sigma_N - P$ (see below).

From Eq. (37), other factors remaining constant, a decrease in P will 'clamp' a fault, making it more stable, and an increase in P will 'unclamp' a fault, potentially resulting in seismicity. The latter change is the accepted mechanism for the widespread occurrence in recent years of seismicity caused by fluid injection (e.g., [95–98]). On this basis, one might conclude that a decrease in the groundwater pressure within the Newdigate fault cannot be the cause of the local seismicity.

However, it is generally accepted that the mechanics of faults, notably whether they are stable or can slip seismically, are determined by the properties of strong patches – asperities – where the opposing fault surfaces are in frictional contact (e.g., [99]). A fault surface consisting, on a microstructural scale, of a fractal size distribution of asperities with a small proportion of the fault surface in contact, in proportion to the normal stress applied to the fault, can mimic the effect, on a macroscopic scale, of a constant coefficient of friction (e.g., [100, 101]). Brown and Scholz [102] showed that natural rock surfaces follow fractal scaling for surface features of height up to ~0.1 m. Laboratory simulations of faulting often include asperities on a microstructural scale (e.g., [103, 104]). Most recently, the view has gained ground that the physics of co-seismic faulting is likewise governed by processes on a microstructural scale (e.g., [105, 106]). For example, McDermott et al. [106] deduced that asperities can be patches of fault with areas of no more than a few square metres, their properties being determined by mineral grains with dimensions of microns. Because these strong patches with fault surfaces in contact occupy only a small proportion of a fault surface, they act as stress concentrations. For example, in the laboratory experiments by Selvadurai and Glaser [104], millimetre-sized asperities with micron-sized heights occupy a very small proportion of the fault area; in one experimental run, a decrease in the mean normal stress across the fault area by ~0.3 MPa caused decreases in the normal stress affecting individual asperities by ~10 MPa.

Figure 5(b) illustrates how a small increase in separation of fault surfaces, Δx, can destabilise a fault through its effect on contact between asperities. Moving from configuration (i) to configuration (ii), two of the three asperities depicted will no longer contribute to fault stability. The third one will experience a significantly reduced normal stress, as a result of the increased separation of the fault surfaces. This will reduce the maximum shear stress that this asperity can sustain, in accordance with Eq. (37), whereas the shear stress that it is required to sustain to keep the fault stable will increase because the other asperities no longer contribute. Overall, it can thus be seen how a small increase in separation of fault surfaces might bring a fault significantly closer to the condition for slip, and might indeed result in coseismic slip.

As others (e.g., [107, 108]) have noted, a general calculation of this 'unclamping' effect for a fault with a general size-distribution of asperities would be very complex; this is thus not attempted here. A simplified calculation is instead presented, assuming that a patch of fault is prevented from slipping by a single asperity. For this patch, of area A, the normal stress and shear stress are σ_N and τ; the single asperity has cross-sectional area a, uncompressed height b, and Young's modulus E. The effect of σ_N compresses this model asperity to height d (**Figure 11**). The tip of this asperity will act as a stress concentration, with normal stress and shear stress $(A / a)\ \sigma_N$ and $(A / a)\ \tau$ and coefficient of friction c. The areas of fault where the wall rocks are not in contact initially contain fluid with pressure P_O (**Figure 11(a)**), the fault being stable with $\Phi = -\Delta\Phi$. The reduction in fluid pressure to $P_O - \Delta P_O$ is followed by poroelastic separation of the fault wall rocks by distance Δx, which brings the fault to the condition for slip ($\Phi = 0$). One may thus state versions of Eq. (37) at the tip of the model asperity for these 'before' and 'after' conditions:

$$\tau\frac{A}{a} - c\left(E\frac{(b-d)}{b} - P_O\right) = -\Delta\Phi, \tag{38}$$

and

$$\tau\frac{A}{a} - c\left(E\frac{(b-d-\Delta x/2)}{b} - P_O + \Delta P_O\right) = 0. \tag{39}$$

Subtraction of Eq. (38) from Eq. (39) gives

$$\Delta\Phi = -c(\Delta P_O - \Delta\sigma_N) = -c\left(\Delta P_O - \frac{E\,\Delta x}{2b}\right), \tag{40}$$

or, substituting Δx from Eq. (27),

$$\Delta\Phi = c\,\Delta P_O\left(\frac{E\alpha}{b\,B_E}\sqrt{\left(\frac{Dt}{\pi}\right)} - 1\right). \tag{41}$$

One may also combine Eqs. (27) and (40) by eliminating ΔP_O, to obtain

$$\Delta x = \frac{2\,\Delta\Phi}{c\left(\frac{E}{b} - \frac{B_E}{2\gamma\alpha}\sqrt{\left(\frac{\pi}{Dt}\right)}\right)}. \tag{42}$$

In the limit of high $D\,t\,/\,b$, at time $t \gg t_s$, where

$$t_s = \frac{\pi b^2 {B_E}^2}{4\gamma^2\alpha^2 E^2 D}, \tag{43}$$

this equation simplifies to

$$\Delta x \approx \frac{2b\,\Delta\Phi}{cE}, \tag{44}$$

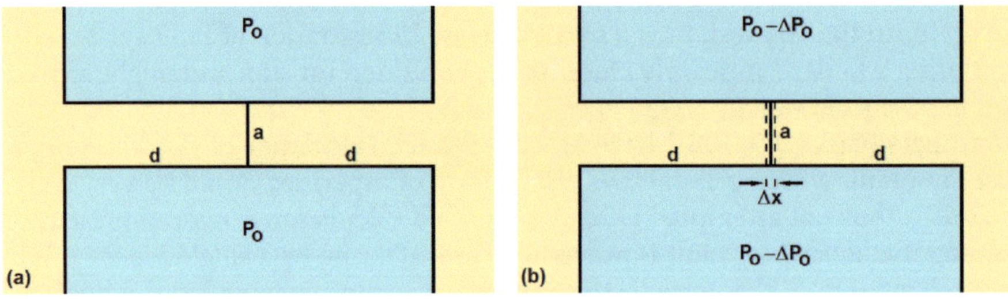

Figure 11.
*Schematic model asperity used to calculate effects of poroelastic fault unclamping. **(a)** Initial state, with fluid pressure within the fault P_O and the asperity (of uncompressed height b and cross-sectional area a) compressed to height d by the normal stress across the fault. **(b)** Modified state, with fluid pressure within the fault reduced to P_O-ΔP_O and the asperity compressed to height d-Δx/2 as a result of the separation Δx between the wall rocks on both sides of the fault caused by their poroelastic compaction.*

indicating that this poroelastic unclamping effect requires Δx / b to be comparable to $\Delta\Phi$ / E. Eq. (44) may also be rearranged, recalling from Eq. (40) that $\Delta\sigma_N = E\,\Delta x / (2\,b)$, to give

$$\Delta\Phi = c\,\Delta\sigma_N. \tag{45}$$

The geomechanical consequences of this model can now be illustrated using the Mohr circle construction (**Figure 12**), for a model fault with c = 0.6. A model stress field is adopted, similar to that deduced by Westaway [11] at 2400 m depth for the Preese Hall case study, with σ_H = 64.488 MPa, σ_V = 54.300 MPa, and σ_h = 37.880 MPa, with hydrostatic groundwater pressure P = 23.544 MPa. As **Figure 12(a)** indicates, the state of stress on a vertical model fault (characterised by effective normal stress of magnitude σ_N' = σ_N-P = 22.332 MPa and shear stress τ = 12.199 MPa), with normal vector oriented at 57° to the maximum principal stress, plots below the Coulomb failure envelope, indicating stability, the differential stress $\Delta\sigma$ being 26.608 MPa. If P within this fault decreases by 10 kPa to 23.534 MPa and this causes, via the poroelastic mechanism described above, in a reduction in the magnitude of σ_N' by 2 MPa to 20.332 MPa, the resulting state of stress is depicted in **Figure 12(b)**. This condition, with τ still 12.199 MPa, now plots on the Coulomb failure envelope, indicating instability. The fault normal vector is now oriented at 60° to the maximum principal stress, reflecting the slight rotation of the stress field in the vicinity of the fault, caused by the poroelastic reduction in σ_N, which accompanies an increase in $\Delta\sigma$ to 28.260 MPa. The state of stress on the model fault has thus effectively shifted left by 2 MPa on the Mohr-Coulomb plot, moving towards the failure envelope by 1.2 MPa, these adjustments being interrelated via Eq. (45) given c = 0.6. This poroelastic adjustment to the stress field thus involves reducing the magnitude of σ_N' keeping τ constant, rotating the near-fault stress field and increasing $\Delta\sigma$. It is different from what occurs during 'fracking' of impermeable rocks, where an increase in P causes a Mohr circle of constant diameter (indicating constant $\Delta\sigma$) to shift leftward until part of its circumference touches the failure envelope.

4.4 Estimation of model parameters

Application of the above theory to the Newdigate case study requires determination of many model parameters, representing properties of key lithologies (Portland Sandstone, calcite 'beef', and Dinantian limestone) and of the Newdigate fault. To facilitate this, it is noted that the elastic moduli that appear in the foregoing analysis are interrelated via standard formulas, such as

$$E \equiv 3B_R(1-2\nu) \tag{46}$$

and

$$\mu \equiv 3B_R\frac{1-2\nu}{2(1-\nu)}. \tag{47}$$

Hydraulic conductivity K and permeability k are also interrelated thus:

$$K \equiv \frac{k\rho_W g}{\eta} \tag{48}$$

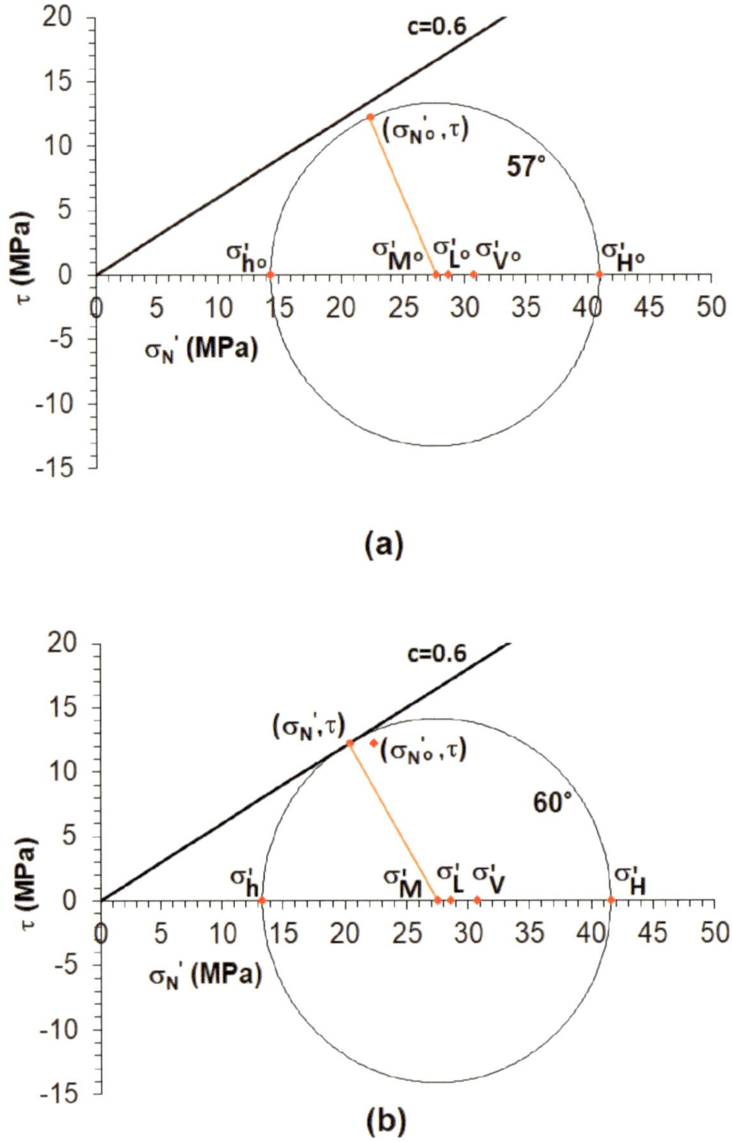

Figure 12.
Mohr circle diagrams representing a model stress field at 2400 m depth, to indicate the sense of changes to the state of stress envisaged as causing the Newdigate seismicity. ***(a)*** *for a model stress field with σ_{HO} = 64.488 MPa, σ_{VO} = 54.300 MPa, and σ_{hO} = 37.880 MPa, with σ_H oriented at 57° to the normal direction to the fault. Hydrostatic groundwater pressure P = 23.544 MPa causes $\sigma_{H'O}$ = 40.944 MPa, $\sigma_{V'O}$ = 30.756 MPa, and $\sigma_{h'O}$ = 14.336 MPa, resulting in $\sigma_{L'O}$ = 28.723 MPa and $\sigma_{M'O}$ = 27.706 MPa. The resolved shear stress and normal stress on the model fault, τ = 12.199 and $\sigma_{N'O}$ = 22.332 MPa, plot below the coulomb failure line for c = 0.6, with (from Eq. (37)) Φ = -1.2 MPa, indicating that the fault is stable under these conditions.* ***(b)*** *for revised conditions consistent with the set of processes in* ***Fig. 5****. Groundwater pressure adjusts by ΔP = -10 kPa, to P_f = 23.534 MPa, and the principal stresses adjust to σ_H = 65.130 MPa and σ_h = 36.870 MPa keeping σ_V = σ_{VO} = 54.300 MPa, with σ_H now oriented at 60° to the fault normal. As a result, σ_{H}' = 41.596 MPa, σ_{V}' = 30.766 MPa, and σ_{h}' = 13.336 MPa, resulting in σ_{L}' = 28.566 MPa and σ_{M}' = 27.466 MPa. The resolved shear stress and normal stress on the fault, τ = 12.199 and σ_{N}' = 20.332 MPa, now plot on the coulomb failure line for c = 0.6, with Φ = 0, indicating that the fault is now frictionally unstable. The calculated increase in Φ by 1.2 MPa, for $\Delta\sigma_N$ = 2 MPa with c = 0.6, is consistent with Eq. (45).*

where g is the acceleration due to gravity. Different formulas for hydraulic diffusivity D, subject to different boundary conditions, have already been noted. However, the considerable uncertainty in model parameters for lithologies in the present study region makes such distinctions moot; D will, therefore, be estimated using Eq. (5).

For the Upper Portland Sandstone at Brockham, Angus [27] reported a 3 m thick layer with ϕ 0.25 and k 200 mD. Lee [109] reported B_R 13 GPa as typical for dry sandstone with ϕ 0.25. Taking B_W 2.15 GPa for water (from [110]), using Eq. (4), B_E for Portland Sandstone with pore space occupied by water is ~6 GPa. At Horse Hill, Xodus [26] reported a 35 foot or 11 m section in Portland Sandstone with permeability up to 20 mD. The water in contact with the Portland reservoirs at both sites, at ~600 m depth, is at ~25 °C (cf. [111]), for which η = 0.9 mPa s [112], with ρ_w 1000 kg m^{-3} and g 9.81 m s^{-2}. Under these conditions the oil at Brockham has η = 11 cP or 11 mPa s [27]. Using Eq. (5), with k 200 mD, B_E ~ 6 GPa, and η 0.9 mPa s, D for Portland Sandstone is ~1.3 m^2 s^{-1}, which can be rounded to 1 m^2 s^{-1}.

Many workers (e.g., [42, 113, 114]) have investigated the aperture, or width, of bedding-parallel fractures (typically filled with 'beef') in shale, as a guide to its hydraulic properties. In a study spanning several shale provinces, Wang [113] found fractures with width between 15 μm and 87 mm. Many of the wider ones could be seen to have opened by multiple increments, each adding a few tens of microns of width, prior to cementation due to growth of calcite. Permeability and fracture aperture can be interrelated by comparing the Darcy equation for laminar fluid flow, Q = (k A / η) dP/dx, and the Poiseulle equation for laminar flow between parallel boundaries, Q = (D W^2 / (12 η)) dP/dx, which is a solution to the more general Navier–Stokes equation for fluid flow (e.g., [115]). Here Q is the volume flow rate, η the viscosity of the fluid, dP/dx the pressure gradient in the direction of flow, k the permeability of the medium, A the cross-sectional area of the flow, and W and D the width of the channel and its length in the direction perpendicular to the flow. Combining these two formulae, equating A to D × W, gives $k \equiv W^2/12$. This formula gives the permeability equivalent to W = 15 μm as ~20 D (~2×10^{-11} m^2). Overall, it is inferred that the ~900 mD value, from Carey et al. [50], might be applicable to calcite 'beef' in the present study area. Identifying a suitable representative value for B_R, the bulk modulus for calcite 'beef', is problematic, because of its strongly anisotropic character. 'Beef' is abundant within mudstones of the Neuquen Basin of Argentina (e.g., [116]). Sosa Massaro et al. [117] estimated a representative vertical Young's modulus and Poisson's ratio for this lithology as ~15 GPa and ~ 0.25; using Eq. (46) these parameters yield a bulk modulus of ~10 GPa. Using Eq. (5), with k 900 mD, K_B ~ 10 GPa, and η 0.9 mPa s, D for calcite 'beef' can be estimated - subject to considerable uncertainty - as ~10 m^2 s^{-1}. The calcite 'beef', reported in the BGS borehole viewer online documentation as 'veins of secondary calcite', in the CF1 borehole occurred at depths of 2327–2329 feet or 709.3–709.9 m TVD, thus ~626 m TVDSS. Although this interval was cored, the core was not analysed for permeability; however, core between 2300 and 2301 feet TVD yielded k 1650 mD. The top Portland in this borehole is at 1753 feet or ~ 534 m TVDSS, the estimated base of the oil reservoir at ~566 m TVDSS (**Figure** 7); this 'beef' layer is thus ~100 m below the top Portland. Based on this information this layer is assigned a nominal thickness h_B of 1 m for the purpose of modelling.

Carbonate rocks such as the Dinantian limestone are likely to be complex, being fractured, so water storage within them will be in part by opening of fractures and in part by opening of pore space. Dinantian limestone typically has low matrix porosity (e.g., [118]), its ability to store and transmit groundwater being largely via fractures. Parameter values for Dinantian limestone include B_R = 50 GPa and ϕ = 0.04, along with Young's modulus E_R = 75 GPa and Poisson's ratio ν = 0.25, from Bell [119]. With this set of values, B_E is ~27 GPa, and α = 1–27/50 = 0.46 (Eq. (22)). Poisson's ratio ν ranges between 0.19 and 0.31 [119] so 0.25 is adopted, for which

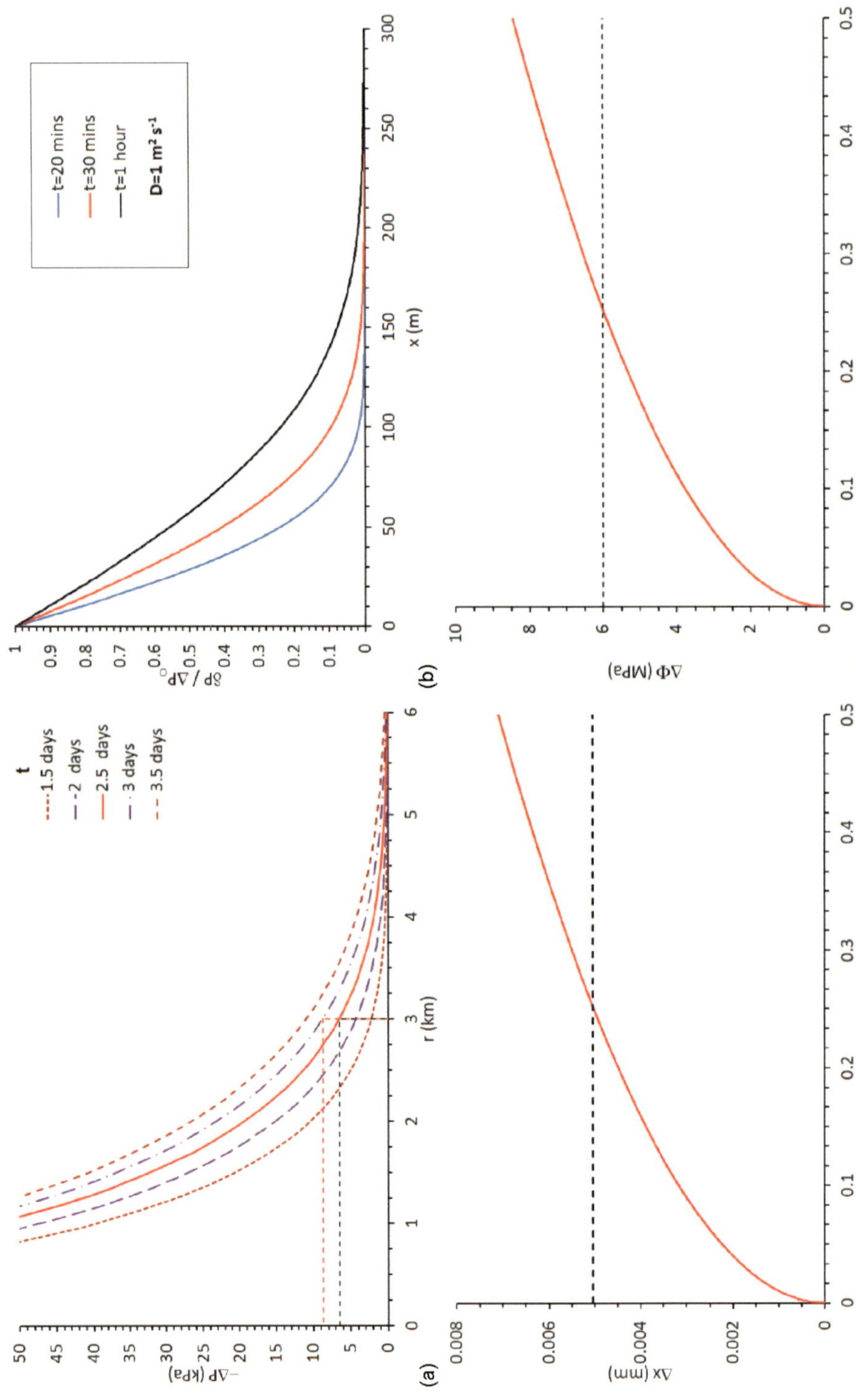

Figure 13.
Modelling of hydraulic consequences of the phase of production from well HH1 starting in February 2019. ***(a)*** *Graphs of -ΔP versus radial distance r within the layer of calcite 'beef' at ~600 m depth, which is inferred to connect the Horse Hill oil reservoir with the Newdigate fault, at different times t after the start of production. Calculations use Eq. (8) with Q = 0.4 l s^{-1}, η 0.9 mPa s, h_B 1 m, and D_B 10 m^2 s^{-1}. Dashed lines indicate, for r = 3 km, ΔP = -6.5 kPa after t = 2.5 days and ΔP = -8.7 kPa after t = 3 days.* ***(b)*** *Graphs of -$\delta P/\Delta P_O$ versus distance x perpendicular to the Newdigate fault, at different times after the pressure variation in (a) reached this fault. Calculations use Eq. (20) with D_D 1 m^2 s^{-1}, for Dinantian limestone.* ***(c)*** *Graph of Δx versus time*

$\mu \equiv B$ (Eq. (47)). No attempt is made here to determine the value of γ; a value of 0.5 will be assumed, consistent with depressurization causing closure of both vertical and horizontal fractures to an equivalent extent. Skempton's coefficient B has been reported as ~0.4 for many limestone samples (e.g., [120–122]). Bell [119] noted a range of values of K for laboratory samples of Dinantian limestone, ranging from 0.07×10^{-9} m s^{-1} to 0.3×10^{-9} m s^{-1}. Lewis et al. [123] reported much higher values ranging from $\sim 10^{-6}$ m s^{-1} to $\sim 10^{-2}$ m s^{-1} in karstified regions, which correspond (using Eq. (48)) to $k \geq 100$ mD. Using the latter set of values, Eq. (2) gives values for D ranging upward from ~3 m^2 s^{-1}. For comparison, Shepley [124] determined an upper bound to D for Dinantian limestone in the Peak District of central England by modelling the hydrology of Meerbrook Sough, a disused mine drainage adit that drains a ~ 40 km^2 area. His analysis reported an upper bound of 50,000 m^2 day^{-1} or ~ 0.6 m^2 s^{-1}. However, this analysis did not reproduce the observed magnitude of seasonal fluctuations in flow, favouring a higher value of D. Overall, it is inferred that that D ~ 1 m^2 s^{-1} is appropriate for karstified Dinantian limestone, subject to considerable uncertainty. For comparison, Hornbach et al. [125] deduced that a poroelastic pressure pulse resulting from large-scale injection of waste water propagated for up to ~40 km through the Ellenburger Formation, a karstified limestone of Ordovician age, in ~6 years, resulting in earthquakes in the vicinity of Dallas, Texas. Using Eq. (30) gives an upper bound to D for the Ellenburger Formation of ~1.4 m^2 s^{-1}, in reasonable agreement. Zhang et al. [126] reported a nominal value of 1 m^2 s^{-1} for this karstified Ordovician limestone.

For an ensemble of faults in different lithologies, Brodsky et al. [107] deduced that the typical asperity height b in length L of a fault scales as

$$b = b_O L^{\zeta}, \tag{49}$$

where ζ = ~0.6 and b_O = $\sim 10^{-3}$ $m^{0.4}$. For example, with L = 100 m, Eq. (49) gives b ~ 16 mm.

To investigate the area of the patch of fault that slipped in each earthquake, and to thus quantify the associated asperity height, seismic moment M_O was first determined using the standard formula

$$\log_{10}(M_o/Nm) = 9.05 + 1.5M_W \tag{50}$$

[127], where M_W is moment magnitude. Next, the radius a of the equivalent circular seismic source was determined from M_o, assuming a nominal coseismic stress drop $\delta\sigma$:

$$M_o = \frac{16\,\delta\sigma(1-\nu)a^3}{3(2-\nu)} \tag{51}$$

(e.g., [128]). The corresponding diameter 2a is equated to fault length L to characterise asperity height. One thus obtains

$$L = 2a = \left(\frac{3(2-\nu)M_o}{2\delta\sigma(1-\nu)}\right)^{1/3}. \tag{52}$$

---

*for the pressure variations depicted in (b). Calculations use Eq. (27) with ΔP_O 8.7 kPa (from (a)), and B_E 27 GPa, α = 0.46, and D again 1 m^2 s^{-1}, for Dinantian limestone. Dashed line indicates Δx = 0.00504 mm, which arises after 0.25 hours or 15 minutes. **(d)** Graph of $\Delta\Phi$ for the patch of the Newdigate fault that slipped in the 14 February 2019 earthquake versus time for the variations in Δx depicted in (c). Calculations use Eq. (44), with the same parameters as for (c) plus b 18.9 mm, c 0.6, and E 75 GPa for the model asperity on the seismogenic patch of fault. Dashed line indicates $\Delta\Phi$ = 6 MPa, which arises after 0.25 hours or 15 minutes.*

4.5 Application of model: Horse Hill

The first topic to be assessed is the possibility that the resumption in February 2019 of production from the Portland reservoir in well HH1 caused the third 'burst' of Newdigate seismicity. This phase of production is inferred to have begun at 08:00 GMT on 11 February 2019 and caused earthquakes starting with the event (M_W 2.5; M_O 6.8 × 10^{12} N m) at 07:43 GMT on 14 February 2019 (located at TQ 22959 41543, after [1], at a probable depth of ~2400 m), event H in **Figure** 7. The time delay in this instance was thus 3 days or ~ 72 hours, for a separation of ~3 km (**Figures 1**, 7), the HH1 well bottom being ~2.85 km from nearest point where the top Portland sandstone is transected by the Newdigate Fault and ~ 3.0 km from an updip projection of the hypocentre to this cutoff. This production took place at ~220 barrels per day or ~ 0.40 l s^{-1}. The pressure drawdown during this phase has not been reported. However, as detailed in the online supplement, during the flow testing of the Portland reservoir in well HH1 in July–August 2018, the developer reported production at a sustained rate of ~190 barrels per day or ~ 0.35 l s^{-1} (~140–160 barrels per day being of oil), with stable bottom hole pressures ~1.4 MPa below the initial reservoir pressure of ~6.3 MPa; scaling in proportion gives a ~ 1.6 MPa pressure drawdown at the HH1 well bottom in February 2019 (i.e., ~1.4 MPa × 220/190). Using Eq. (52), with ν = 0.25 and $\delta\sigma$ = 10 MPa, gives L = 134 m; using Eq. (49) gives b = 18.9 mm.

Figure 13(a) shows the predicted pressure variation in the 'beef' layer that is inferred to hydraulically connect the Horse Hill Portland reservoir and the Newdigate fault. With the chosen parameter values, the resulting pressure decrease at 3 km distance is calculated using Eq. (8) as ~6.5 kPa after 2.5 days and ~ 8.7 kPa after 3 days. Taking h_P = 11 m, η = 0.9 mPa s, D = 1.3 m^2 s^{-1}, and k_P = 35 mD (somewhat higher than the 20 mD reported by Xodus [26], with r_W = 88.9 mm (appropriate for 7-inch diameter casing), using Eq. (14) the pressure decline at the well bottom, 2–4 weeks after the start of production, would be ~1.6 MPa, as expected. If η were adjusted to 11 mPa s, to reflect oil rather than water, k_P ~ 430 mD would be required to match this pressure variation.

Figure 13(b) shows the variation in fluid pressure inside the Dinantian limestone alongside the seismogenic part of the Newdigate fault, as a result of the imposed pressure variation in the 'beef'. **Figure 13(c)** shows the corresponding variation in Δx, the poroelastic separation of the opposing faces of this fault, and **Figure 13(d)** shows the corresponding variation in $\Delta\Phi$. The upper limit to $\Delta\Phi$ is taken as c $\delta\sigma$, or 6 MPa, the coseismic stress drop, $\delta\sigma$ = 10 MPa, also being the increase in shear stress during a complete earthquake cycle; this value of $\Delta\Phi$ corresponds (using Eq. (44)) to Δx = 0.005 mm. These conditions are taken as indicating the condition for an earthquake to occur on the specified patch of fault. It is evident that with the specified combination of parameter values, these changes will occur very rapidly, in just 15 minutes after the idealised step onset of the pressure variation in the fault. In reality, **Figure 13(a)** indicates a gradual onset of this pressure variation, rather than an abrupt step, so the calculated changes Δx and $\Delta\Phi$ will occur gradually, leading to an earthquake at the calculated time, roughly three days after the start of production from the well. This modelling demonstrates that, with the parameter values adopted, production from the Portland reservoir by well HH1 has been readily able to cause seismicity on the Newdigate fault.

4.6 Application of model: Brockham

The second topic assessed will be the possibility that the ~4 m^3 of production from well BRX2Y on 23 March 2018, inferred to have started at 08:00 BST, caused earthquakes starting with the event ($M_W \sim 2.8$; $M_O \sim 1.3 \times 10^{13}$ N m) at 12:10 BST on 1 April 2018 (located at TQ 21992 41976, after [1], also at a probable depth of ~2400 m), event B in **Figure 7**, with hypocentre close to the western or WNW limit of the seismogenic part of the Newdigate fault. The time delay in this instance was 9 days ~4 hours or ~ 220 hours. In a straight line, the BRX2Y well bottom is ~7.2 km NNW of the point, updip from this hypocentre, on the top Portland cutoff of the Newdigate fault (between circa TQ 18440 48310 and circa TQ 22100 42200). However, as already noted, it is possible that the high throw on the Brockham Fault along this direct path blocks any high-permeability connection, in which case the actual path length between these end points, through localities farther west with low fault offsets (**Figure 8**), and around the western end of the Leigh Fault (**Figure 1**), might be >1 km longer. Using Eq. (52), with $\nu = 0.25$ and $\delta\sigma = 10$ MPa, $M_O \sim 1.3 \times 10^{13}$ N m for the 1 April 2018 earthquake gives L = 166 m; using Eq. (49) gives b = 21.5 mm.

Figure 14(a) shows the predicted variation in fluid pressure in the 'beef' layer that is inferred to hydraulically connect the Brockham Portland oil reservoir and the Newdigate fault, calculated using Eq. (15). With the chosen set of parameter values, following 4 m^3 of production the resulting pressure decrease at r = 8 km distance is predicted to be ~50 Pa after 9 days for $D_B = 10$ m^2 s^{-1}. Assuming $h_P = 3$ m, $\eta = 0.9$ mPa s, $D_P = 1.3$ m^2 s^{-1}, and $k_P = 200$ mD, with $r_W = 88.9$ mm (again appropriate for 7-inch diameter casing), using Eq. (14) the pressure decline at the well bottom at the end of production would be ~530 kPa if the production took 4 hours or ~ 1.0 MPa if completed in 2 hours. Within an hour of this idealised ~50 Pa step pressure variation affecting the Dinantian limestone, Δx would be ~0.06 μm (**Figure 14(b)**). This would be sufficient to unclamp the Newdigate fault by ~60 kPa or ~ 1% of δσ (**Figure 14(c)**). With D_B adjusted to 20 m^2 s^{-1}, the pressure decrease at r = 8 km would peak, circa 9 days after this pulse of production, at ~150 Pa (**Figure 14(d)**). Within three quarters of an hour of this idealised ~150 Pa step pressure variation affecting the Dinantian limestone, Δx would be ~0.14 μm (**Figure 14(e)**). This would be sufficient to unclamp the Newdigate fault by ~150 kPa or ~ 2.5% of δσ (**Figure 14(f)**). Evidently, for such a small change in the state of stress to have caused the observed seismicity, the patch of this fault that slipped on 1 April 2018 must have already been within a very small proportion of δσ of the Mohr-Coulomb condition for slip.

4.7 Application of model: interference between the wells

The final topic assessed will be pressure interference between the wells. Like the path from the Brockham well to the Newdigate fault, the most permeable connection, through 'beef', between the two wells will exceed the straight line distance; separation r = 10 km is adopted. Again calculated using Eq. (29), with $D_B = 10$ m^2 s^{-1} the 4 m^3 of production from well BRX2Y would cause a maximum pressure decline in the vicinity of well HH1 of ~50 Pa after a ~ 1 month delay (**Figure 15(a)**). With D_B increased to 20 m^2 s^{-1}, this production would cause a maximum pressure decline of ~90 Pa after a ~ 15 day delay (**Figure 15(b)**). Variations in HH1 bottom hole pressure of this order, developing and dissipating on timescales of weeks or months in response to attempts at production from well BRX2Y, would be extremely difficult, if not impossible, to recognise given the >1 MPa pressure variations caused by the production from this well. This is consistent with the statement by the OGA [5] that no pressure variation at HH1 was detectable in response to the pulses of production from BRX2Y.

A second test is possible, given that ~20 years of production at Brockham (followed by two years of shut-in, during 2016–2018) resulted in a bottom-hole

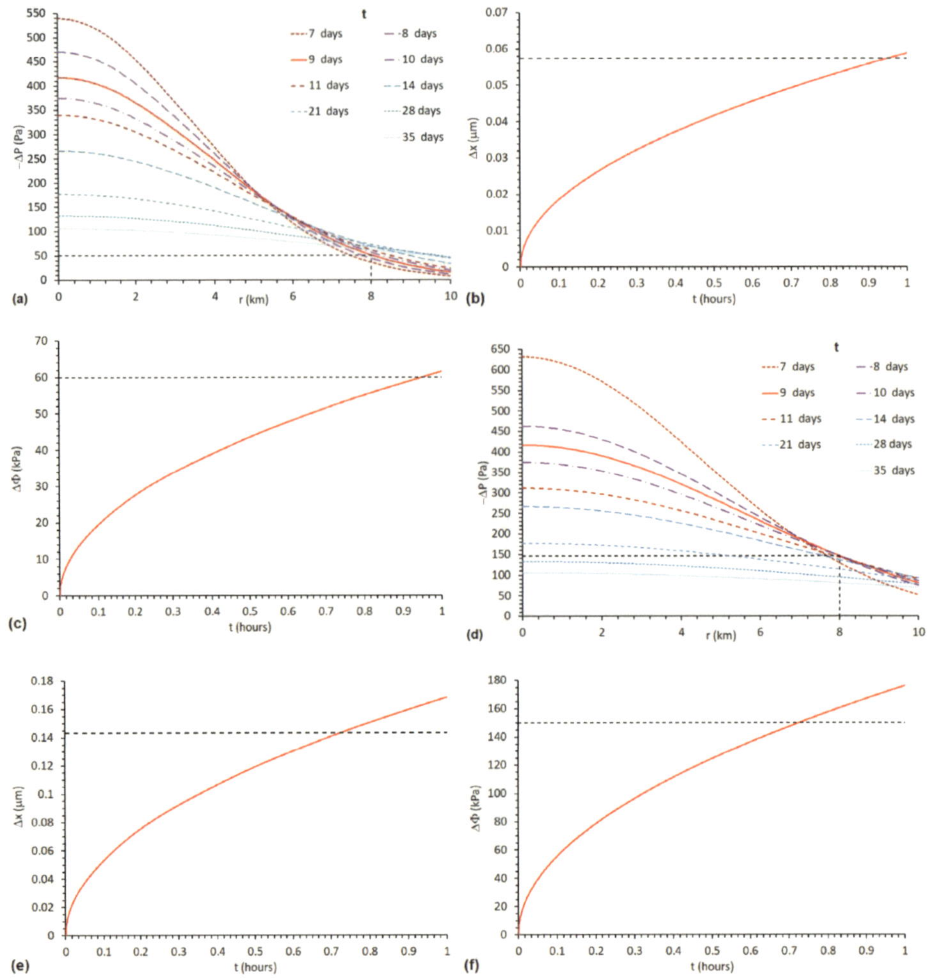

Figure 14.
Modelling of hydraulic consequences of the pulse of production from well BRX2Y on 23 march 2018. ***(a)*** *Graphs of -ΔP versus radial distance r within the layer of calcite 'beef' at ~600 m depth, which is inferred to connect the Brockham oil reservoir with the Newdigate fault, at different times t after the start of the pulse of production from well BRX2Y on 23 march 2018, time t being measured from the start of production. Calculations using Eq. (15) assume 4 m^3 produced volume, calculated as Q 1.39 × 10^{-4} m^3 s^{-1} for Δt 8 hours, η 0.9 mPa s, h_B 1 m, k_B 900 mD, and D_B 10 m^2 s^{-1}. Dashed line indicates ΔP = -51 Pa for r = 8 km after t = 9 days.* ***(b)*** *Graph of Δx versus time following a step pressure reduction within the Newdigate fault. Calculations use Eq. (27) with ΔP_O = 51 Pa (cf. (a)), with K_D 27 GPa, and D_D again 1 m^2 s^{-1}, for Dinantian limestone. Dashed line indicates Δx = 0.057 μm, which arises after 0.95 hours or 57 minutes.* ***(c)*** *Graph of $\Delta\Phi$ for the patch of the Newdigate fault that slipped in the 1 April 2018 earthquake versus time for the variations in Δx depicted in (b). Calculations use Eq. (44), with the same parameters as for (b) plus b 21.5 mm, c 0.6, and E 75 GPa for the model asperity on the seismogenic patch of fault. Dashed line indicates $\Delta\Phi$ = 60 kPa, which arises after 0.95 hours or 57 minutes.* ***(d)*** *Graphs as for (a), except D_B is now 20 m^2 s^{-1}. Dashed line now indicates Δ_P = -146 Pa for r = 8 km after t = 9 days.* ***(e)*** *Graph as for (b), except D_B is now 20 m^2 s^{-1}. Dashed line now indicates Δx = 0.143 μm, which arises after 0.72 hours or 43 minutes.* ***(f)*** *Graph as for (c), except D_B is now 20 m^2 s^{-1}. Dashed line indicates $\Delta\Phi$ = 150 kPa, which arises after 0.72 hours or 43 minutes.*

pressure of ~3.4 MPa, roughly 3 MPa below hydrostatic, whereas the initial bottom-hole pressure in well HH1 was ~6.3 MPa, roughly hydrostatic. The pressure draw-down at Brockham evidently had no significant effect on the pressure at HH1. It might thus be inferred that the two wells are not hydraulically connected, in contrast with the arguments in the present study. To explore this issue, Eq. (8) is used to calculate the effect of twenty years of production at the steady rate of 6×10^{-5} m^3 s^{-1} to obtain the 36,900 m^3 produced (**Figure 15(c)**). Taking into

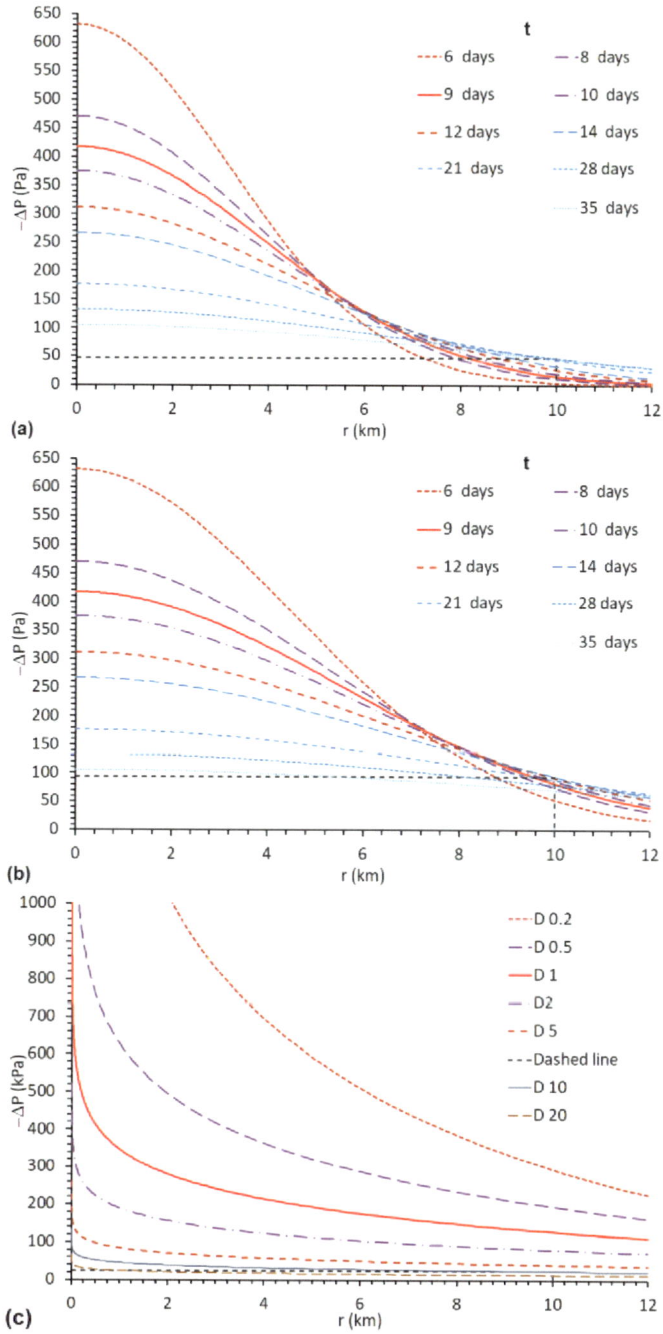

Figure 15.
Modelling of pressure interference of production from well BRX2Y on well HH1. ***(a)*** *Graphs of pressure variations -Δ_P in 'beef' adjoining well HH1 caused by the pulse of production from well BRX2Y on 23 March 2018, time t being measured from the start of production. Calculations using Eq. (15) assume 4 m^3 produced volume, calculated as Q 1.39 × 10^{-4} m^3 s^{-1} for Δt 8 hours, η 0.9 mPa s, h_B 1 m, k_B 900 mD, and D_B 10 m^2 s^{-1}. Horizontal dashed line indicates Δ_P = -47 Pa for r = 10 km after t = 35 days.* ***(b)*** *Graphs as for (a), except D_B is now 20 m^2 s^{-1}. Horizontal dashed line now indicates Δ_P = -93 Pa for r = 10 km after t = 35 days.* ***(c)*** *Graphs illustrating the pressure drawdown caused by 20 years of production from well BRX2Y. Calculations using Eq. (8) assume Q 6 × 10^{-5} m^3 s^{-1}, η 0.9 mPa s, and h_B 1 m, for variable k_B and D_B. For D_B = 10 m^2 s^{-1}, k_B = 900 mD; for other values of D_B, k_B is adjusted in proportion (cf. Eq. (5)). Horizontal dashed line now indicates ΔP = -23.7 kPa for r = 10 km with D_B = 10 m^2 s^{-1}. Using Eq. (8), with k_P = 20 mD, h_P = 2 m, r_w = 0.0889 m, and D_P = 1 m^2 s^{-1}, gives a predicted bottom-hole pressure decline at well BRX2Y of ~2.9 MPa, roughly as observed.*

account the dependence of D_B on permeability, it is evident that for $D_B \sim 10$–$20\ m^2\ s^{-1}$, as envisaged in the present study, the predicted pressure decline at HH1 was only a few tens of kilopascals, thus not significant in relation to the measured pressure. The large difference in bottom-hole pressure between the two wells in 2018 is thus not evidence against the proposed model.

The production test from well HH1 that started on 11 February lasted until late June 2019, some 140 days or 20 weeks, albeit with some intermittency (see supplement). Using Eq. (8), the resulting pressure decline is depicted in **Figure 16(a)** for $D_B = 10\ m^2\ s^{-1}$ and in **Figure 16(b)** for $D_B = 20\ m^2\ s^{-1}$. At r = 10 km, in the vicinity of well BRX2Y, the maximum pressure decline is ~40 kPa (**Figure 16(a)**) or ~ 60 kPa (**Figure 16(b)**). The production from well HH1 thus influenced bottom hole pressure in well BRX2Y by three orders of magnitude more than for the effect

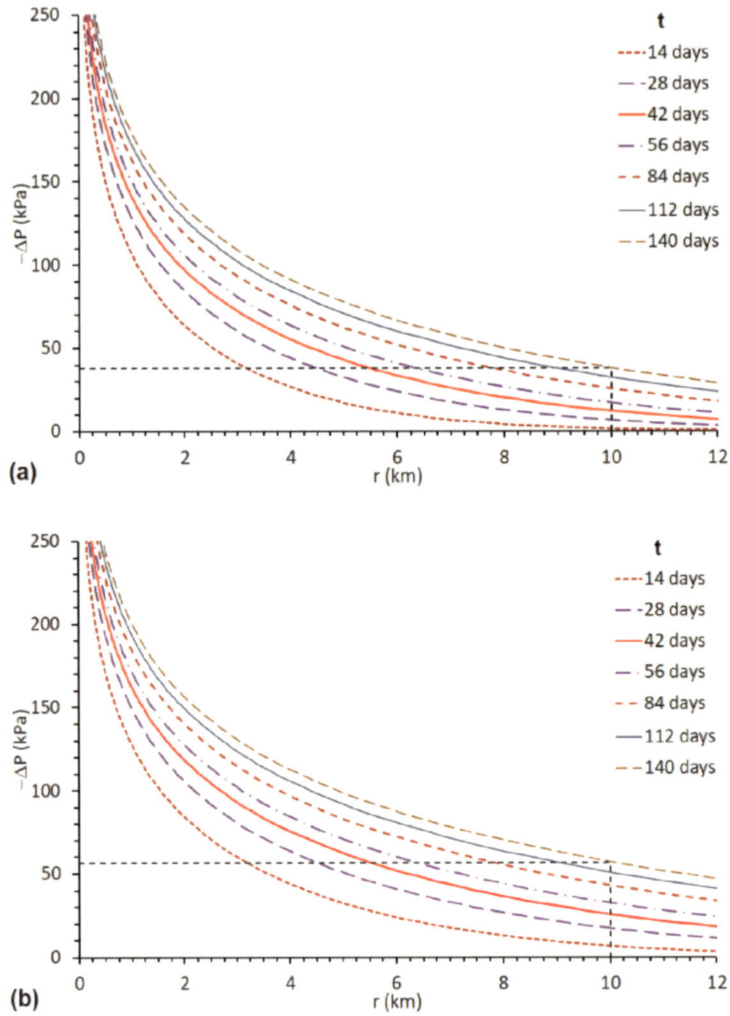

Figure 16.
Modelling of pressure interference of production from well HH1 on well BRX2Y. ***(a)*** *Graphs of pressure variations -Δ_P in 'beef' adjoining well BRX2Y caused by the phase of production from well HH1 starting in February 2019, time t being measured from the start of production. Calculations using Eq. (8) assume Q $4 \times 10^{-3}\ m^3\ s^{-1}$, η 0.9 mPa s, h_B 1 m, k_B 900 mD, and D_B $10\ m^2\ s^{-1}$. Horizontal dashed line indicates ΔP = -37.8 kPa for r = 10 km after t = 140 days.* ***(b)*** *Graphs as for (a), except D_B is now $20\ m^2\ s^{-1}$. Horizontal dashed line now indicates ΔP = -56.6 kPa for r = 10 km after t = 140 days.*

of well BRX2Y on well HH1. The OGA [6] made no mention of any pressure data for well BRX2Y that might be used to test this deduction.

5. Discussion

The proposed mechanism for the Newdigate seismicity depends on a pressure drop within the Dinantian limestone alongside the seismogenic strand of the Newdigate fault zone, as a result of depressurization of the water within this fault, caused by oil production from neighbouring wells (**Figure 5**). For the production from BRX2Y to have caused seismicity by this mechanism, the seismogenic fault must already have been extremely close (maybe within ~60 kPa; see above) to the Mohr-Coulomb failure condition. It can be inferred that the same mechanism, operating during the previous production from this well, contributed to creating this state of stress by progressively depressurizing the Dinantian limestone. To test this possibility, one may use Eq. (36), noting the 2.7 MPa depressurization of the reservoir over ~20 years and estimating from the previous analyses (e.g., **Figure 15 (d)**) that the resulting value of ΔP (and, thus, δP) within this limestone would be ~10 kPa. With B_E = 27 GPa, α = 0.46, H = 70 m, and ΔV = 36,900 m^3 to balance the production, depressurization of a ~ 6000 km^2 area would be indicated; if roughly equidimensional, this would have a radius of ~40 km. However, in reality, it is to be expected that such depressurization would be largely cancelled by recharge of water into the Dinantian limestone from other directions, which is not incorporated into the model. For example, if after this cumulative production, x_M were ~ 6 km and ΔP_O were ~ 1 kPa, then substituting Eq. (21) into Eq. (41), and taking b ~ 0.1 m (obtained for L = 1.5 km, the length of the seismogenic part of the Newdigate Fault, using Eq. (49)), $\Delta\Phi$ would be ~6 MPa. Notwithstanding the approximations made in the model, it is thus indeed plausible that the cumulative production at Brockham brought this fault to the condition for shear failure, assuming that it was already critically stressed before this production began.

In principle, testing of the proposed mechanism is possible, given the predicted vertical compaction in the Dinantian limestone (Eq. (35)). Such compaction will cause subsidence of the Earth's surface, and so is in principle observable using multiple techniques, including interferometric synthetic-aperture radar (InSAR) and repeated gravity and GPS measurements. A combined dataset of this type has been analysed for a region of southeast England, including the northern Weald Basin, by Aldiss et al. [129]. At the October 2018 workshop attention was also drawn to an InSAR-derived surface deformation map of Britain by GVL [130], spanning October 2015 to October 2017. However, the predicted subsidence, resulting from the two decades of production at Brockham, will be only a small fraction of 1 mm (~36,900 m^3 / ~200 $km^2 \approx$ 0.2 mm), even if none of the fluid withdrawal from this limestone were recharged. The Aldiss et al. [129] analysis revealed vertical crustal motions at ~1 mm a^{-1}, caused by processes such as extraction from or recharge of shallow groundwater reservoirs. Such rates make it impossible to resolve the much smaller effect expected from compaction of the Dinantian limestone at Newdigate.

Much has been made by participants in the OGA [5] workshop regarding the extent to which the Newdigate earthquake 'swarm' might fit the standard criteria identified by Davis and Frohlich [131] for establishing whether instances of seismicity are anthropogenic (e.g., [54]). UKOG [8] have argued that this set of criteria is inapplicable as they relate to seismicity caused by fluid injection, which is not the causal mechanism in this case. However, familiarity with the literature in this field (e.g., [132]) indicates that these criteria are widely used irrespective of the geomechanical cause of any particular anthropogenic earthquake. Verdon et al. [6]

proposed a different approach to assessing anthropogenic seismicity. This approach appears problematic, since it replaces the objective (yes / no) criteria recommended by Davis and Frohlich [131] with subjective numerical scores. The development of a conceptual geomechanical model for the Newdigate earthquakes supersedes the other Davis and Frohlich [131] criteria; nonetheless, an appraisal of this seismicity in terms of these criteria is included in the online supplement.

Although fluid injection is nowadays recognised as a widespread cause of induced seismicity, it is worth noting that reductions in fluid pressure caused by fluid extraction has been linked to seismicity for far longer. The first instance recognised is by Pratt and Johnson [133], for earthquakes accompanying oil production near Houston, Texas. Other case studies subsequently recognised include those by Calloi et al. [134], Kovach [135], Rothé and Lui [136], Simpson and Leith [137], Pennington et al. [138], Wetmiller [139], Grasso and Wittlinger [140], Doser et al. [141], Ottemöller et al. [142], Dahm et al. [143], and Hornbach et al. [144], whereas works discussing physical mechanisms for such seismicity, including compaction of reservoir rocks, include those by Yerkes and Castle [145], Simpson et al. [146], Segall [147], Segall and Fitzgerald [88], Ottemöller et al. [142], and Soltanzadeh and Hawkes [90]. Some of the above works (e.g., [138]) have recognised the significance of processes required in the present conceptual model (e.g., compaction of limestone and failure of asperities), and others (e.g., [148–150]) have recognised that highly permeable connections can cause seismicity at significant distance from the source of the causative change in fluid pressure. However, no previous case study known to the present author has proposed a geometry between the source of depressurization and the seismogenic fault that resembles the present conceptual model (**Figure 5**).

Hicks et al. [1] considered and rejected the possibility that the Newdigate seismicity was caused by compaction, as a result of depressurization caused by oil production, from one of the oil reservoirs in the area. They reached this conclusion on the basis of the strike-slip focal mechanisms (**Figure 1**), because in their view compaction would be expected to cause dip-slip earthquakes; they thus argued that these strike-slip earthquakes must be natural. They cited in support of this conclusion work on the Groningen seismicity in the Netherlands, where gas field depletion has caused many earthquakes, almost all with normal-faulting focal mechanisms (e.g., [151]). Compaction was initially thought to be the cause of the Groningen seismicity [152]. However, these events were later reinterpreted as caused by the combined effects of compaction and poro-elastic changes to the state of stress on faults [153], the normal faulting focal mechanisms thus reflecting the extensional stress state in the Netherlands. There is therefore no contradiction between this work and the occurrence at Newdigate of strike-slip earthquakes, the focal mechanism orientation again consistent with the local state of stress. Differences between the Newdigate and Groningen case studies concern, for the latter, the much larger scale (reservoir area $\sim$ 900 km^2; surface subsidence $\sim$30 cm; cumulative production $\sim 10^{10}$ m^3 at reservoir pressure; [154]), and the lower hydraulic diffusivity and elastic moduli (D 0.38 m^2 s^{-1} and E $\sim$ 8–25 GPa; [153, 155]). Regarding the geomechanics, the principal differences are that the analysis for Groningen has neglected consideration of fault asperities and has assumed that the elastic moduli of the fault wall rocks vary over time following the start of compaction, relaxing from short timescale to long timescale values (this effect being assumed to occur over a characteristic timescale, specified as 7.3 years for no clear reason other than to make the model work). In the present study, the time-dependence of the response develops as a result of the time required for poroelastic compaction in the Dinantian limestone to create fault opening Δx (Eq. (24)) comparable to the typical fault asperity height b. This seems a more physically realistic approach, avoiding any arbitrary timescale parameter and in keeping with modern ideas (noted above) that coseismic slip on faults is governed by interactions of asperities. Nonetheless, as is evident, the present analysis makes many simplifying assumptions, not least regarding the geometry of the fluid flow and its variations, for example: recharge of the Dinantian limestone is neglected; and smooth variations in the pressure of the fluid being drawn from the Newdigate Fault (as in **Figure 13(a)**) are approximated as step changes (as in **Figure 13(b)**). Furthermore, as Segall [87] noted, the choice of boundary conditions for analysis of the combined pressure and strain response in the Dinantian limestone is one end member of a range of possibilities. The present analysis is essentially a 'proof-of-concept', to demonstrate that it is plausible that oil production caused the Newdigate seismicity and to shed light on the combination of hydraulic properties that makes this possible; it does not, of course, prove that the production caused this seismicity.

The remainder of this discussion will concentrate on geomechanical issues. Following each of the Newdigate earthquakes, the spatially averaged shear stress on the patch of fault that slipped will reduce by a value equal to the coseismic stress drop $\Delta\sigma$, moving the state of stress away from the Coulomb failure condition. The decline in seismicity in late 2019 suggests that it had reached some form of limiting state; the conditions governing this decline will now be assessed. To determine the area of the patch of fault that slipped in each earthquake, seismic moment M_O was first calculated from M_W using Eq. (50). For most of the Newdigate earthquakes, M_W values are unavailable from Hicks et al. [1], only M_L has been determined. Nonetheless, Deichmann [156] has shown that M_L can serve as an equivalent proxy of earthquake 'size' as M_W, provided it is appropriately calibrated. Hicks et al. [1] used the Luckett et al. [157] M_L calibration, which in the UK has superseded the more familiar Ottemöller and Sergeant [158] version. No calibration against M_W of the Luckett et al. [157] M_L formula has been reported; nonetheless, for the Newdigate events for which both M_W and M_L have been determined, both these measures are in close agreement, so M_L is used as here a proxy for M_W. Next, the radius a of the equivalent circular seismic source is determined from M_o, using Eq. (51). The source area $A = \pi a^2$ and the mean slip $u = M_o / (\mu A)$ are then calculated. For an ideal circular earthquake source, $u = 16 (1 - \nu) a \delta\sigma / (3 (1 - \nu) \pi \mu)$ and the maximum slip u' is $8 (1 - \nu) a \delta\sigma / ((1 - \nu) \pi \mu)$ (e.g., [128]), so $u' = 3 u / 2$. Coseismic stress drop $\delta\sigma$ is set as 10 MPa (a plausible upper bound), Poisson's ratio ν is set as 0.25, and shear modulus $\mu \equiv B_R$ for $\nu = 0.25$ (Eq. (47)); as before $B_R = 50$ GPa, from Bell [119]. This task was carried out for the complete Hicks et al. [1] earthquake dataset, plus the additional events listed in **Table 4**. The cumulative seismic moment thus obtained was $\sim 1.5 \times 10^{14}$ N m, equivalent to a single event of $M_W \sim 3.4$, with cumulative area of fault rupture $\sim 2.1 \times 10^5$ m^2 and maximum coseismic slip in the largest earthquake $\sim$3.4 cm.

Taking 1.5 km as the length of the seismogenic fault, from **Figures 1** and 7, and 70 m as the thickness of the Dinantian limestone, the area of fault in this lithology is $\sim 10^5$ m^2. Calculated on this basis, the total area of coseismic ruptures exceeds the area of the fault, and would be even greater if lower $\delta\sigma$ were adopted. Thus, either patches of fault slipped more than once, or that the seismicity propagated into the overlying and/or underlying lithologies, although the compact hypocentre 'cloud' (**Figure 1**) indicates no clear propagation in any direction. Nonetheless, the calculations indicate that the eight largest earthquakes (with $M_W \geq 1.96$) have source diameters larger than the estimated 70 m thickness of the limestone; evidently, these events either ruptured outside this layer or ruptured patches of fault that are elongated horizontally. Assuming the latter explanation, and that the overall earthquake population was distributed to produce constant overall coseismic slip across the seismogenic fault, this amount is determined as was $\sim 1.5 \times 10^{14}$ N m /

(50 GPa × 1500 m × 70 m) or ~2.9 cm, roughly equal to the maximum slip in the largest individual earthquake (~3.4 cm for M_W 3.25) (the value would be ~3.3 cm, in better agreement, if the thickness of this lithology were taken as 60 m). It is thus possible that the earthquake swarm was indeed 'self-limiting', and that once the full extent of the seismogenic fault had slipped by this distance, the fault was effectively 'de-stressed' ($\delta\sigma$ = 10 MPa having reduced $\Delta\Phi$ by ~6 MPa) and the activity died out, consistent with its observed decline and near cessation in late 2019 (**Table 4**). This seismogenic patch of fault is bounded to the east by the eastern end of the Newdigate Fault, but has no obvious boundary to the west, although it is noted that at its limit the downthrow of the top Portland sandstone is ≤~60 m (**Figure 7**), comparable to the thickness of the Dinantian limestone; the state of stress must be different, for some reason yet to be resolved, farther west along this fault. Further analysis of this aspect is evidently warranted, given the possibility that significant seismicity might resume (two very small earthquakes having occurred in the spring of 2020; **Table 4**), as a result of pressure changes and fluid withdrawal arising from the planned increase in production at the Horse Hill site. A simple approach to mitigation, suggested by the present study, would be to ensure that oil production is balanced by reinjection to minimise the spatial extent of subsurface pressure changes. Suitably targeted reinjection could ensure that pressure decreases are confined to the Portland reservoir and do not propagate to the 'beef' and thence to the Newdigate Fault or other faults.

As already noted, the proposed physical mechanism, whereby a decrease in the fluid pressure within a fault can destabilise the fault, is the opposite of what might be termed the 'standard' effect, familiar from many studies of 'fracking': the unclamping that will occur when the fluid pressure within a fault increases and reduces the effective normal stress. The possibility of these two opposite responses to a fluid pressure change in a fault can be seen from inspection of Eq. (41). In the limit of D = 0 this adjusts to the conventional expression $\Delta\Phi = -c\ \Delta P_O$, whereby a pressure decrease by ΔP_O would cause fault 'clamping' and a pressure increase would be necessary for 'unclamping', but if D is large enough, a pressure decrease by ΔP_O can outweigh this effect, causing fault 'unclamping' and coseismic slip. The 'standard' effect of an increase in fluid pressure causing fault unclamping is to be expected if the fault is in impermeable rocks, where the fluid pressure only acts within the fault and not within the adjoining rock volume (e.g., [11, 159]). In general, for faults within permeable rocks, one can expect these two effects to counteract each other; whether the poroelastic effect will predominate or not will depend on the conditions in each case.

In this context, it is noteworthy that much of the seismicity associated with fluid injection in the USA occurs as a result of pressure increases in faults in impermeable basement rocks, rather than in the permeable rocks into which the injection takes place. A significant case study illustrating this issue is provided by the attempt to control seismicity in the Rangely, Colorado, oilfield by varying the reservoir pressure [160]. In this instance earthquakes occurred on patches of a fault where it transects both the reservoir and deeper crustal basement, their frequency of occurrence in both locations increasing with fluid pressure. The fault unclamping effect of increasing fluid pressure was the same in the reservoir rocks as in the impermeable basement rocks. In this case, evidently, the direct effect of fluid pressure on the Mohr-Coulomb failure condition for the patch of fault in the reservoir rocks outweighed the poroelastic effect (cf. Eq. (40)), as might be expected given the relatively low permeability (~1 mD; [160]) of the reservoir rocks, orders-of-magnitude smaller than is expected for Dinantian limestone (see above). A second case study highlighting the complexity of this issue was documented by Hornbach et al. [144] (see also, e.g., [126, 161]), at Azle near Fort Worth, Texas, where earthquake activity began in 2013 in a locality that had experienced both injection (of industrial wastewater) and production (of brine, oil and natural gas). The injection was initially suspected as the cause, on account of its very large volume, but Hornbach et al. [144] deduced that the pressure reduction caused by oil and gas production was the most important individual factor. In other instances, for example that discussed by Justinic et al. [162], authors have emphasised the proximity of hypocentres to injection points to highlight the anthropogenic cause, when many hypocentres are in fact rather deeper and indicate earthquakes within the underlying impermeable basement. Hincks et al. [163] have noted that fluid injection into faults or fractures in basement or near the contact with basement at the base of permeable sediments is statistically much more likely to result in seismicity than injection well above basement. Consideration of poroelasticity provides a natural explanation for this empirical observation.

6. Conclusions

The seismicity at Newdigate, Surrey, during 2018–2019, has been reassessed, amending aspects of the Hicks et al. [1] analysis. First-order correction of their seismic velocity model, which was too slow for the local stratigraphy, adjusts the hypocentres ~400 m deeper than previously thought, to depths of ~2400 m, placing them within the Palaeozoic 'basement' beneath the Weald Basin rather than within its Jurassic sedimentary sequence. These earthquakes involved mainly right-lateral slip on a steeply north dipping fault, part of the Newdigate fault zone (**Figure 2**).

Oil was produced during 2018–2019 in this vicinity from the Upper Portland Sandstone by the Brockham-X2Y and Horse Hill-1 wells. The correlation between phases of production from this reservoir and 'bursts' of earthquake activity (**Figure 4**) warrants consideration of potential geomechanical mechanisms. A conceptual model that can account for this causal connection is indicated schematically in **Figure 5**. It is thus suggested that the seismicity occurred within a thin (estimated ~70 m thick) layer of permeable Dinantian limestone, hydraulically connected to the Portland reservoir via permeable strands of the Newdigate fault zone and by the highly permeable calcite 'beef' fabric within the Portland sandstone. It is hypothesised that past oil production at Brockham depressurized the Portland reservoir around this well and drew groundwater from the Dinantian limestone, causing it to compact and 'unclamp' the seismogenic fault but not sufficiently to reach the Mohr-Coulomb failure criterion to initiate coseismic slip. The resumption of production at Brockham in March 2018 caused a negative pressure pulse to propagate through the hydraulic connection to the Dinantian limestone, which, it is suggested, reached the failure threshold, initiating the first 'burst' of Newdigate seismicity in April 2018. Likewise, negative pressure pulse following resumption of production from the Portland reservoir at Horse Hill in February 2019 initiated a subsequent 'burst' of seismicity. This mechanism requires hydraulic diffusivity ~10–20 $m^2\ s^{-1}$ in the calcite 'beef' and ~ 1 $m^2\ s^{-1}$ in the Dinantian limestone; it predicts unclamping of fault patches by many megapascals as a result of the Horse Hill production in February 2019 and by up to ~0.1 MPa as a result of the Brockham production in March 2018. At other times, the complexity of production patterns (e.g., from both BRX2Y and HH1 in summer 2018) and the absence of pressure data prevent any detailed conclusions being drawn, although the general correlation of seismicity with production from the Portland reservoir (**Figure 4**) is compelling. The proposed 'unclamping' effect requires consideration of the roughness of the seismogenic fault, determined by the height of its asperities and their response to compaction of the adjoining limestone. Such behaviour is particularly significant in this instance because of the high permeability of the Dinantian limestone; in impermeable rocks a reduction in pore pressure would cause fault clamping rather than unclamping. In principle this model is fully testable, but required data, notably the history of pressure variations in the oil wells, is not currently in the public domain. The recognition that this instance of seismicity is arguably caused by human activity, and the role of highly permeable hydraulic connections extending for many kilometres, has significant implications for regulation to mitigate the potential nuisance from future seismicity caused by oil production in the Weald Basin, and may also inform the understanding of anthropogenic seismicity in other settings.

Acknowledgements

Imagery and metadata for seismic lines C79-36, TWLD90-15 and TWLD90-21 were kindly provided by Malcolm Butler from the UK Onshore Geophysical Library / 'Beneath Britain' archive. This research did not receive any specific grant from any funding agency.

Declaration of interests

The author declares that he has no competing financial interests or personal relationships that could have appeared to influence the work reported in this paper.

A. Supplementary data

Provided online: https://eprints.gla.ac.uk/227098/

Author details

Rob Westaway
James Watt School of Engineering, University of Glasgow, Glasgow, UK

*Address all correspondence to: robert.westaway@gla.ac.uk

References

[1] Hicks, S., Verdon, J., Baptie, B., Luckett, R., Mildon, Z., Gernon, T., 2019. A shallow earthquake swarm close to hydrocarbon activities: discriminating between natural and induced causes for the 2018–19 Surrey, UK earthquake sequence, Seismological Research Letters, 90, 2095–2110.

[2] Hayhurst, R., 2018. Oil company says "We're not to blame for Surrey earthquake" – but local concerns remain. Drill or Drop? Magazine. https://drillordrop.com/2018/04/04/oil-company-says-were-not-to-blame-for-surrey-earthquake-but-local-concerns-remain/

[3] BBC, 2018. Seventh tremor strikes in Surrey 'quake swarm'. British Broadcasting Corporation, London. https://www.bbc.co.uk/news/uk-england-44727326

[4] Gilfillan, S., Haszeldine, S., McGuire, B., Selley, R., 2018. Surrey quake fears. Letter to the Editor, The Times, 6 August 2018.

[5] OGA, 2018. OGA Newdigate Seismicity Workshop –3 October 2018. Summary and conclusion. UK Oil & Gas Authority, Aberdeen. https://www.ogauthority.co.uk/media/5174/2018_11_23-newdigate-workshop-summary-finalv3.pdf

[6] Verdon, J.P., Baptie, B.J., Bommer, J. J., 2019. An improved framework for discriminating seismicity induced by industrial activities from natural earthquakes. Seismological Research Letters, 90, 1592–1611.

[7] Horse Hill, 2018. FACTS ABOUT HORSE HILL. What we ARE doing & what we ARE NOT doing. Horse Hill Developments Ltd., London. http://www.horsehilldevelopments.co.uk/ul/FACTS%20about%20HH%20Final%20Draft%20181018.pdf

[8] UKOG, 2019. Why earth tremors in Surrey should not be blamed on oil exploration. Statement from UK Oil & Gas Plc, in response to "unscientific" claims made by Dr Cavanagh, Dr Gilfillan and Professor Haszeldine. UK Oil & Gas Plc, London. https://www.ukogplc.com/ul/Technical%20Response%20to%20Edinburgh%20University%20120419.pdf

[9] Davies, R., Foulger, G., Bindley, A., Styles, P., 2013. Induced seismicity and hydraulic fracturing for the recovery of hydrocarbons. Marine and Petroleum Geology, 45, 171–185.

[10] Hitzman, M.W., Clarke, D.D., Detournay, E., Dieterich, J.H., Dillon, D. K., Green, S.J., Habiger, R.M., McGuire, R.K., Mitchell, J.K., Shemeta, J.E., Smith, J.L., 2013. Induced Seismicity Potential in Energy Technologies. The National Academies Press, Washington, DC, 263 pp. http://www.nap.edu/catalog/13355/induced-seismicity-potential-inenergy-technologies.

[11] Westaway, R., 2017. Integrating induced seismicity with rock mechanics: a conceptual model for the 2011 Preese Hall fracture development and induced seismicity. In: Rutter, E.H., Mecklenburgh, J., Taylor, K.G. (eds). Geomechanical and Petrophysical Properties of Mudrocks. Geological Society, London, Special Publications, 454, 327–359.

[12] Butler, M., Pullan, C.P., 1990. Tertiary structures and hydrocarbon entrapment in the Weald Basin of southern England. In: Hardman, R.F.P., Brooks, J. (eds), Tectonic Events Responsible for Britain's Oil and Gas Reserves. Geological Society, London, Special Publications, 55, 371–391.

[13] Pullan, C.P., Butler, M., 2018. Paleozoic gas potential in the Weald Basin of southern England. In: Monaghan, A.A., Underhill, J.R.,

Hewett, A.J., Marshall, J.E.A. (eds), Paleozoic Plays of NW Europe. Geological Society, London, Special Publications, 471, 333–363.

[14] Dines, H.G., Edmunds, F.H., 1933. The geology of the country around Reigate and Dorking: memoir for 1:63,360 geological map sheet 286 (England and Wales). H.M.S.O., London, 204 pp.

[15] Gallois, R.W., Worssam, B.C., 1993. The geology of the country around Horsham: memoir for 1:50,000 geological map sheet 302 (England and Wales). H.M.S.O., London, 130 pp.

[16] Trueman, S., 2003. The Humbly Grove, Herriard, Storrington, Singleton, Stockbridge, Goodworth, Horndean, Palmers Wood, Bletchingley and Albury Fields, Hampshire, Surrey and Sussex, UK Onshore. In: Gluyas, J., Hitchens, H. M. (eds)., United Kingdom Oil and Gas Fields Commemorative Millennium Volume. Geological Society, London, Memoir 20, pp. 929–941.

[17] DECC, 2013. The Hydrocarbon Prospectivity of Britain's Onshore Basins. Department of Energy and Climate Change, London, 93 pp. https://www.ogauthority.co.uk/media/1695/uk_onshore_2013.pdf

[18] Andrews, I.J., 2014. The Jurassic shales of the Weald Basin: geology and shale oil and shale gas resource estimation. British Geological Survey for Department of Energy and Climate Change, London, UK, 79 pp.

[19] Chadwick, R.A., 1986. Extension tectonics in the Wessex Basin, southern England. Journal of the Geological Society, London, 143, 465–488.

[20] Chadwick, R.A., Kenolty, N., Whittaker, A., 1983. Crustal structure beneath southern England from deep seismic reflection profiles. Journal of the Geological Society, London, 140, 893–911.

[21] Hawkes, P.W., Fraser, A.J., Einchcomb, C.C.G., 1998. The tectono-stratigraphic development and tectonic history of the Weald and Wessex Basins, Southern England. In: Underhill, J.R. (ed.), The Development, Evolution and Petroleum Geology of the Wessex Basin. Geological Society, London, Special Publications, 133, 33–69.

[22] Karner, G.D., Lake, S.D., Dewey, J. F., 1987. The thermal and mechanical development of the Wessex Basin, southern England. In: Coward, M.P., Dewey, J.F., Hancock, P.L. (eds), Continental Extensional Tectonics. Geological Society, London, Special Publications, 28, 517–536.

[23] Lake, S.D., Karner, G.D., 1987. The structure and evolution of the Wessex Basin, southern England: an example of inversion tectonics. Tectonophysics, 137, 347–356, 358–378.

[24] Stoneley, R., 1982. The structural development of the Wessex Basin. Journal of the Geological Society, London, 139, 543–552.

[25] Wigley, P., 2015. Exploration in the UK Weald Basin: Déjà vu. Search and Discovery, 70182, 8 pp. http://www.searchanddiscovery.com/pdfz/documents/2015/70182wigley/ndx_wigley.pdf.html

[26] Xodus, 2018. 2018 Competent Person's Report to UK Oil & Gas Investments PLC. Xodus Group, London, 112 pp. https://www.ukogplc.com/ul/UKOG%202018%20CPR%20060618.pdf

[27] Angus, 2018. Brockham Portland & Kimmeridge Reservoirs. Addendum to the Field Development Plan. Angus Energy, Plc., London, 32 pp. https://www.whatdotheyknow.com/request/523935/response/1277768/attach/4/brockfdpadd%20Redacted.pdf?cookie_passthrough=1 (partly redacted version

released under a Freedom of Information request)

[28] Europa, 2004. Application for Production Licence. Appendix B – Geotechnical Information. Europa Oil & Gas Ltd., London, 38 pp. https://ukogl.org.uk/map/php/pdf.php?subfolder=industry_reports&filename=41436.pdf

[29] Angus, 2018. OGA – Surrey earthquakes. Angus Energy, Plc., London. https://www.ogauthority.co.uk/media/5160/7c-angus-maps-for-oga-meeting.pdf

[30] Busby, J.P., Smith, N.J.P., 2001. The nature of the Variscan basement in southeast England: evidence from integrated potential field modelling. Geological Magazine, 138, 669–685.

[31] Webster, T., 1826. Observations on the Purbeck and Portland Beds. Transactions of the Geological Society, London, Series 2, 2, 37–44.

[32] Cobbold, P.R., Rodrigues, N., 2007. Seepage forces, important factors in the formation of horizontal hydraulic fractures and bedding-parallel fibrous veins ('beef' and 'cone-in-cone'). Geofluids, 7, 313–322.

[33] Zanella, A., Cobbold, P.R., Boassen, T., 2015. Natural hydraulic fractures in the Wessex Basin, SW England: widespread distribution, composition and history. Marine and Petroleum Geology, 68, 438–448.

[34] Buckland, W., De la Beche, H.T., 1835. On the geology of the neighbourhood of Weymouth and the adjacent parts of the coast of Dorset. Transactions of the Geological Society, London, Series 2, 4, 1–46.

[35] Cobbold, P.R., Zanella, A., Rodrigues, N., Løseth, H., 2013. Bedding-parallel fibrous veins (beef and cone-in-cone): Worldwide occurrence and possible significance in terms of fluid overpressure, hydrocarbon generation and mineralization. Marine and Petroleum Geology, 43, 1–20.

[36] Meng QingFeng, Hooker, J., Cartwright, J., 2017. Early overpressuring in organic-rich shales during burial: evidence from fibrous calcite veins in the Lower Jurassic Shales-with-Beef Member in the Wessex Basin, UK. Journal of the Geological Society, London, 174, 869–882.

[37] Tarney, J., Schreiber, B.C., 1977. Cone-in-cone and beef-in-shale textures from DSDP Site 330, Falkland Plateau, South Atlantic. Deep Sea Drilling Project Initial Reports, 36, 865–870.

[38] Al Duhailan, M.A., Sonnenberg, S. A., Longman, M., 2015. Analyzing beef fractures: Genesis and relationship with organic-rich shale facies. SPE Unconventional Resources Technology Conference, 20-22 July 2015, San Antonio, Texas, paper URTEC-2151959-MS, doi: 10.15530/URTEC-2015-2151959

[39] Lash, G.G., Engelder, T., 2005. An analysis of horizontal microcracking during catagenesis: Example from the Catskill delta complex. AAPG Bulletin, 89, 1433–1449.

[40] Parnell, J., Honghan, C., Middleton, D., Haggan, T., Carey, P., 2000. Significance of fibrous mineral veins in hydrocarbon migration: fluid inclusion studies. Journal of Geochemical Exploration, 69, 623–627.

[41] Zhang Bo, Yin CongYuan, Gu ZhiDong, Zhang JinJiang, Yan ShuYu, Wang Yang, 2015. New indicators from bedding-parallel beef veins for the fault valve mechanism. Science China: Earth Sciences, 58, 1320–1336.

[42] Maher Jr, H.D., Ogata, K., Braathen, A., 2017. Cone-in-cone and beef mineralization associated with Triassic growth basin faulting and shallow shale

diagenesis, Edgeøya, Svalbard. Geological Magazine, 154, 201–216.

[43] Al Duhailan, M.A., Sonnenberg, S. A., 2014. The curious case of hydrocarbon-expulsion fractures: Genesis and impact on the Bakken Shales. Search and Discovery, 80398, 30 pp.

[44] GeoSierra, 2019. Newdigate Seismicity and Link to Horse Hill HH-1 Well Activities. GeoSierra LLC, Norcross, Georgia. http://www.geosierra.com/files/132555955.pdf

[45] Hesselbo, S.P., Jenkyns, H.C.A., 1995. A comparison of the Hettangian to Bajocian successions of Dorset and Yorkshire. In :Taylor, P.D. (ed.), Field geology of the British Jurassic. Geological Society, London, pp. 105–150.

[46] Lang, W.D., 1914. The geology of Charmouth cliffs, beach and fore-shore. Proceedings of the Geologists' Association, 25, 293–360.

[47] Lang, W.D., Spath, L.F., Richardson, W.A., 1923. Shales-with-'Beef', a sequence in the Lower Lias of the Dorset coast. Quarterly Journal of the Geological Society, London, 79, 47–99.

[48] Howett, F., 1964. Stratigraphy and structure of the Purbeck inliers of Sussex (England). Quarterly Journal of the Geological Society, London, 120, 77–113.

[49] Bisdom, K., Baud, E., Estrada, S., Sanz-Perl, Y., Gauthier, B., Bertotti, G., 2016. Coupled stress-fluid pressure modelling of stimulated rock volume in shale - impact of natural fractures and beef. 78th EAGE Conference and Exhibition, Vienna, Austria, 30 May - 2 June 2016. doi: 10.3997/2214-4609.201601164

[50] Carey, J.W., Lei Zhou, Rougier, E., Mori, H., Viswanathan, H., 2015. Fracture-permeability behavior of shale. Journal of Unconventional Oil and Gas Resources, 11, 27–43.

[51] Bond, C.E., Gibbs, A., Shipton, Z.K., Jones, S., 2007. What do you think this is? "Conceptual uncertainty" in geoscience interpretation. GSA Today, 17 (11), 4–10.

[52] Cordier, J.P., 1985. Calculation of Interval Velocities. In: Velocities in Reflection Seismology. Seismology and Exploration Geophysics, vol 3. Springer, Dordrecht, the Netherlands, pp. 101–114. https://doi.org/10.1007/978-94-017-3641-1_10

[53] Dix, C.H., 1955. Seismic velocities from surface measurements. Geophysics, 20 (1), 68–86.

[54] Baptie, B., Luckett, R., 2018. The Newdigate earthquake sequence, 2018. British Geological Survey Internal Report OR/18/059, 20 pp. https://earthquakes.bgs.ac.uk/research/NewdigateEarthquakesReport.pdf

[55] Waldhauser, F., Ellsworth, W.L., 2000. A double-difference earthquake location algorithm: method and application to the northern Hayward Fault, California. Bulletin of the Seismological Society of America, 90, 1353–1368.

[56] Kingdon, A., Fellgett, M.W., Williams, J.D.O., 2016. Use of borehole imaging to improve understanding of the *in-situ* stress orientation of Central and Northern England and its implications for unconventional hydrocarbon resources. Marine and Petroleum Geology, 73, 1–20.

[57] Fellgett, M.W., Kingdon, A., Williams, J.D.O., Gent, C.M.A., 2017. State of stress across UK regions. British Geological Survey GeoAnalytics and Modelling Directorate Open Report OR/17/048, 60 pp. http://nora.nerc.ac.uk/id/eprint/517414/1/OR17048.pdf

[58] Evans, C.J., Brereton, N.R., 1990. *In situ* crustal stress in the United Kingdom from borehole breakouts. In: Hurst, A., Lovell, M.A., Morton, A.C. (eds), Geological Applications of Wireline Logs. Geological Society, London, Special Publications, 48, 327–338.

[59] McKenzie, D.P., 1969. The relationship between fault plane solutions for earthquakes and the directions of the principal stresses. Bulletin of the Seismological Society of America, 59, 591–601.

[60] Hainzl, S., 2004. Seismicity patterns of earthquake swarms due to fluid intrusion and stress triggering. Geophysical Journal International, 159, 1090–1096.

[61] Dake, L.P., 1998. Fundamentals of Reservoir Engineering. Developments in Petroleum Science series, volume 8, 17th edition. Elsevier, London, 498 pp.

[62] Guo BoYun, Sun Kai, Ghalambor, A., eds., 2008. Well Productivity Handbook. Elsevier, London, 334 pp.

[63] Horse Hill, 2018. Horse Hill-1 rig-less intervention and well testing programme. Report HHDL-HH1-RIWTP-R0. Horse Hill Developments Ltd., London, 81 pp. https://brockhamoilwell.files.wordpress.com/2019/04/disclosure-201808357-2.pdf

[64] Townend, J., Zoback, M.D., 2000. How faulting keeps the crust strong. Geology, 28, 399–402.

[65] Zoback, M.L., Zoback, M.D., 2007. Lithosphere stress and deformation. Treatise on Geophysics, 6, 253–273.

[66] Bense, V.F., Gleeson, T., Loveless, S.E., Bour, O., Scibek, J., 2013. Fault zone hydrogeology. Earth-Science Reviews, 127, 171–192.

[67] Caine, J.S., Evans, J.P., Forster, C.B., 1996. Fault zone architecture and permeability structure. Geology, 24, 1025–1028.

[68] Evans, J.P., Forster, C.B., Goddard, J.V., 1997. Permeability of fault-related rocks, and implications for hydraulic structure of fault zones. Journal of Structural Geology, 19, 1393–1404.

[69] Haines, T., Michie, E.A.H, Neilson, J.E., Healy, D., 2016. Permeability evolution across carbonate hosted normal fault zones. Marine and Petroleum Geology, 72, 62–82.

[70] Lunn, R.J., Shipton, Z.K., Bright, A. M., 2008. How can we improve estimates of bulk fault zone hydraulic properties? In: Wibberley, C.A.J., Kurtz, W., Imber, J., Holdsworth, R.E., Collettini, C. (eds), The Internal Structure of Fault Zones: Implications for Mechanical and Fluid-Flow Properties. Geological Society, London, Special Publications, 299, 231–237.

[71] Agosta, F., Prasad, M., Aydin, A., 2007. Physical properties of carbonate fault rocks, Fucino Basin (central Italy): implications for fault seal in platform carbonates. Geofluids, 7, 19–32.

[72] Barton, C.A., Zoback, M.D., Moos, D., 1995. Fluid flow along potentially active faults in crystalline rock. Geology, 23, 683–686.

[73] Westwood, R.F., Toon, S.M., Styles, P., Cassidy, N.J., 2017. Horizontal respect distance for hydraulic fracturing in the vicinity of existing faults in deep geological reservoirs: a review and modelling study. Geomechanics and Geophysics for Geo-Energy and Geo-Resources, 3, 379–391.

[74] Wilson, M.P., Worrall, F., Davies, R.J., Almond, S., 2018. Fracking: How far from faults? Geomechanics and Geophysics for Geo-Energy and Geo-Resources, 4, 193–199.

[75] Zimmerman, R.W., 2018. Fluid Flow in Porous Media. World Scientific, Singapore, 205 pp.

[76] Gurevich, B., Schoenberg, M., 1999. Interface conditions for Biot's equations of poroelasticity. Journal of the Acoustic Society of America, 105, 2585–2589. doi: 10.1121/1.426874.

[77] Detournay, E., Cheng, A.H.-D., 1993. Fundamentals of poroelasticity. Chapter 5. In: Fairhurst, C. (ed.), Comprehensive Rock Engineering: Principles, Practice and Projects, Vol. II, Analysis and Design Method. Pergamon Press, Oxford, pp. 113-171.

[78] Costain, J.K., 2017. Groundwater recharge as the trigger of naturally occurring intraplate earthquakes. In: Landgraf, A., Kübler, S., Hintersberger, E., Stein, S. (eds), Seismicity, Fault Rupture and Earthquake Hazards in Slowly Deforming Regions. Geological Society, London, Special Publications, 432, 91–118.

[79] Theis, C.V., 1935, The relationship between the lowering of the piezometric surface and the rate and duration of discharge of a well using ground-water storage. Transactions American Geophysical Union, 16, 519–524.

[80] Turuntaev, S.B., Riga, V.Y., 2017. Non-linear effects of pore pressure increase on seismic event generation in a multi-degree-of-freedom rate-and-state model of tectonic fault sliding. Nonlinear Processes in Geophysics, 24, 215–225.

[81] Masina, E., 2019. A review on the Exponential-Integral special function and other strictly related special functions. 37 pp. https://arxiv.org/pdf/1907.12373.pdf

[82] Schurman, G., 2017. The Exponential Integral. Part I - Derivation and Solution. http://www.appliedbusinesseconomics.com/files/gvsexpint01.pdf

[83] Harris, F.E., 1957. Tables of the Exponential Integral Ei(x). Mathematical Tables and Other Aids to Computation 11, 9–16. http://ftp.math.utah.edu/pub/tex/bib/toc/mathcomp1950.html#11(57):January:1957

[84] Stegun, I.A., Zucker, R., 1976. Automatic computing methods for special functions. Part III. The Sine, Cosine, Exponential Integrals, and Related Functions. Journal of Research of the National Bureau of Standards-B. Mathematical Sciences, 80B (2), 291–311. https://pdfs.semanticscholar.org/a1a6/7bdae39125c2369577d7afefff90ba91940b.pdf

[85] Barry, D.A., Parlange, J.-Y., Li, L., 2000. Approximation for the exponential integral (Theis well function). Journal of Hydrology, 227, 287–291.

[86] Engelder, T., Fischer, M.P., 1994. Influence of poroelastic behavior on the magnitude of minimum horizontal stress, S_h, in overpressured parts of sedimentary basins. Geology, 22, 949–952.

[87] Segall, P., 1992. Induced stresses due to fluid extraction from axisymmetric reservoirs. Pure and Applied Geophysics, 139, 535–560.

[88] Segall, P., Fitzgerald, S.D., 1998. A note on induced stress changes in hydrocarbon and geothermal reservoirs. Tectonophysics, 289, 117–128.

[89] Yehya, A., Yang, Z., Rice, J.R., 2018. Effect of fault architecture and permeability evolution on response to fluid injection. Journal of Geophysical Research: Solid Earth, 123, 9982–9997.

[90] Soltanzadeh, H., C.D. Hawkes, 2008. Semi-analytical models for stress change and fault reactivation induced by reservoir production and injection. Journal of Petroleum Science and Engineering, 60, 71–85. doi: 10.1016/j.petrol.2007.05.006.

[91] Van Wees, V.J., Buijze, L., van Thienen-Visser, K., Nepveu, M., Wassing, B., Orlic, B., Fokker, P., 2014. Geomechanics response and induced seismicity during gas field depletion in the Netherlands. Geothermics, 52, 206–219.

[92] Abramowitz, M., Stegun, I.A., 1972. Handbook of Mathematical Functions with Formulas, Graphs, and Mathematical Tables, 10th ed. Dover, New York, 1040 pp.

[93] Weisstein, E.W., 2019. Erfc. From MathWorld - A Wolfram Web Resource. http://mathworld.wolfram.com/Erfc.html

[94] Hettema, M., Papamichos, E., Schutjens, P., 2002. Subsidence delay: Field observations and analysis. Oil & Gas Science and Technology, 57, 443–458.

[95] Ellsworth, W.L., 2013. Injection-induced earthquakes. Science, 341, 1225942, 7 pp. doi: 10.1126/science.1225942.

[96] Keranen, K.M., Weingarten, M., Abers, G.A., Bekins, B.A., Ge SheMin, 2014. Sharp increase in central Oklahoma seismicity since 2008 induced by massive wastewater injection. Science, 345, 448–451.

[97] Walsh, F.R., Zoback, M.D., 2015. Oklahoma's recent earthquakes and saltwater disposal. Science Advances, 1, e1500195, 9 pp., doi: 10.1126/sciadv.1500195.

[98] Weingarten, M., Ge SheMin, Godt, J.W., Bekins, B.A., Rubinstein, J.L., 2015. High-rate injection is associated with the increase in U.S. mid-continent seismicity. Science, 348, 1336–1340.

[99] Reiter, M., 1999. Stress analyses of a simple fault asperity. In: Vail Rocks 1999, The 37th U.S. Symposium on Rock Mechanics, 7-9 June 1999, Vail, Colorado. American Rock Mechanics Association paper ARMA-99-0391, 8 pp.

[100] Archard, J.F., 1957. Elastic deformation and the laws of friction. Proceedings of the Royal Society of London, Series A, Mathematical and Physical Sciences, 243, 190–205.

[101] Mitchell, E.K., Fialko, Y., Brown, K.M., 2013. Temperature dependence of frictional healing of Westerly granite: experimental observations and numerical simulations. Geochemistry, Geophysics, Geosystems, 14, 567–582.

[102] Brown, S.R., Scholz, C.H., 1985. Broad bandwidth study of the topography of natural rock surfaces. Journal of Geophysical Research, 90, 12,575-12,582.

[103] Harbord, C.W.A., Nielsen, S.B., De Paola, N., Holdsworth, R.E., 2017. Earthquake nucleation on rough faults. Geology, 45, 931–934.

[104] Selvadurai, P.A., Glaser, S.D., 2017. Asperity generation and its relationship to seismicity on a planar fault: a laboratory simulation. Geophysical Journal International, 208, 1009–1025.

[105] Acosta, M., Passelègue, F.X., Schubnel, A., Violay, M., 2018. Dynamic weakening during earthquakes controlled by fluid thermodynamics. Nature Communications, 9, 3074, 9 pp., doi: 10.1038/s41467-018-05603-9

[106] McDermott, R.G., Ault, A.K., Evans, J.P., Reiners, P.W., 2017. Thermochronometric and textural evidence for seismicity via asperity flash heating on exhumed hematite fault mirrors, Wasatch fault zone, UT, USA. Earth and Planetary Science Letters, 471, 85–93.

[107] Brodsky, E.E., Kirkpatrick, J.D., Candela, T., 2016. Constraints from

fault roughness on the scale-dependent strength of rocks. Geology, 44, 19–22.

[108] Chen XiaoFeng, Carpenter, B.M., Reches, Z., 2020. Asperity failure control of stick–slip along brittle faults. Pure and Applied Geophysics, 177, 3225–3242.

[109] Lee, M.W., 2005. Proposed moduli of dry rock and their application to predicting elastic velocities of sandstones. Scientific Investigations Report 2005–5119. U.S. Geological Survey, Reston, Virginia, 18 pp. https://pubs.usgs.gov/sir/2005/5119/pdf/SIR-2005-5119.pdf

[110] Engineering ToolBox, 2004. Bulk Modulus and Fluid Elasticity. https://www.engineeringtoolbox.com/bulk-modulus-elasticity-d_585.html

[111] Busby, J., Kingdon, A., Williams, J., 2011. The measured shallow temperature field in Britain. Quarterly Journal of Engineering Geology and Hydrogeology, 44, 373–387.

[112] Engineering ToolBox, 2004. Water - Dynamic and Kinematic Viscosity. https://www.engineeringtoolbox.com/water-dynamic-kinematic-viscosity-d_596.html

[113] Wang QiQi, 2016. Characterization of bedding-parallel fractures in shale - Morphology, size distribution and spatial organization. Master of Science thesis, The University of Texas at Austin, 324 pp.

[114] Meng QingFeng, Hooker, J., Cartwright, J., 2018. Displacive widening of calcite veins in shale: Insights into the force of crystallization. Journal of Sedimentary Research, 88, 327–343.

[115] Zimmerman, R.W., Bodvarsson, G. S., 1996. Hydraulic conductivity of rock fractures. Transport in Porous Media, 23, 1–30.

[116] Rodrigues, N., Cobbold, P.R., Loseth, H., Ruffet, G., 2009. Widespread bedding-parallel veins of fibrous calcite ('beef') in a mature source rock (Vaca Muerta Fm, Neuquén Basin, Argentina): evidence for overpressure and horizontal compression. Journal of the Geological Society, London, 166, 695–709.

[117] Sosa Massaro, A., Espinoza, D.N., Frydman, M., Barredo, S., Cuervo, S, 2017. Analyzing a suitable elastic geomechanical model for Vaca Muerta Formation. Journal of South American Earth Sciences, 79, 472–488.

[118] Newson, M.D., 1973. The Carboniferous Limestone of the UK as an aquifer rock. The Geographical Journal, 139 (2), 294–305.

[119] Bell, F.G., 1981. A survey of the physical properties of some carbonate rocks. Bulletin of the International Association of Engineering Geology, 24, 105–110.

[120] Hart, D.J., 2000. Laboratory measurements of poroelastic constants and flow parameters and some associated phenomena. Ph.D. thesis, The University of Wisconsin – Madison, 130 pp.

[121] Hasanov, A.K., 2014. Reservoir transport and poroelastic properties from oscillating pore pressure experiments. Master of Science thesis, Colorado School of Mines, 231 pp.

[122] Hassanzadegan, A., Guerizec, R., Reinsch, T., Blöcher, G., Zimmermann, G., Milsch, H., 2016. Static and dynamic moduli of Malm carbonate: A poroelastic correlation. Pure and Applied Geophysics, 173, 2841—2855.

[123] Lewis, M.A., Cheney, C.S., Ó Dochartaigh, B.É., 2006. Guide to

Permeability Indices. British Geological Survey, Information Products Programme, Open Report CR/06/160N. British Geological Survey, Keyworth, Nottingham, 29 pp.

[124] Shepley, M.G., 2007. Analysis of flows from a large Carboniferous Limestone drainage adit, Derbyshire, England. Quarterly Journal of Engineering Geology and Hydrology, 40, 123–135.

[125] Hornbach, M.J., Jones, M., Scales, M., DeShon, H.R., Magnani, M.B., Frohlich, C., Stump, B., Hayward, C., Layton, M., 2016. Ellenburger wastewater injection and seismicity in North Texas. Physics of the Earth and Planetary Interiors, 261, 54–68.

[126] Zhang YiPeng, Person, M., Rupp, J., Ellett, K., Celia, M.A., Gable, C.W., Bowen, B., Evans, J., Bandilla, K., Mozley, P., Dewers, T., Elliot, T., 2013. Hydrogeologic controls on induced seismicity in crystalline basement rocks due to fluid injection into basal reservoirs. Groundwater, 51, 525–538.

[127] Hanks, T.C., Kanamori, H., 1979. A moment magnitude scale. Journal of Geophysical Research, 84, 2348–2350.

[128] Westaway, R., Younger, P.L., 2014. Quantification of potential macroseismic effects of the induced seismicity that might result from hydraulic fracturing for shale gas exploitation in the UK. Quarterly Journal of Engineering Geology and Hydrogeology, 47, 333–350.

[129] Aldiss, D., Burke, H., Chacksfield, B., Bingley, R., Teferle, N., Williams, S., Blackman, D., Burren, R., Press, N., 2014. Geological interpretation of current subsidence and uplift in the London area, UK, as shown by high precision satellite-based surveying. Proceedings of the Geologists' Association, 125, 1–13.

[130] GVL, 2018. United Kingdom Relative Deformation Map. Geomatic Ventures Limited, Nottingham. https://mangomap.com/geomatic-ventures-limited/maps/72883/united-kingdom-relative-deformation-map?preview=true#

[131] Davis, S.D., Frohlich, C., 1993. Did (or will) fluid injection cause earthquakes? - Criteria for a rational assessment. Seismological Research Letters, 64 (3–4), 207–224.

[132] Foulger, G.R., Wilson, M.P., Gluyas, J.G., Julian, B.R., Davies, R.J., 2018. Global review of human-induced earthquakes. Earth-Science Reviews, 178, 438–514.

[133] Pratt, W.E., Johnson, D.W., 1926. Local subsidence of the Goose Creek oil field. Journal of Geology, 34, 577–590.

[134] Calloi, P., DePanfilis, M., DeFilippo, D., Marcelli, L., Spadea, M. C., 1956. Terrimoti della Val Padana del 15–16 Maggio 1951. Annali di Geofisica, 9, 63–105 (with summary in English).

[135] Kovach, R.L., 1974. Source mechanisms for Wilmington oil field, California, subsidence earthquakes. Bulletin of the Seismological Society of America, 64, 699–711.

[136] Rothé, G.H., Lui, C.Y., 1983. Possibility of induced seismicity in the vicinity of the sleepy hollow oil field, southwestern Nebraska. Bulletin of the Seismological Society of America, 73, 1357–1367.

[137] Simpson, D.W., Leith, W., 1985. The 1976 and 1984 Gazli, USSR, earthquakes were induced? Bulletin of the Seismological Society of America, 75, 1465–1468.

[138] Pennington, W.D., Davis, S.D., Carlson, S.M., Dupree, J., Ewing, T.E., 1986. The evolution of seismic barriers and asperities caused by the depressuring of fault planes in oil and

gas fields of South Texas. Bulletin of the Seismological Society of America, 76, 939–948.

[139] Wetmiller, R.J., 1986. Earthquakes near Rocky Mountain House, Alberta, and their relationship to gas production facilities. Canadian Journal of Earth Science, 23, 172–181.

[140] Grasso, J.R., Wittlinger, G., 1990. Ten years of seismic monitoring over a gas field area. Bulletin of the Seismological Society of America, 80, 450–474.

[141] Doser, D.I., Baker, M.R., Mason, D. B., 1991. Seismicity in the War-Wink gas field, Delaware Basin, West Texas, and its relationship to petroleum. Bulletin of the Seismological Society of America, 81, 971–986.

[142] Ottemöller, L., Nielsen, H., Atakan, K., Braunmiller, J., Havskov, J., 2005. The 7 May 2001 induced seismic event in the Ekofisk oil field, North Sea. Journal of Geophysical Research, 110, B10301, 15 pp., doi: 10.1029/2004jb003374.

[143] Dahm, T., Cesca, S., Hainzl, S., Braun, T., Krüger, F., 2015. Discrimination between induced, triggered, and natural earthquakes close to hydrocarbon reservoirs: A probabilistic approach based on the modeling of depletion-induced stress changes and seismological source parameters. Journal of Geophysical Research, 120, 2491–2509.

[144] Hornbach, M.J., DeShon, H.R., Ellsworth, W.L., Stump, B.W., Hayward, C., Frohlich, C., Oldham, H. R., Olson, J.E., Magnani, M.B., Brokaw, C., Luetgert, J.H., 2015. Causal factors for seismicity near Azle, Texas. Nature Communications, 6, 7728, 11 pp., doi: 10.1038/ncomms7728

[145] Yerkes, R.F., Castle, R.D., 1976. Seismicity and faulting attributed to fluid extraction. Engineering Geology, 10, 151–167.

[146] Simpson, D.W., Leith, W.S., Scholz, C.H., 1988. Two types of reservoir-induced seismicity. Bulletin of the Seismological Society of America, 78, 2025–2040.

[147] Segall, P., 1989. Earthquakes triggered by fluid extraction. Geology, 17, 942–946.

[148] Holland, A.A., 2013. Earthquakes triggered by hydraulic fracturing in south-central Oklahoma. Bulletin of the Seismological Society of America, 103, 1784–1792.

[149] Igonin, N., Verdon, J.P., Kendall, J-M., Eaton, D.W., 2019. The importance of pre-existing fracture networks for fault reactivation during hydraulic fracturing. Earth and Space Science Open Archive, 10500976, 27 pp. doi: 10.1002/essoar.10500976.1. https://www.essoar.org/doi/pdf/10.1002/essoar.10500976.1

[150] Schultz, R., Stern, V., Novakovic, M., Atkinson, G., Gu, Y.J., 2015. Hydraulic fracturing and the Crooked Lake sequences: Insights gleaned from regional seismic networks. Geophysical Research Letters, 42, 2750–2758.

[151] Willacy, C., van Dedem, E., Minisini, S., Li JunLun, Blokland, J.-W., Das, I., Droujinine, A., 2019. Full-waveform event location and moment tensor inversion for induced seismicity. Geophysics, 84 (2), KS39–KS57. doi: 10.1190/GEO2018-0212.1

[152] Bourne, S.J., Oates, S.J., van Elk, J., Doornhof, D., 2014. A seismological model for earthquakes induced by fluid extraction from a subsurface reservoir. Journal of Geophysical Research: Solid Earth, 119, 8991–9015. doi: 10.1002/2014JB011663.

[153] van Thienen-Visser, K., Sijacic, D., van Wees, J.-D., Kraaijpoel, D., Roholl,

J., 2016. Groningen field 2013 to present: Gas production and induced seismicity. Report R10425. TNO, Utrecht, the Netherlands, 92 pp. https://www.nlog.nl/sites/default/files/final_tno%20report_2016_r10425.pdf

[154] van Thienen-Visser, K., Breunese, J., 2015. Induced seismicity of the Groningen gas field: History and recent developments. The Leading Edge, 34 (6), 664–671.

[155] van Thienen-Visser, K., Nepveu, M., van Kempen, B., Kortekaas, M., Hettelaar, J., Peters, L., van Gessel, S., Breunese, J., 2014, Recent developments of the Groningen field in 2014 and, specifically, the southwest periphery of the field. Report R11703. TNO, Utrecht, the Netherlands, 96 pp. https://www.nlog.nl/sites/default/files/final%20tno%20report%20ekl.pdf

[156] Deichmann, N., 2006. Local magnitude, a moment revisited. Bulletin of the Seismological Society of America, 96, 1267–1277.

[157] Luckett, R., Ottemöller, L., Butcher, A., Baptie, B., 2019. Extending local magnitude M_L to short distances. Geophysical Journal International, 216, 1145–1156.

[158] Ottemöller, L., Sargeant, S., 2013. A local magnitude scale M_L for the United Kingdom. Bulletin of the Seismological Society of America, 103, 2884–2893.

[159] Hackston, A., Rutter, E., 2016. The Mohr–Coulomb criterion for intact rock strength and friction – a re-evaluation and consideration of failure under polyaxial stresses. Solid Earth, 7, 493–508.

[160] Raleigh, C.B., Healy, J.H., Bredehoeft, J.D., 1976. An experiment in earthquake control at Rangely, Colorado. Science, 191, 1230–1237.

[161] Kim Won-Young, 2013. Induced seismicity associated with fluid injection into a deep well in Youngstown, Ohio. Journal of Geophysical Research Solid Earth, 118, 3506–3518.

[162] Justinic, A.H., Stump, B., Hayward, C., Frohlich, C., 2013. Analysis of the Cleburne, Texas earthquake sequence from June 2009 to June 2010. Bulletin of the Seismological Society of America, 103, 3083–3093.

[163] Hincks, T., Aspinall, W., Cooke, R, Gernon, T., 2018. Oklahoma's induced seismicity strongly linked to wastewater injection depth. Science, 359, 1251–1255.

Section 2

Site Effects Evaluation

Chapter 5

S-Wave Site Amplification Factors from Observed Ground Motions in Japan: Validation of Delineated Velocity Structures and Proposal for Empirical Correction

Eri Ito, Kenichi Nakano, Shigeki Senna and Hiroshi Kawase

Abstract

We first derived site amplification factors (SAFs) from the observed strong motions by the Japanese nationwide networks, namely, K-NET and KiK-net of National Institute of Earthquake Research and Disaster Resilience and Shindokei (Instrumental Seismic Intensity) Network of Japan Meteorological Agency by using the so-called generalized spectral inversion technique. We can use these SAFs for strong motion prediction at these observation sites, however, we need at least observed weak motion or microtremor data to quantify SAF at an arbitrary site. So we tested the capability of the current velocity models in Japan whether they can reproduce or not the observed SAFs at the nearest grid of every 250 m as the one-dimensional theoretical transfer functions (TTF). We found that at about one-half of the sites the calculated 1D TTFs show more or less acceptable fit to the observed SAFs, however, the TTFs tend to underestimate the observed SAFs in general. Therefore, we propose a simple, empirical method to fill the gap between the observed SAFs and the calculated TTFs. Validation examples show that our proposed method effectively predict better SAFs than the direct substitute of TTFs at sites without observed data.

Keywords: site effect, generalized spectral inversion, strong motion, theoretical transfer function, velocity structure, seismological bedrock

1. Introduction

The quantitative strong motion prediction with a source- and site-specific scheme is very important for mitigation of earthquake disaster and seismic design of important structures. It is especially true in Japan where large mega-thrust earthquakes are expected to occur within the coming 30 years. That is why we have a couple of nation-wide strong-motion networks in which a considerable number of strong motion records have been accumulated since 1996 [1].

There are several ways to simulate strong-motions as waveforms on the surface at a target site located at an arbitrary position. One is a theoretical Green's function method (TGF) in which wave generation at the source, propagation from the source to the site, and amplification near the site are represented by the numerical modeling for the whole process from the source to the site. In this method we need a physical model of the medium to represent the wave propagation in the whole path. In other words, we need to calculate the theoretical Green's function for a point source on the fault surface. The other is an empirical method in which we use observed ground motions of a small earthquake as a substitute for the Green's function and sum up all the contributions from the elemental sources on the fault surface. It is called the empirical Green's function method (EGF). If there are no appropriate small earthquake records to be used as the empirical Green's function, we first generate synthetic waveforms based on many records of small earthquakes. It is called the statistical Green's function method (SGF). Because the frequency range for the theoretical approach with coherent nature is limited to the lower end, usually below 1 Hz or lower, while the effective frequency range of EGF or SGF with inherent nature of stochasticity should be higher than that, a hybrid scheme with TGF and EGF or TGF and SGF are used naturally, as has been used in the current national project for strong motion predictions with specific sources [2].

After the deployment of the dense national strong motion observation networks, namely K-NET, KiK-net, and JMA Shindokei (Instrumental Seismic Intensity) network in Japan, a significant number of data has been accumulated. We can use these data to construct a model of SGF in a broadband frequency range. As long as we can generate the SGF for an arbitrary size of a small earthquake at an arbitrary location of a site in the frequency range of engineering interest, namely from 0.1 Hz to 20 Hz, we need not use a hybrid scheme. Thus we have been analyzing these strong-motion data in Japan by using the generalized spectral inversion technique (GIT) initially developed in 1980's [3, 4] to delineate statistical properties of the three major terms, namely, the source term, path term, and site term [5–8]. The novelty of our approach is that the hypothesized (i.e., extracted) seismological bedrock spectra at a reference site, YMGH01, are used as a reference to calculate site amplification factors at all the other observed sites. Such a separation of observed spectra into three major terms is sufficient to generate SGF at these observed sites. However, strong-motion simulations for the whole region near the seismogenic fault would be still difficult by SGF because we cannot estimate the site term at an arbitrary location other than the observed sites used in GIT. Thus, we need to develop a method to evaluate the site term at an arbitrary location as precisely as possible.

When we look at the site term as a function of frequency evaluated at K-NET, KiK-net, and JMA Shindokei network, we found that they show strong spectral fluctuations from 1 to 10 as a normal range of fluctuations and from 1 to 50 at tens of extraordinary sites with various peak frequencies. Several attempts have been made to correlate the primary characteristics of the observed site amplification factor (SAF) with a site proxy or proxies such as the S-wave velocity (Vs) averaged over top xx m, Vs_xx (e.g., Vs30) or the depth to the layer with the S-wave velocity higher than y.y km/s, Z_y.y (e.g., Z1.0) [9–13], trying to reproduce primary characteristics of SAF such as the fundamental peak frequency f_0 and its peak amplitude A_0. Unfortunately, these extracted characteristics are not sufficient to reproduce synthetic seismograms needed in the SGF summation. We should find a different strategy.

In what follows, we first introduce the fundamental characteristics of the observed SAF in the horizontal component (hSAF) derived from GIT [7, 8]. Then, we show comparisons of these hSAFs with the 1D theoretical ones calculated from the recently established unified velocity model (UVM) of the National Research Institute for Earth Science and Disaster Resilience (NIED) in the Kanto and Tokai regions. Next, we obtain the modification ratios to reduce the gap between them at the observation points and propose a scheme to evaluate hSAF at an arbitrary point by using the theoretical hSAF and the interpolated modification ratios, named FMR as the frequency modification ratio and AMR as the amplitude modification ratio. Finally, we propose an interpolation scheme to get hSAF in every 250 m grid point and validate the scheme at selected sites.

2. Observed SAF from GIT

In this section, we briefly introduce the observed horizontal SAF (hSAF) and vertical SAF (vSAF) derived from GIT [7, 8]. Here we introduce only their basic aspects because we are using their results as a starting point.

They restricted events and sites with the Japan Meteorological Agency (JMA)'s magnitude $M_{JMA} \geq 4.5$; source depth ≤ 60 km; hypocentral distance ≤ 200 km; peak ground acceleration ≤ 2 m/s^2; and a number of observation sites triggered simultaneously for one event ≥ 3. These selection criteria resulted in 150,468 event-station pairs at 2,593 sites for 1,734 events. Only a relatively short duration of acceleration record from the onset of the S-wave was analyzed (5 s if $4.5 < M_{JMA} \leq 6$; 10 s if $6 < M_{JMA} \leq 7$; 15 s if $7 < M_{JMA} \leq 8$). A Parzen window of 0.1 Hz was used for a minimum level of smoothing. Note that the mainshock of the 2011 Off the Pacific Coast of Tohoku earthquake was excluded because the durations of those records are extraordinarily long.

As mentioned above, the most important feature of their GIT is that they determined the S-wave velocity structure at the reference site (YMGH01) using the transfer function (the spectral ratio and the phase difference) between the surface and the borehole 200 m below and that the observed Fourier spectra on the surface were deconvolved (divided by the amplification factor) to obtain the hypothesized outcrop spectra on the seismological bedrock with Vs of 3,450 m/s. The resultant S-wave velocity profile determined by the matching of the theoretical transfer function to the observed transfer function is quite similar to the original P-S logging data published by NIED [1], only with higher bedrock velocity of 3,450 m/s from 3,100 m/s. After the determination of the velocity profile, Nakano et al. [7] corrected (divided) all the observed spectra at YMGH01 by the calculated 1D S-wave site amplification factor on the surface with respect to the outcrop motion on the bedrock (=twice of the input) and used as the reference spectra in the subsequent GIT analyses, as if they were observed at YMGH01. Thus, their separated site terms, hSAF and vSAF, are considered to be the site amplification factors with respect to the outcrop seismological bedrock, on which there is virtually no site effect. Nakano et al. [7] successfully separated the source spectra and path terms as evidenced by their correspondence to the ω^{-2} source spectra shapes and Q values similar to the previous studies in Japan.

Figure 1 shows examples of the observed hSAF at four sites in the Tokai region. We can see significant differences from site to site. Although we did not show vSAF here, the amplitude and its fluctuation of vSAF is much smaller than hSAF, especially below 3 to 4 Hz, that is, below the fundamental peak frequency of vSAF. That is why the earthquake horizontal-to-vertical spectral ratio, eHVSR, which is equal to hSAF/vSAF, tends to be similar to hSAF until the fundamental peak frequency of vSAF. However, to get hSAF from eHVSR, we need to correct vSAF, as recently proposed by Ito et al. [14]. Please note that precisely speaking, vSAF in this paper should be referred to as vSAF* as in [14] because we use the same reference condition for both hSAF and vSAF as the seismological bedrock spectra in the horizontal component so that we need to have correction due to the vertical-to-horizontal spectral ratio on the seismological bedrock on top of the vertical-to-vertical (i.e., P-wave) site amplification.

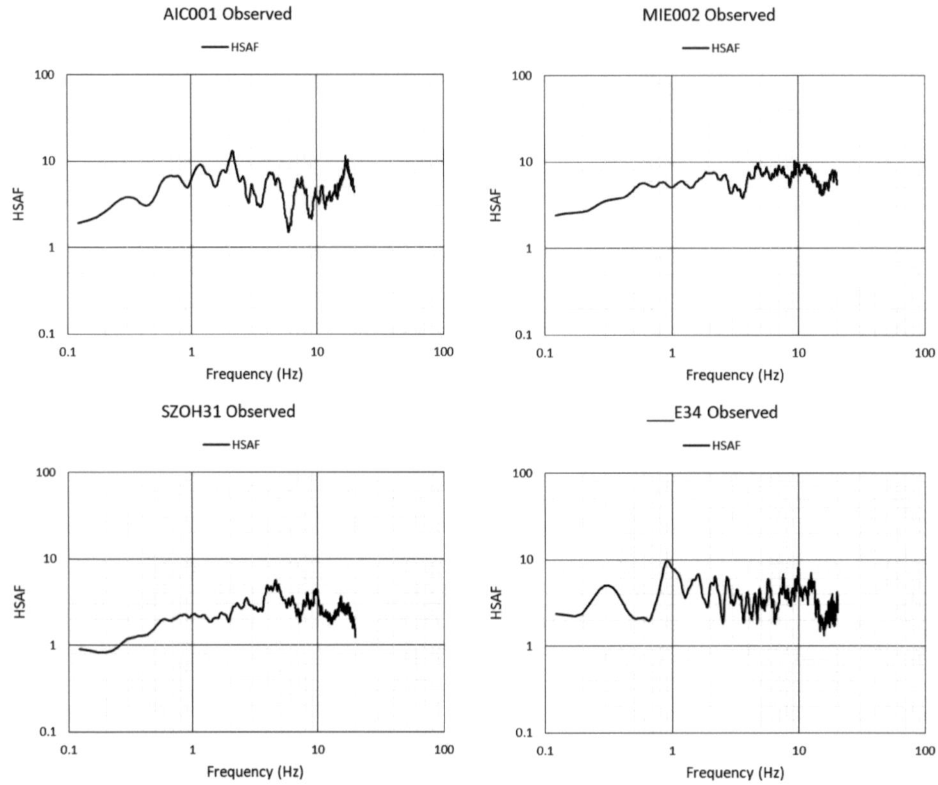

Figure 1.
Observed horizontal site amplification factor, hSAF, extracted from strong motions at K-NET, KiK-net, and JMA Shindokei network by GIT after [7].

3. Unified velocity model of NIED

To theoretically reproduce the observed hSAF and vSAF from GIT we need a velocity structure at each site from the seismological bedrock to the surface because they are the spectral ratios with respect to the outcrop spectra on the seismological bedrock. Note again that the seismological bedrock here is the surface of the crust on which we can assume no site amplification, whose S-wave velocity should be equal to or higher than 3 km/s. We have been delineating velocity structures in the deeper- and shallower-parts separately, primarily because we need to use different methods to explore the velocity structures in different depths. The boundary between them is the so-called engineering bedrock, whose Vs would be in between 350 m/s to 450 m/s. This is so beause we are using plenty of borehole information to constrain velocity structures in the shallower-parts, for which engineers need to gather information for their construction works. They usually need information only down to the layer with Vs in between 350 m/s to 450 m/s. However, it is apparent that higher-mode contributions of reverberated S- and P-waves within the whole basin above the seismological bedrock should show up in the frequency range higher than the fundamental peak frequency [15]. Therefore, we need a unified velocity model that integrates both shallower- and deeper-parts.

To that end, the National Research Institute for Earth Science and Disaster Resilience (NIED) has developed a unified velocity model (UVM) by integrating shallower- and deeper-parts of the structures above the seismological bedrock in the Kanto and Tokai regions [16–18]. The procedure to develop the model was based on the "concept of creating a subsurface structure model" released by the government agency, the Headquarters for Earthquake Research Promotion (HERP) [19]. Details of the procedure can be found in the papers referenced above, but the following is a brief description of the procedure.

An initial model of the shallow structure from the ground surface to the engineering bedrock was created based on existing studies and continuously collected SPT values in the boring data. Meanwhile, an initial model of the deep structure from the engineering bedrock down to the seismological bedrock was created based on the velocity models developed in existing studies by HERP. Then, an initial UVM was created by connecting them at the engineering bedrock. Next, the initial UVM was adjusted by using S-wave velocity structures at the strong-motion sites in the regions and those of spatially uniform and dense array microtremor explorations conducted as a part of Japan's national Strategic Innovation Promotion project. Finally, the adjusted UVM was verified by using earthquake data at the strong-motion sites of NIED.

Examples of the important features of the resultant UVM are shown in **Figures 2** and **3**. **Figure 2** shows the depth to the seismological bedrock with the Vs of 3.1 km/s (Z3.1) in the Tokai region, while **Figure 3** shows the depth to the engineering bedrock with the Vs of 0.35 km/s (Z0.35), which is the interface between the shallower- and deeper-parts of the UVM in the Tokai region [18]. A similar map can be seen for the Kanto region in Senna et al. [16, 17].

Please note that in the following sections when we calculate theoretical one-dimensional (1D) S-wave SAF, we use the following Q values for intrinsic and scattering attenuation:

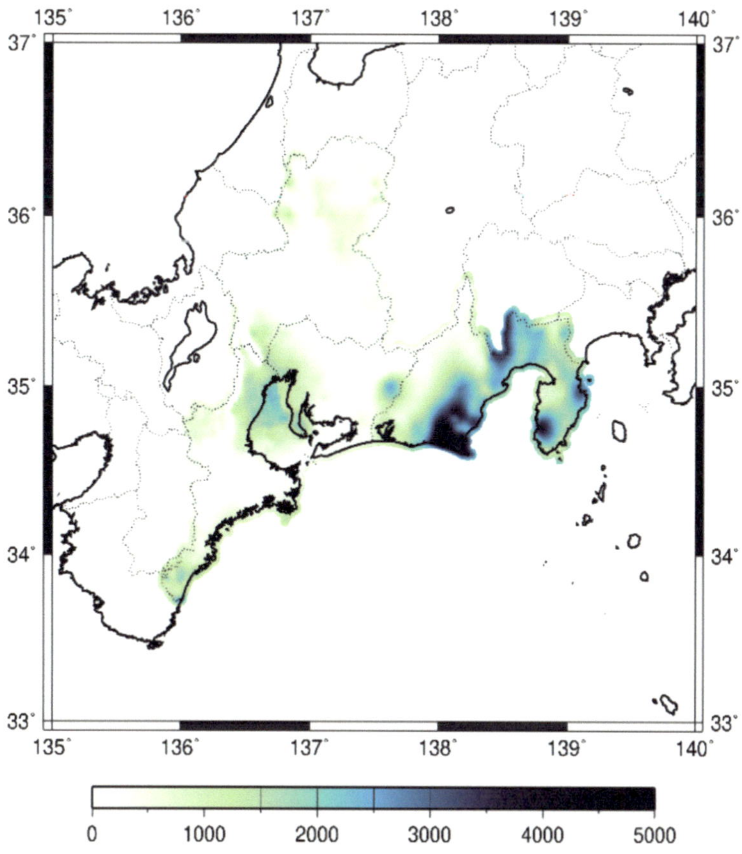

Figure 2.
Depth contour in meters to the seismological bedrock with Vs of 3.1 km/s after [18].

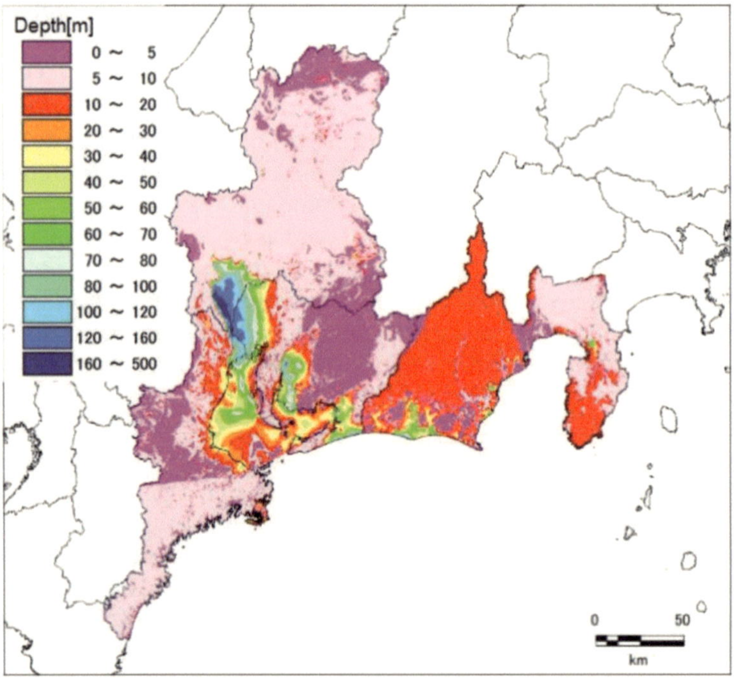

Figure 3.
Depth contour of the engineering bedrock with Vs of 350 m/s after [18].

$$\begin{aligned} Q &= Q_0 * 0.5 \quad &\text{if } f \le 0.5\text{Hz} \\ Q &= Q_0 * f \quad &\text{if } 0.5\text{Hz} < f < 5\text{Hz} \\ Q &= Q_0 * 5 \quad &\text{if } 5\text{Hz} \le f \end{aligned} \tag{1}$$

The Q_0 values for the deeper part are listed in **Table 1**, while those for the shallower part we assume Q_0 = Vs/10.

To connect the bottommost layer of the shallower part Lsb with the topmost layer of the deeper part Ldt, we prioritize the shallower part if the depth Lsb is deeper than Ldt. If there is a gap between the two depths and Vs of Lsb is equal to or larger than Vs of Ldt, we extend Lsb down to Ldt. If Vs of Lsb is much smaller than

No.	Vp (m/s)	Vs (m/s)	Density (kg/m^3)	Q_0
1	1600	350	1850	70
2	1600	400	1850	80
3	1700	450	1900	90
4	1800	500	1900	100
5	1800	550	1900	110
6	2000	600	1900	120
7	2000	650	1950	130
8	2100	700	2000	140
9	2100	750	2000	150
10	2200	800	2000	160
11	2300	850	2050	170

No.	Vp (m/s)	Vs (m/s)	Density (kg/m^3)	Q_0
12	2400	900	2050	180
13	2400	950	2100	190
14	2500	1000	2100	200
15	2500	1100	2150	220
16	2600	1200	2150	240
17	2700	1300	2200	260
18	3000	1400	2250	280
19	3200	1500	2250	300
20	3400	1600	2300	320
21	3500	1700	2300	340
22	3600	1800	2350	360
23	3700	1900	2350	380
24	3800	2000	2400	400
25	4000	2100	2400	420
26	5000	2700	2500	540
27	4600	2900	2550	580
28	5500	3100	2600	620
29	5500	3200	2650	640

Table 1.
Assumed layer profiles for the deeper part of UVM after [18].

Vs of Ldt, then we insert three layers with a gentle gradient of increasing Vs. **Table 1** shows the parameters of layers assumed commonly in the deeper part. The bedrock S-wave velocity of UVM, 3,200 m/s, is close enough to that of the hypothesized bedrock of 3,450 m/s in GIT so we can compare both SAFs directly.

4. Observed and theoretical SAF

As mentioned in the previous section, the UVM of NIED is considered to be the most reliable velocity model for the strong motion simulation because the UVM combines all the available geophysical information to date related to the velocity structures from the ground surface to the seismological bedrock as densely sampled as possible. However, the actual S-wave SAF at a specific site, as shown in **Figure 1**, can be significantly different from the theoretical one calculated by a simple 1D S-wave multi-reflection theory for a stack of layers [20–22]. To see the difference, we plot in **Figure 4** comparisons of the 1D theoretical hSAF with the observed hSAF at the same four sites in **Figure 1**. We calculate the theoretical hSAF as the 1D soil response on the surface of a combined velocity structure of the shallower- and deeper-parts with respect to the outcrop of the seismological bedrock motion (i.e., twice of the input at the bottom of the deeper part). Except for the site SZOH31, the theory tends to underestimate the observation.

The major reasons for discrepancy are twofold; one is due to an inevitable inaccuracy of the derived velocity structure, and the other is due to a too simplistic assumption of the 1D horizontally-flat layered model. In the former there are two possible causes; one is the inaccuracy of the referenced values to delineate the

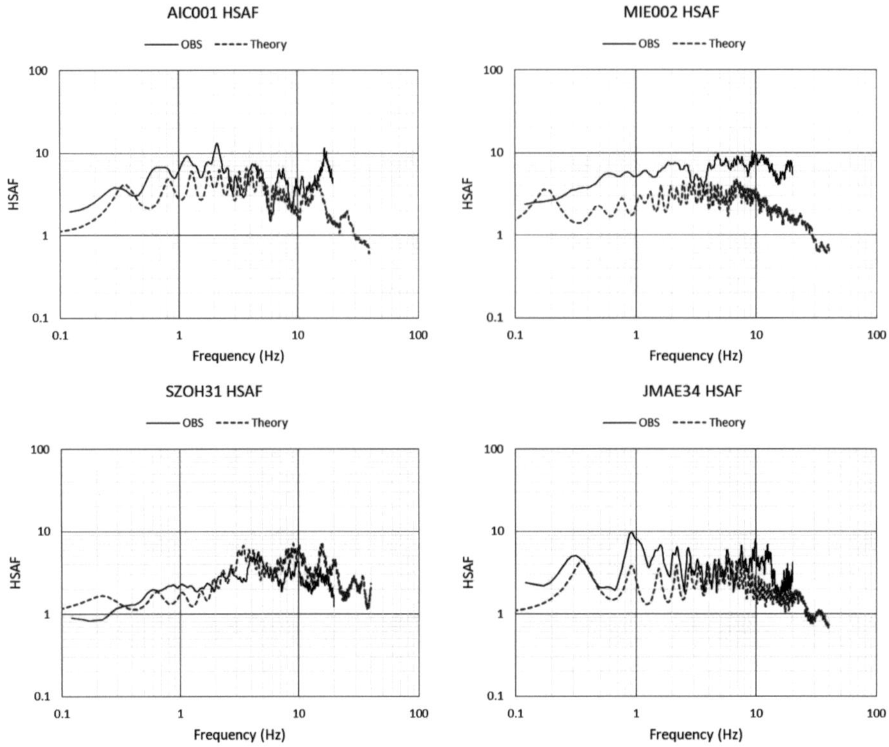

Figure 4.
Observed hSAF extracted by GIT and 1D theoretical hSAF from UVM after [18]. Fundamental characteristics are well reproduced, however, the theoretical hSAF tends to underestimate the observed hSAF.

structure including the assumed Q-values, and the other is the rapid spatial variations within the 250 m grid. In the latter there are two possible causes; one is the additional amplification due to the basin-induced surface waves generated at the edge of two or three-dimensional (2D or 3D) basins (see for example [23–25]) and the other is the topographic effects near the surface of irregular shapes such as a hill, a valley, or a cliff (see Kawase [26]).

To account for the effects of the basin-induced surface waves inside sedimentary basins, Nakano [8] and Nakano et al. [27] proposed to use an empirical ratio called the whole-wave-to-S-wave ratio (WSR), where the spectral ratios of the whole duration with respect to the S-wave portion with relatively short duration (5 to 15 s depending on the source magnitude as used in GIT) are averaged over all the observed events at a site. They found that the WSR tends to be close to unity irrespective of frequency for a site on hard rock, whereas it can easily exceed 10 in the lower frequency range for a site inside a soft sedimentary basin. Even for such a site, WSR will converge to unity in a higher frequency range above a few Hz. Because the spatial variation of WSR at one specific frequency highly correlates with that of the basin depths, as seen in Nakano et al. [27], Nakano [8] proposed a scheme to interpolate WSRs to make it possible to calculate a scenario-type hazard map with much higher spatial density than those of the strong motion observation sites. This WSR correction is a simple, empirical way to account for the additional amplification due to basin-induced surface waves.

However, other than the contributions of basin-induced surface waves through the modeling of WSR on top of hSAF as proposed by Nakano [8], it is difficult to account for the physical cause of the discrepancy between observation and theory at every sites as seen in **Figure 4**. We have been attempting to fill the gap by inverting the velocity structures from observed horizontal-to-vertical spectral ratios (HVSRs) of earthquakes under the diffuse field assumption [28–30], which is quite successful to reproduce observed HVSRs (and consequently also hSAF). This approach can provide us an equivalent 1D structure that will reproduce the observed hSAF at the target site quite precisely, however, the method is valid only for a site with either earthquake or microtremor records. We need a different strategy to evaluate hSAF as precisely as possible at an arbitrary site without any records. Because the velocity structure in the UVM of NIED is obtained with the spatial density of the 250 m grid, we want to reflect the fundamental characteristics of the theoretical transfer function of that structure, yet the resultant hSAF for strong motion simulation should be close enough to the observed hSAF.

5. Frequency and amplitude modification ratio

To overcome the difficulty to obtain better velocity models through physical parametrization of the underground structure at the sites without observed records, we would like to propose a different but simple approach here.

Suppose that a theoretical 1D hSAF from the UVM at a certain site deviates from the true hSAF by one of the aforementioned causes or their combination, the difference of the theoretical hSAF would be manifested in both the frequency axis and the amplitude axis. If we have stiffer or thinner layers in reality than the assumed profile in UVM, then the frequency characteristics would be all shifted towards the higher side. Or if we have a stronger velocity gradient within layers in reality, then the peak amplitudes would be all shifted towards the higher side. Thus, we have two different ways to adjust the theoretical hSAF to make it closer to the observed hSAF, one in the frequency axis and the other in the amplitude axis. Here is a simple way of correction for the theoretical hSAF, $HSAF_{the}$:

$$HSAF_{mod}(f) = AMR * HSAF_{the}(f / FMR), \quad (2)$$

where FMR is the frequency modification ratio and AMR is the amplitude modification ratio. $HSAF_{mod}$ is the resultant hSAF after both modifications as a function of frequency f. We need to determine FMR and AMR to make RES, the residual between $HSAF_{mod}$ and the observed hSAF ($HSAF_{obs}$), minimum:

$$RES(FMR, AMR) = \sum_{f=fmin}^{fmax} \frac{\left|\log 10(HSAF_{mod}) - \log 10(HSAF_{obs})\right|}{f} \quad (3)$$

where $fmin$ and $fmax$ are the minimum and maximum frequencies of interest and we set them 0.12 Hz and 20 Hz, respectively. Here we use frequency f as a weight because the higher the frequency the denser the data in the linear space.

5.1 Grid search scheme

Because the calculation of Eq. (2) is quite easy, we use the grid search to obtain the best FMR and AMR combination. However, after several experiments, we found that the evaluation function in Eq. (2) seems too weak to determine FMR and AMR in a reasonable range because there is a trade-off between them. Therefore, we introduce the correlation function between $HSAF_{mod}$ and $HSAF_{obs}$ as an additional constraint. Then the target function to be maximized, GOF, becomes.

$$GOF = RES_{min} / RES(FMR, AMR) * COR(FMR) / COR_{max} \quad (4)$$

where RES_{min} is the minimum residual in the searching range, $RES(FMR, AMR)$ is the residual shown in Eq. (2) as a function of FMR and AMR, $COR(FMR)$ is the correlation coefficient between $HSAF_{mod}$ and $HSAF_{obs}$, which is a function of only FMR, not a function of AMR, and COR_{max} is the maximum correlation coefficient in the searching range. Thus 1.0 is the maximum of GOF.

We set the searching range for FMR depending on the original correlation without frequency modulation, which is $COR(1.0)$, as follows:

$$\begin{gathered}\text{If } 0.6 \leqq COR(1.0),\ 0.80 \leq FMR \leq 1.25 \\ \text{If } 0.4 \leqq COR(1.0) < 0.6,\ 0.67 \leq FMR \leq 1.50 \\ \text{If } COR(1.0) < 0.4,\ 0.50 \leq FMR \leq 2.00\end{gathered} \tag{5}$$

We use these searching ranges because if the correlation of the original model is sufficiently high, we should not modify its frequency characteristics so much. For AMR we set the searching range to be $0.333 \leq \text{AMR} \leq 3.00$ irrespective of $COR(1.0)$ because there are a few tens of sites with those amplitude differences as high as 3 times or as low as 1/3 and AMR does not alter $COR(1.0)$. To efficiently search the best FMR and AMR with the precision of two digits, we employ the two-step grid search; first with every 0.1 increments, then with every 0.01 increments around the best FMR and AMR in the first step.

5.2 Evaluated *FMR* and *AMR*

Figure 5 shows examples of the resultant $HSAF_{mod}$ in comparison with $HSAF_{obs}$ and $HSAF_{the}$ at the same K-NET, KiK-net, or JMA Shindokei network sites shown in **Figures 1** and **4**. As we can see, $HSAF_{mod}$ determined by the grid search are quite close to $HSAF_{obs}$ in both frequency fluctuations and amplitudes.

Figure 6 shows the resultant optimal values of FMR and AMR at all the 546 sites used. We can see a very weak correlation between them. As for the searching range of AMR, namely 1/3 to 3, looks sufficient. On the other hand, we see a significant concentration of sites near the searching range boundary, 1/2 or 2 for FMR. This means that we could obtain better residuals and correlations if we search the optimal FMR in the frequency range wider than those limits. However, when we increase the searching range for FMR too much, we will see some cases where the reverberated fluctuations by the 1D resonance within the sediments seem to be shifted to the next overtone.

5.3 Correlation and residual improvement

If no improvement to the matching with the observed hSAF is achieved, our correction method does not have any merit. Therefore, we need to check if we can see significant improvement or not.

Figure 7 shows the distributions of the obtained correlation and the averaged residuals between $HSAF_{mod}$ and $HSAF_{obs}$. We can see most of the site shows residuals less than 1.5 and correlations higher than 0.5. **Table 2** shows the percentage of the sites in different categories in terms of the goodness of fit to $HSAF_{obs}$. When we compare the matching seen in **Figure 5** and the distribution of these data in **Figure** 7, we can see that the average residual is more important than the correlation in terms of the quality of the modified hSAF because the correlation can be deteriorated easily by small fluctuations at different frequencies.

Figure 8 shows significant improvements in the correlations from the original ones to the modified ones. There is no data with decreased correlations, however,

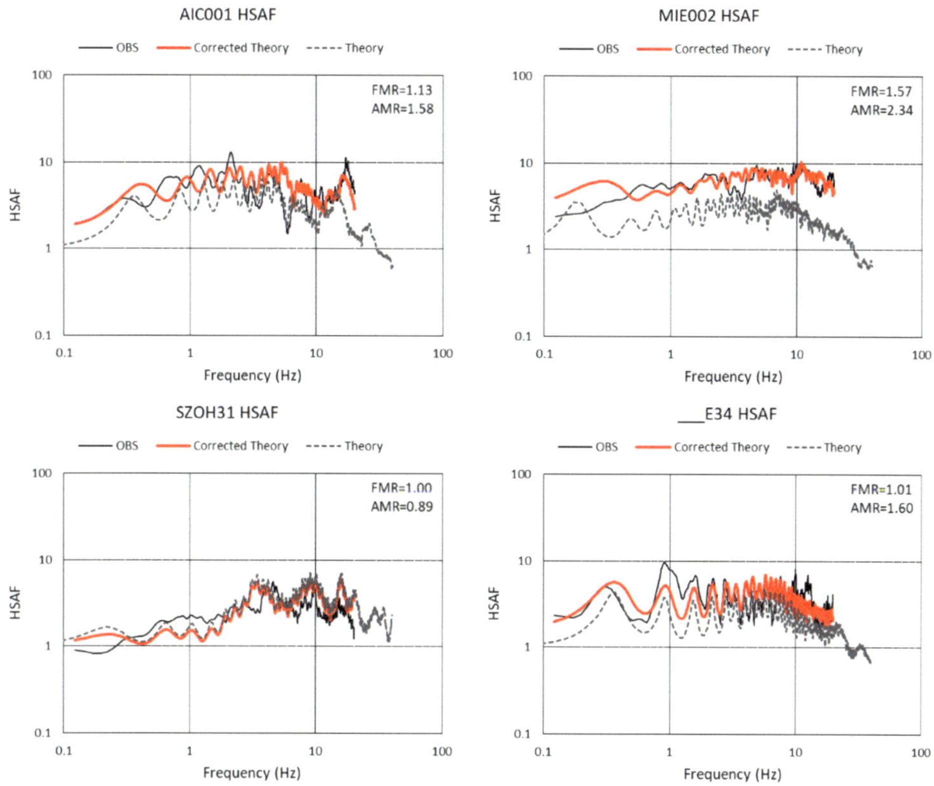

Figure 5.
Modified hSAF ($HSAF_{mod}$) *by using* FMR *and* AMR *in comparison with observed hSAF* ($HSAF_{obs}$) *extracted by GIT and 1D theoretical hSAF* ($HSAF_{the}$) *from the velocity structure taken from UVM after [18] at the same sites in* ***Figures 1*** *and* ***4****. The optimal values of* FMR *and* AMR *for each site are shown in the upper-right corner.*

there still remain 7 sites with the correlations less than zero, which was decreased from 84. **Figure 9** also shows correlation improvements but as a function of *FMR*. We can see a clear concentration of *FMR* near the boundaries at 0.8 and 1.25, the boundaries of the searching range if the original correlation is larger than 0.6 as shown in Eq. (4). We can see a smaller improvement in this *FMR* range for higher correlation sites, in comparison to those with lower correlations.

5.4 Spatial interpolation

Now we have more than 500 sites in the Kanto and Tokai regions where we have determined modification ratios for frequency and amplitude, *FMR* and *AMR*. There are various ways to utilize these ratios for the prediction of site amplifications with much denser spatial resolutions based on the UVM in the 250 m grid. One way is to establish relationships of these modification ratios with respect to a site proxy or proxies such as Vs30 or Z1.0 as mentioned in the introduction. As seen in a lot of previous studies for site effects based on such relationships, however, we need to accept large deviations from the average relationships from site to site because it is the nature of the site amplification. We also face the possibility of the inaccurate choice of a site proxy in UVM for an arbitrary site used for modification ratios.

Therefore, we decide to use a direct spatial interpolation scheme as Nakano [8] proposed for WSR. In this scheme, we employ first GMT's "surface" function [31, 32] in which the curvature minimization scheme is used together with the

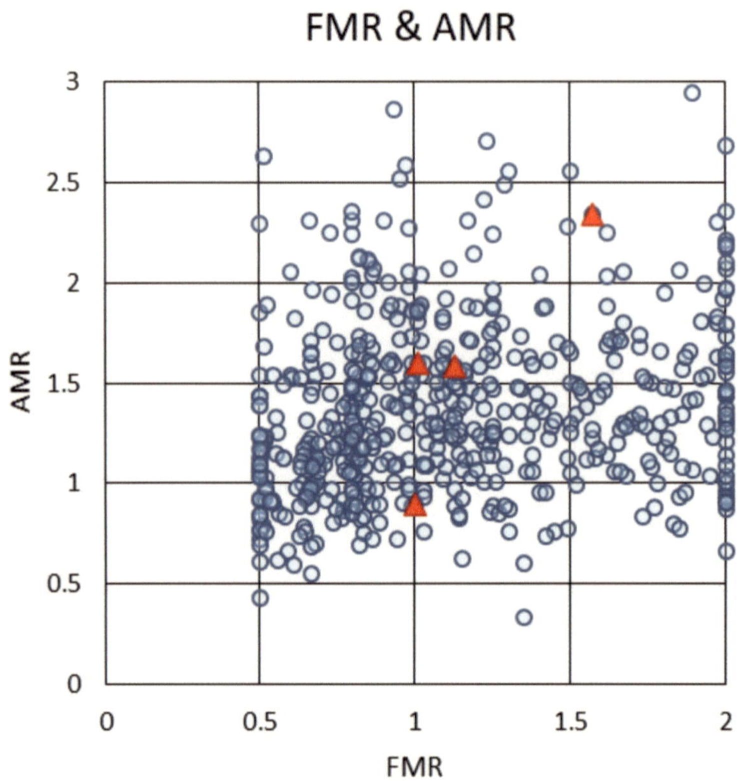

Figure 6.
Resultant values of FMR *and* AMR *after the grid search for 546 sites in the Kanto and Tokai regions. Red triangles are those for the sites in* ***Figure 5****.*

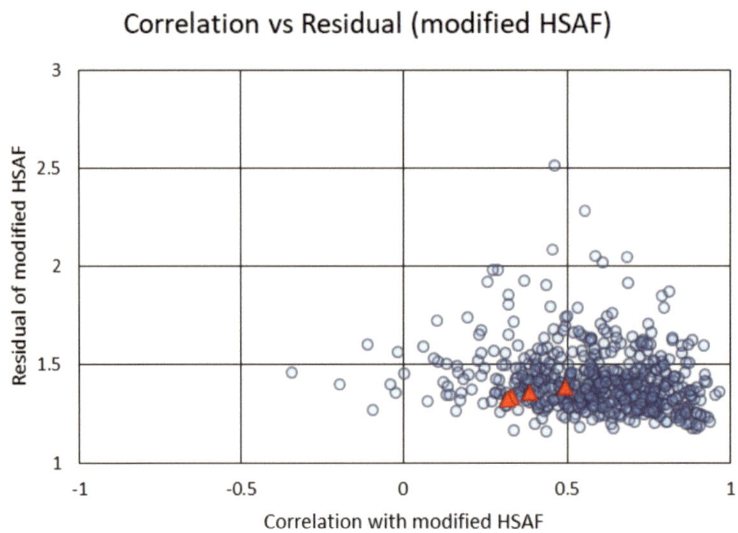

Figure 7.
Distributions of the correlation of $HSAF_{mod}$ *versus the averaged residual (the averaged spectral ratio between* $HSAF_{mod}$ *and* $HSAF_{obs}$*) for the same 546 sites shown in* ***Figure 6.*** *Triangles are values at the sites shown in* ***Figure 5****.*

smoothing constraint of an elastic shell with the tension factor of 0.25. In this Step-1 interpolation, we use the 3 km equal-spaced grid. Then in Step-2, we use the 250 m grid to interpolate further by using the modified Shepard's method [33].

Residual	Before	After	Correlation	Before	After
<1.25	5	43	<0.0	84	7
1.25–1.50	175	389	0.0–0.4	227	86
1.50–2.00	286	108	0.4–0.6	116	186
2.00–3.00	77	6	0.6–0.8	85	193
3.00<	3	0	0.8–1.0	34	74
Total	546	546	Total	546	546

Table 2.
Residual and correlation improvement in terms of the number of sites in each range before and after the correction.

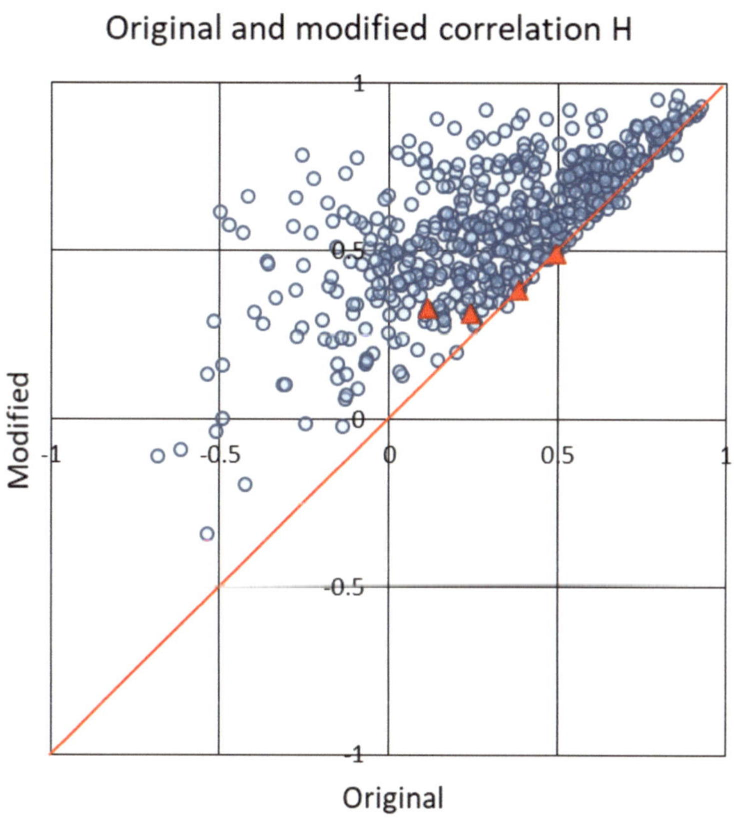

Figure 8.
Improvement from the original correlations of $\mathrm{HSAF_{the}}$ *to the modified ones of* $\mathrm{HSAF_{mod}}$*. Triangles are values at the sites shown in* ***Figure 5****.*

Figure 10 shows the comparison of interpolated values in Step-1 with those used as targets for (a) *FMR* and (b) *AMR*. Interpolation in *AMR* is better because its spatial variation is smoother than *FMR* as shown later. Although we see tens of sites in **Figure 10a** away from the 1:1 line, the average deviation from unity for *FMR is* about 10% (1.1 or 0.9 times) and 91% of the interpolated values are within the range between 1/1.25 and 1.25 times of the original *FMR*. The number of outliers is misleading because we use the "blockmedian" command of GMT to refer to only the median value if we have plural sites in the same 3 km grid.

Similarly, **Figure 11** shows the comparison of interpolated values in Step-2 with those used as targets from Step-1 for (a) *FMR* and (b) *AMR*. In Step-2 the

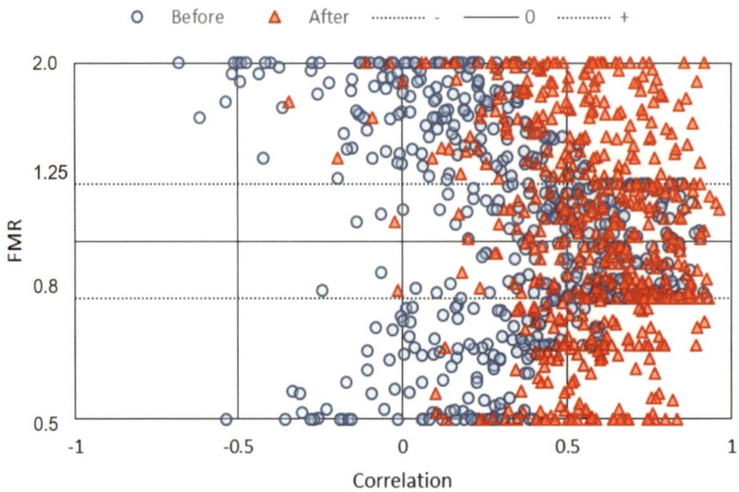

Figure 9.
Improvement from the original correlations of $HSAF_{the}$ *(before) to those of* $HSAF_{mod}$ *(after) as a function of* FMR*. Data after the correction are concentrated near the boundaries of the searching range (0.8 and 1.25) when the original correlation is more than 0.6 because of the setting used. Note that* FMR *in the vertical axis of this figure is common for both before and after the correction.*

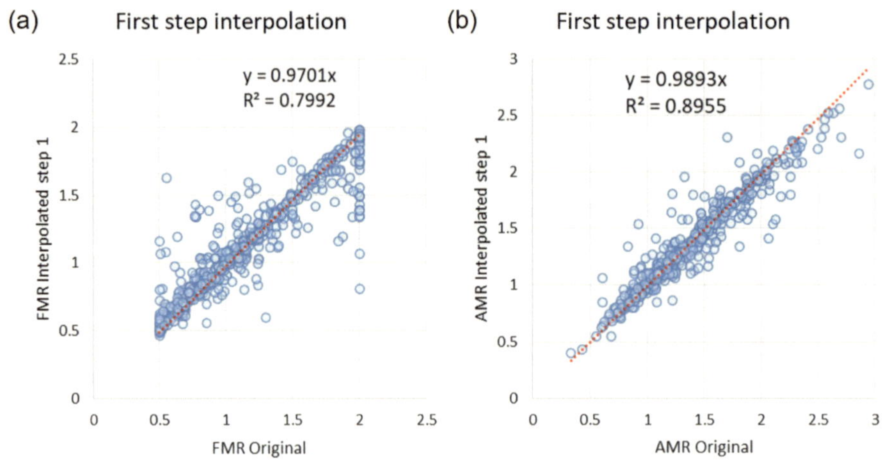

Figure 10.
Comparison of the original FMR *and* AMR *used as targets of interpolations and those of interpolated values in Step-1 at strong motion observation sites. Red broken lines are linear regression lines whose inclinations and coefficients of determination* R^2 *are listed inside. (a) FMR and (b) AMR.*

interpolation is performed from the 3 km grid in Step-1 to the 250 m grid. The linearity of interpolation in Step-2 is much higher than that in Step-1 in the case of *FMR*, whereas that of Step-2 is as high as that of Step-1 in the case of *AMR*.

We can see the spatial stability of the interpolation scheme as a gross picture shown in **Figure 12** for *FMR* and *AMR*. These are the results of Step-2 with a spatial resolution of 250 m. Apparently, *AMR* is much smoother than *FMR* in terms of spatial variability so that the interpolation for *AMR* should be much easier and precise than *FMR*. On average, the Kanto region needs smaller correction values in *FMR* than those in the Tokai region, although it needs relatively larger correction values in AMR inside the whole soft-sedimentary areas in the north of Tokyo Bay.

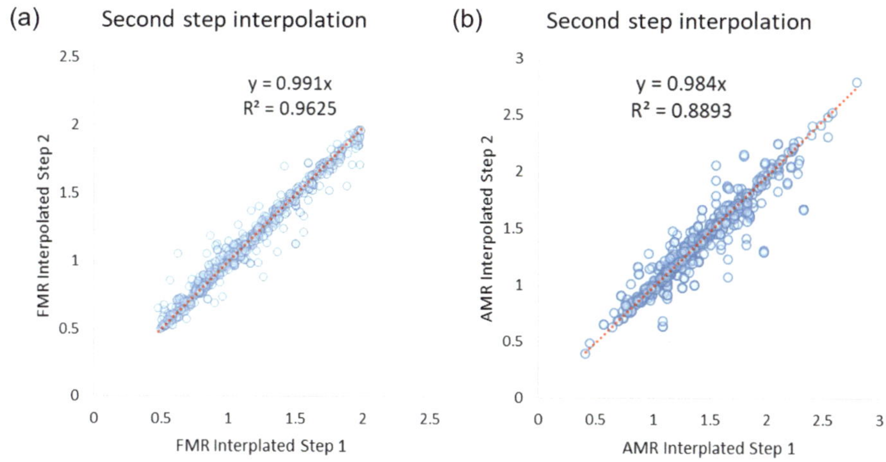

Figure 11.
Comparison of Step-1 FMR *and* AMR *used as targets of interpolations in Step-2 interpolation and those of interpolated values in Step-2 at strong motion observation sites. Red broken lines are linear regression lines whose inclinations and coefficients of determination* R^2 *are listed inside. (a) FMR and (b) AMR.*

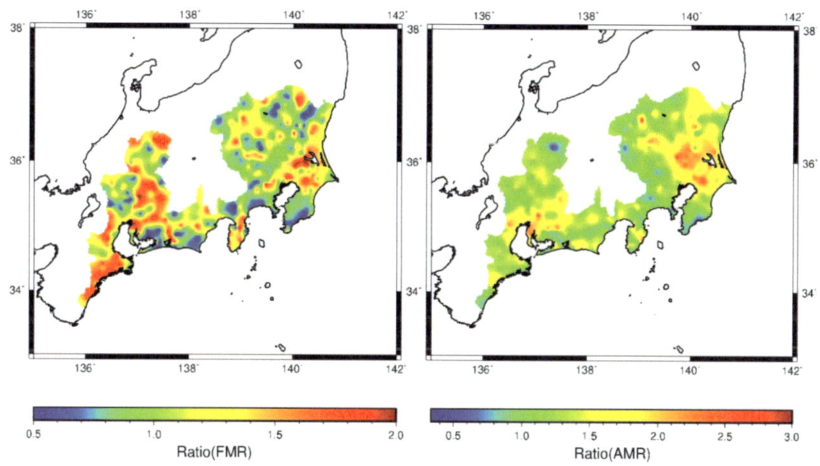

Figure 12.
Interpolated contour maps of FMR *(left) and* AMR *(right) after the Step-2 interpolation with the spatial resolution of 250 m.*

5.5 Validation

So far we show that a simple two-step scheme of interpolation works to calculate modification ratios for both frequency and amplitude, namely, *FMR* and *AMR* from 546 strong motion stations in the Kanto and Tokai regions in a grid as small as 250 m. Because the UVM in these regions has a spatial resolution of 250 m, we can directly use these interpolated modification ratios once we calculate 1D theoretical hSAF at any of these grid points. To validate the method, we take four sites shown in **Figure 5** out from the control points used for interpolation and let the program interpolate the modification ratios there and see how it works.

Figure 13 shows contour maps of *FMR* and *AMR* without four points used as examples in **Figure 5**. We can see smooth interpolation is achieved at these four points. **Figure 14** shows the correspondence of original and interpolated values of *FMR* and *AMR* for the cases with and without four points. In case of *FMR*, the

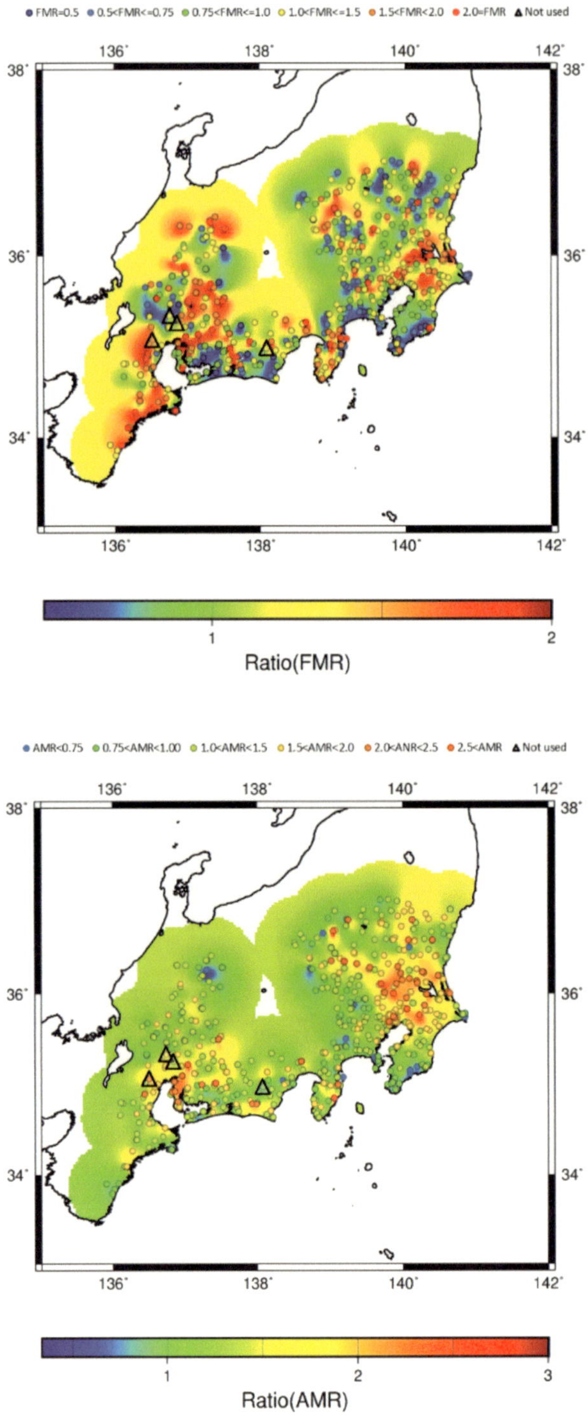

Figure 13.
Interpolated contour maps of FMR *(top) and* AMR *(bottom) after the Step-2 interpolation without four points shown by triangles. Inside the triangles interpolated values at these sites are shown by color-coded circles.*

original and interpolated values are close to the 1:1 line and the pure interpolation values at three sites out of four are close to the original ones. We should note that the worst site of the interpolated *FMR*, AIC001, and the JMA site ___E34 were very close to each other as shown in **Figure 13**. In case of *AMR*, at three sites the

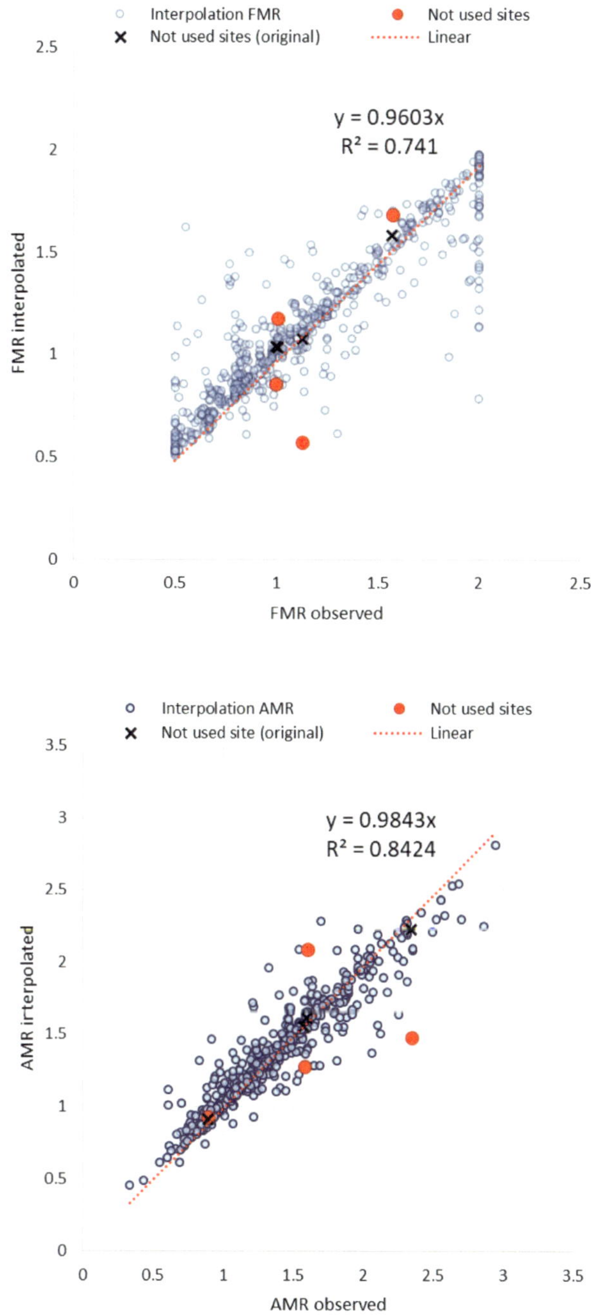

Figure 14.
Comparison of the original FMR *and* AMR *used as targets of interpolations and those of interpolated values in Step-2 with and without four example sites in* **Figure** *5. Black crosses are original values at four sites and red circles are interpolated values without referring to those original values, that is to say, purely interpolated values. Except for one site for* FMR *(AIC001), our interpolation scheme works.*

interpolation values without four sites were not as good as those with four sites. Still the deviation from the original values are within the range of 1.5 or 1/1.5 times.

Finally, we compare the corrected hSAF at those four sites not used in the spatial interpolation as references but used as the targets of the interpolation with the observed hSAF in **Figure 15**. Although the corrections by the original *FMR* and *AMR* seen in **Figure 5** are much better than the corrections by these

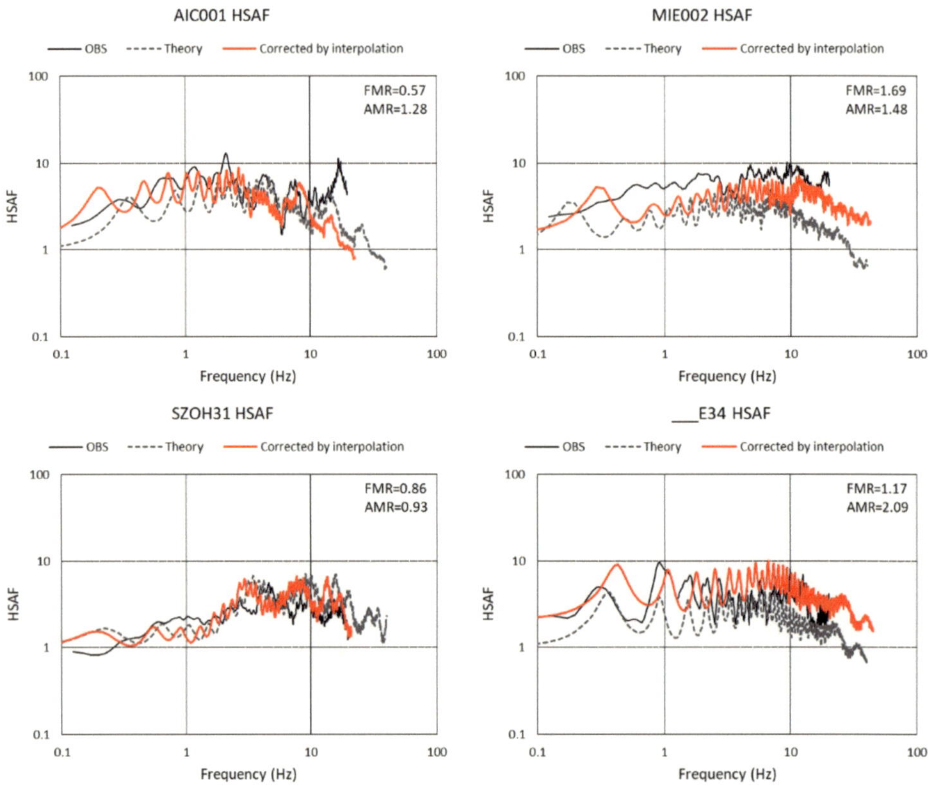

Figure 15.
Modified hSAF ($HSAF_{mod}$ *) by using* FMR *and* AMR *after interpolation (without using these four site) in comparison with observed hSAF (* $HSAF_{obs}$ *) extracted by GIT and 1D theoretical hSAF (* $HSAF_{the}$ *) from the velocity structure taken from UVM at the same sites in* ***Figures 1, 4*** *and* ***5****. The values of* FMR *and* AMR *by interpolation for each site are shown in the upper-right corner.*

interpolated *FMR* and *AMR* in this figure, the interpolated corrections still make theoretical hSAF closer to the observed hSAF.

6. Conclusions

In order to evaluate an equivalent 1D S-wave site amplification factor at an arbitrary point, we propose an empirical method of correction on to the theoretical site amplification factor calculated from the unified velocity model of NIED for the Kanto and Tokai regions where the shallower- and deeper parts of the velocity structure are combined. First, we check how well the current unified velocity model in Japan can reproduce horizontal site amplification factors derived from the observed strong motions in the form of the equivalent 1D S-wave theoretical transfer functions at the nearest grid of every 250 m. The observed site amplification factors were obtained by GIT relative to the reference spectra extracted as the outcrop motions on the seismological bedrock. To be consistent with these observed site amplification factors, the theoretical transfer functions are calculated relative to the outcrop motions (twice of the input) on the seismological bedrock. We find that at about one-half of the sites the calculated 1D amplification factors show more or less acceptable fit to the observed ones, however, they tend to underestimate the observed amplifications in general. Therefore, we propose a simple, empirical method to fill the gap between the observed site amplification factors and the calculated ones based on the frequency and amplitude modification ratios. Once we obtain these modification ratios at 546 observed sites, we can interpolate them in space to obtain the modification ratios at an arbitrary point. Validation examples show that our proposed method effectively predict better site amplifications than the direct substitute of theoretical amplification factors at a site without observed data.

In the future investigation, we will apply the proposed correction method to the sites where we have observed records of either earthquakes or microtremors but we do not include them in the delineation of the modification ratios in order to further validate the effectiveness of the proposed method.

Acknowledgements

This study has used the strong-motion observation records from the K-NET and KiK-net of the National Research Institute for Earth Science and Disaster Resilience (doi: 10.17598/NIED.0004) as well as the seismic intensity (Shindokei) network of the Japan Meteorological Agency (https://www.jma.go.jp/jma/indexe.html, last accessed November 2020). A part of this study was supported by the Japan Society for the Promotion of Science (JSPS) Kakenhi Grant-in-Aid for Basic Research (B) Number 19H02405. Continuous support from Hanshin Consultants Co., Ltd. is highly appreciated.

Conflict of interest

The authors declare that they have no conflict of interest.

Author details

Eri Ito[1*], Kenichi Nakano[2], Shigeki Senna[3] and Hiroshi Kawase[1]

1 DPRI, Kyoto University, Kyoto, Japan

2 HAZAMA ANDO CORPORATION, Tsukuba, Japan

3 National Research Institute for Earth Science and Disaster Resilience, Tsukuba, Japan

*Address all correspondence to: ito@sere.dpri.kyoto-u.ac.jp

References

[1] Aoi S, Obara K, Hori S, Kasahara K, and Okada Y: New strong-motion observation network: KiK-net. Eos Transactions, AGU. 2000:81: 329.

[2] NIED: J-SHIS [Internet]. 2020. Available from: http://www.j-shis.bosai.go.jp/en/ [Accessed: 2020-11-24]

[3] Andrew D J: Objective determination of source parameters and similarity of earthquakes of different size, Earthquake Source Mechanics (eds. S. Das, J. Boatwright and C. H. Scholz), American Geophysical Union, Washington, D.C., 1986. doi: 10.1029/GM037p0259.

[4] Iwata T and Irikura K: Source parameters of the 1983 Japan-Sea earthquake sequence, J. Phys. Earth, 1986:36, 155-184.

[5] Kawase H and Matsuo H: Amplification characteristics of K-NET, KiK-net, and JMA Shindokei network sites based on the spectral inversion technique, 13th World Conf. on Earthquake Engineering, Vancouver, Canada, 2004:Paper No. 454.

[6] Kawase H: Site effects derived from spectral inversion method for K-NET, KiK-net, and JMA strong-motion network with special reference to soil nonlinearity in high PGA records, Bull. Earthq. Res. Inst., Univ. Tokyo, 2006:81, 309-315.

[7] Nakano K, Matsushima S and Kawase H: Statistical properties of strong ground motions from the generalized spectral inversion of data observed by K-NET, KiK-net, and the JMA Shindokei Network in Japan, Bull. Seism. Soc. Am., 2015:105: 2662-2680, doi:10.1785/0120140349.

[8] Nakano K: Strong-motion prediction method based on the spectral inversion, Ph.D. Dissertation, Graduate School of Engineering, Kyoto University, 2020 (in Japanese).

[9] Borcherdt R D: Effects of local geology on ground motion near San Francisco Bay, Bull. Seism. Soc. Am., 1970:60, No.1, 29-61.

[10] Matsuoka M and Midorikawa S: GIS-based seismic hazard mapping using the digital national land information, Proc. 9th Japan Earthq. Eng. Symp., 1994:E-331-E-336 (in Japanese with English abstract).

[11] Kawase H and Matsuo H: Relationship of S-wave velocity structures and site effects separated from the observed strong motion data of K-NET, KiK-net, and JMA Network, J. Japan Assoc. Earthq. Eng., 2004:4(4), 126-145 (in Japanese with English abstract).

[12] Zhao J: Geometric spreading functions and modeling of volcanic zones for strong-motion attenuation models derived from records in Japan, Bull. Seism. Soc. Am. 2010:100(2): 712-732.

[13] Zhu C, Pilz M, and Cotton F: Which is a better proxy, site period or depth to bedrock, in modelling linear site response in addition to the average shear-wave velocity?, Bull. Earthq. Eng., 2020:18, 797-820, doi: 10.1007/s10518-019-00738-6.

[14] Ito E, Nakano K, Nagashima F, and Kawase H: A method to directly estimate S-wave site amplification factor from Horizontal-to-Vertical spectral ratio of earthquakes (eHVSRs), Bull. Seism. Soc. Am., 2020:110(6): 2892-2911, doi:10.1785/0120190315.

[15] Kawase H, Nagashima F, Nakano K, and Mori Y: Direct evaluation of S-wave

amplification factors from microtremor H/V ratios: Double empirical corrections to "Nakamura" method, Soil Dyn. Earthq. Eng. 2018:126, doi: 10.1016/j.soildyn.2018.01.049.

[16] Senna S, Maeda T, Inagaki Y, Suzuki H, Matsuyama H and Fujiwara H: Modeling of the subsurface structure from the seismic bedrock to the ground surface for a broadband strong motion evaluation, J. Disaster Res., 2014:8, No.5, 889-903.

[17] Senna S, Wakai A, Yatagai A, Jin K, Matsuyama H, Suzuki H and Fujiwara H: Modeling of the subsurface structure from the seismic bedrock to the ground surface for a broadband strong motion evaluation in Japan, Proc. of 7th Int. Conf. of Earthquake and Geotechnical Engineering, Malta, 2019.

[18] Wakai A, Senna S, Jin K, Yatagai A, Suzuki H, Inagaki W, Matsuyama H, and Fujiwara H: Modeling of subsurface velocity structures from seismic bedrock to ground surface in the Tokai region, Japan, for broadband strong ground motion prediction, Journal of Disaster Research 2019:14, No.9, 1140-1153.

[19] The Headquarters for Earthquake Research Promotion: Modeling concept of subsurface structures from seismic bedrock to ground surface, https://www.jishin.go.jp/evaluation/strong_motion/underground_model/ (in Japanese) [accessed November 20, 2020].

[20] Sezawa K and Kanai K: Possibility of free oscillations of strata excited by seismic waves, Bull. Earthq. Res. Inst. 1930:8, 1-11.

[21] Haskell N A: Crustal reflection of plane P and SV waves, J. Geophy. Res. 1962:67, 4751-4767.

[22] Tsai N C: A note on the steady-state response of an elastic half-space, Bull. Seism. Soc. Am., 1970:60, No. 3, 795-808.

[23] Bard P-Y and Bouchon M: Seismic response of sediment-filled valleys, Part I: The case of incident SH waves, Bull. Seism. Soc. Am. 1980:70, 1263-1286.

[24] Kawase H, and Aki K: A study on the response of a soft basin for incident S, P, and Rayleigh waves with special reference to the long duration observed in Mexico City, Bull. Seism. Soc. Am., 1989:79, 1361-1382.

[25] Kawase H: The cause of the damage belt in Kobe: "The basin-edge effect", Constructive interference of the direct S-wave with the basin-induced diffracted/Rayleigh waves, Seismo. Res. Lett., 1996:67, 25-34.

[26] Kawase H and Aki K: Topography effect at critical SV-wave incidence: possible explanation of damage pattern by the Whittier Narrows, California, earthquake of 1 October 1987, Bull. Seism. Soc. Am. 1990:80, 1-22.

[27] Nakano K, and Kawase H: Source parameters and site amplifications estimated by generalized inversion technique: Focusing on the 2018 Hokkaido Iburi–Tobu earthquake, Earth Planets Space 2019:71, 66, doi: 10.1186/s40623-019-1047-1.

[28] Kawase H, Sánchez-Sesma F J, and Matsushima S: The optimal use of horizontal-to-vertical spectral ratios of earthquake motions for velocity inversions based on diffuse-field theory for plane waves. Bull. Seism. Soc. Am. 2011:101, 2011-2014.

[29] Nagashima F, Matsushima S, Kawase H, Sánchez-Sesma F J, Hayakawa T, Satoh T, and Oshima M: Application of horizontal-to-vertical (H/V) spectral ratios of earthquake ground motions to identify subsurface structures at and around the K-NET

site in Tohoku, Japan, Bull. Seism. Soc. Am. 2014:104, 2288-2302, doi: 10.1785/0120130219.

[30] Nagashima F, Kawase H, and Matsushima S: Estimation of horizontal seismic bedrock motion from vertical surface motion based on horizontal-to-vertical spectral ratios of earthquake motions, Proc. of 16th World Conf. Earthq. Eng., Santiago Chile, 9-13 January 2017, Paper N°3685.

[31] Smith F H W and Wessel P: Gridding with continuous curvature splines in tension, GEOPHYSICS, 1990:55, Issue 3, doi:10.1190/1.1442837.

[32] Wessel P, Smith W H F, Scharroo R, Luis J F, and Wobbe F, Generic Mapping Tools: Improved version released, EOS Trans. AGU, 94, 409-410, 2013.

[33] Renka, R : Algorithm 790: CSHEP2D: Cubic Shepard Method for Bivariate Interpolation of Scattered Data, ACM Transactions on Mathematical Software, 1999:25, No. 1, 70-73, doi:10.1145/305658.305737.

Chapter 6

Assessing Seismic Site Response at Areas Characterized by a Thick Buried Low-Velocity Layer

Daniela Farrugia, Pauline Galea and Sebastiano D'Amico

Abstract

Earthquake ground motion is dependent on various factors, including local ground conditions. Whilst many studies have characterized the effect of having outcropping "soft" geological layers which have the ability to amplify ground motion, there is minimal literature on the effect of having such layers embedded between two harder layers. This situation creates a seismic wave velocity inversion. The Maltese islands (Central Mediterranean) present a good opportunity for the study of velocity inversion as almost half of the islands are characterized by a thick buried layer of clay. The results presented in this chapter are a combination of studies which have been conducted on the Maltese islands, using non-invasive geophysical prospecting techniques in areas characterized by a thick buried low-velocity layer, to characterize the response of earthquake ground shaking in such geological situations.

Keywords: low-velocity layers, site response analysis, shear-wave velocity profiling, V_{S30}, seismic building codes

1. Introduction

It is presently well-established that earthquake ground shaking at a particular site is a function of a set of physical parameters and phenomena including the soil conditions beneath the site. Sites characterized by geological layers with a significant impedance contrast are prone to increasing the amplitude of ground motion, changing the frequency content and also lengthening the duration of the earthquake.

Various studies have investigated the effect of outcropping soft geological layers on earthquake ground motion both experimentally and theoretically [1–4]. However, whilst it is usually assumed that the shear-wave velocity (V_S) increases with depth, soft geological layers with low-velocity "sandwiched" between higher velocity layers can also be present. Such situations create what is known as a velocity inversion.

The presence of a buried thick low-velocity layer is not uncommon, especially in sedimentary environments encompassing clay deposits. However, the effects of such stratigraphies on site effects are not often studied or documented. Moreover, seismic building codes such as the Eurocode 8 (EC8), suggest the use of the average shear-wave velocity in the upper 30 m (V_{S30}) to classify sites in different categories.

IntechOpen

Hypothetically this approach is not viable for sites characterised by buried low-velocity layers because the first 30 m could be characterised by high velocity geological material.

The Maltese islands (Central Mediterranean) present the perfect case for studying the effect of buried low-velocity layers. The islands are characterised by a four-layer sequence of limestones and clays [5] (**Figure 1**). While the eastern half of the archipelago is characterised by limestone layers, the western half exhibits limestone plateaux and hillcaps, ranging in thickness from thickness from 2 m up 162 m, covering a clay layer which can be up to 75 m thick [6].

The research presented in this chapter was motivated by the following problems related to the Maltese islands:

- the lack of information about average V_S values of different lithotypes making up the Maltese islands;
- the lack of site response studies on a national scale;
- the lack of knowledge about the effect of the buried clay on site response;
- the V_{S30} might not be the right proxy for site response;
- the lack of a national annex for the Eurocode8.

Even though this study is based and intended for the Maltese islands, results from it can still shed light on the effect of low-velocity layers in any context globally. To tackle the above-mentioned issues, a two-fold process was taken: firstly V_S profiles were obtained using ambient-noise techniques at various sites around

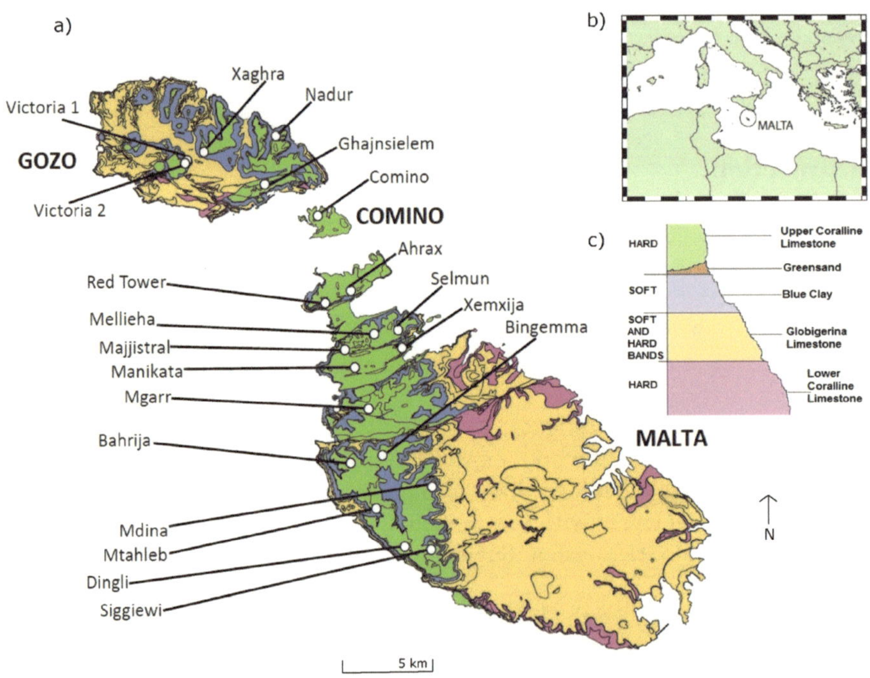

Figure 1.
(a) A geological map of the Maltese islands with the location of the studied sites; (b) the location of the Maltese islands in the Central Mediterranean; (c) schematic showing the geological formations of the Maltese islands.

the islands characterised by buried clay and secondly, these profiles were used as an input to the equivalent-linear site response analysis programme SHAKE2000 to determine the site response.

2. The geology of the Maltese islands

The Maltese archipelago, which consists of three main islands (Malta, Gozo and Comino) and covers an area of around 316 km^2, was formed as marine sediments during the Oligocene and Miocene epochs. The islands lie in the Sicily Channel (**Figure 1b**) on a relatively stable plateau of the African foreland, known as the Pelagian Platform, about 200 km south of the convergent segment of the Europe-Africa plate boundary that runs through Sicily.

Geologically, the islands are made up of four main strata of lime-rich sedimentary rocks, with the composition and texture of each layer depending among other things, on the grain size of the sediment and depth of deposition [7]. Starting from the oldest and the bottom-most layer, the formations are: the Lower Coralline Limestone (LCL), the Globigerina Limestone (GL), the Blue Clay (BC) and the Upper Coralline Limestone (UCL) (**Figure 1c**).

The compact LCL forms most of Malta's southern and south-western coastline along with some inland outcrops associated with faults. The base of the LCL cannot be seen above sea-level and it is exposed through a thickness of up to 140 m. It is non-homogeneous and composed of five different 'facies' according to depositional environment of the sediments [7, 8]. The GL covers large areas of central and southern Malta and Gozo [9]. It is a chalky and soft yellowish fine-grained limestone, which is further subdivided into three layers separated by two thin hardground conglomerate layers. Its thickness can vary from as little as 20 m to over 200 m [5].

The BC layer is the softest in the layer package making it easily erodible. It is mostly found beneath the UCL which is the youngest of the layers. The latter can have variable characteristics, ranging from fractured and friable to highly compact. These two formations are absent in the central and eastern parts of Malta, whereas the western half of Malta and some areas in Gozo retain the full sedimentary sequence. In limited areas, on the uppermost part of the BC layer, one can find a thin layer (between 1 m and 11 m), known as Greensand Formation which is made up of bioclastic limestones rich in the mineral glauconite.

3. Derivation of shear-wave velocities in Maltese rocks using joint inversion of H/V and ESAC curves

Twenty sites have been chosen for this investigation (14 in Malta, 5 in Gozo and 1 in Comino, shown in **Figure 1a**), all of which are characterised by the full sedimentary sequence i.e. the Blue Clay is embedded between the Upper Coralline Limestone above and the Globigerina Limestone below. Since the sites all have similar stratigraphy, any spatial geophysical variations within a particular stratum can also be investigated.

At each site, single-station ambient noise measurements were conducted jointly with geophone array measurements. The sites were chosen not to have any major topographical slopes or irregularities so as to fulfil the 1-D assumption of the array methods. For reasons of clarity, the more detailed results in the next sub-sections are presented only for eight representative sites with a range of stratigraphical characteristics.

3.1 Single-station measurements

Single-station measurements were used to obtain the Horizontal-to-Vertical Spectral Ratio (H/V) for ambient seismic noise. The H/V curve is known to give a peak that matches the S-wave resonance frequency of a site, f_0, which is linked to the S-wave velocity (V_S) of the sedimentary layer and its thickness H by:

$$f_0 = \frac{V_s}{4H}$$

Time-series of 20 minutes each, sampled at 128 Hz, were recorded using the Micromed Tromino™ and analysed using the software Grilla™ to obtain H/V curves in the frequency range of 0.5–64 Hz. The time-series were divided into 60 non-overlapping windows, each 20 s long, as suggested by the SESAME guidelines [10]. Before the analysis, windows which contain any spurious signals were removed to reduce the standard deviation. The H/V curve was obtained by averaging the horizontal spectra using the geometric mean and dividing the mean by the vertical spectrum for each time window. The curves for each window were then averaged to get the final H/V curve [11].

Figure 2 shows the H/V curves obtained at eight representative sites. All curves exhibit a peak between 1 and 2 Hz, the amplitude of which varies between 2 and 5. Previous studies utilising H/V analysis in such areas have also obtained this peak [12–15], which is presumably associated with the boundary separating the BC and the GL [16]. This peak is immediately followed by a drop below 1 in the H/V spectrum over a wide frequency range. This feature has been attributed to the presence of a buried low-velocity layer by [17, 18] and is also evident and consistent in all previous studies of areas of similar lithostratigraphy on the islands.

3.2 Array measurements

The passive seismic array measurements were conducted using Micromed SoilSpy Rosina™ seismic digital acquisition system equipped with 4.5 Hz vertical geophones. The noise signals detected are interpreted as plane Rayleigh waves in their fundamental and higher propagation modes. The number of geophones used varied between 17 and 42 which were placed either in an L- or C-shaped configuration with a regular interstation distance of 5 m, for the majority of the cases. This decision depended on the space available and the expected thickness of the shallow layers. On average the total length of the array was around 150 m and the depth of exploration exceeded 100 m only at a couple of locations. The recordings, each 20 minutes long and sampled at 256 Hz, were analysed using the Extended Spatial

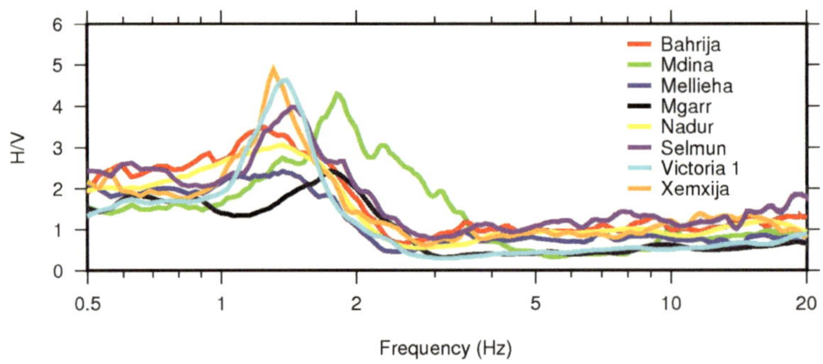

Figure 2.
The H/V curves obtained at eight of the investigated sites.

Auto-Correlation (ESAC) technique and the curves automatically picked by the provided code [19–22].

The ESAC method outputs an effective (or apparent) dispersion curve which in the presence of higher modes, will include a combination of the dispersion curves relative to the relevant modal components [23]. The results for the eight chosen sites are shown in **Figure 3**.

The curves, except in the Mellieha case, exhibit "normal" dispersion characteristics in the low frequency range whereby the effective Rayleigh-wave phase velocity decreases with increasing frequency. At higher frequencies, this trend changes to an inversely dispersive one. This represents an increase in velocity with increasing frequency or decreasing depth. This shape of the effective curves is indicative of the presence of higher modes of surface waves and has been attributed to the presence of a stiff layer overlying a softer one (i.e. UCL and BC, in this case) by various authors (e.g. [23, 24]). In Mellieha, only an inversely dispersive curve was obtained suggesting that the combined thickness of the UCL and BC layers is too high for the GL to be adequately sampled with the given array configuration.

3.3 Data inversion

The H/V and effective dispersion curves were jointly inverted using a Genetic Algorithm (GA; [25]) approach to obtain one-dimensional V_S profiles. The range of allowed values of the most important parameters in the inversion, namely layer thickness and shear-wave velocity, for each site inversion were guided by previous knowledge of the site geology, from geological maps or previous publications. However, the shear-wave velocity in each layer was allowed to vary over a wide range of values, and no indication of where the low-velocity layer is found was given. This was done to assess the ability of the GA to correctly identify and characterize the shear-wave velocity inversion. Up to 10 higher modes were taken into account.

Initially 100 models were randomly generated on which genetic operators (cross-over, mutation and elite selection) are applied for the selection and creation of a second generation of models. The processes were repeated through 150 iterations. For each site, ten separate inversions were run outputting 10 different best-fitting profiles, one from each inversion run. The one profile with the least misfit value from the best 10 profiles was then chosen as the representative profile for the site. The other 9 profiles are useful to estimate the variability and robustness of the final result. **Figures 4** and **5** shows the results of the joint inversion for the eight sites. The V_{S30} was also calculated for each site and is displayed in the figure.

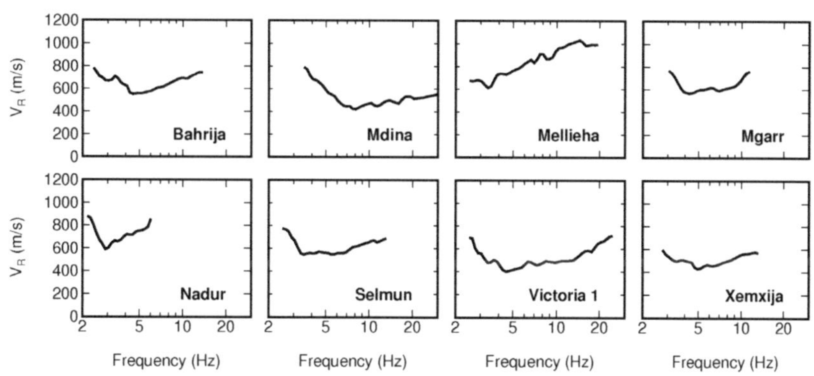

Figure 3.
The effective dispersion curves (Rayleigh-wave vs. frequency) obtained at eight of the investigated sites.

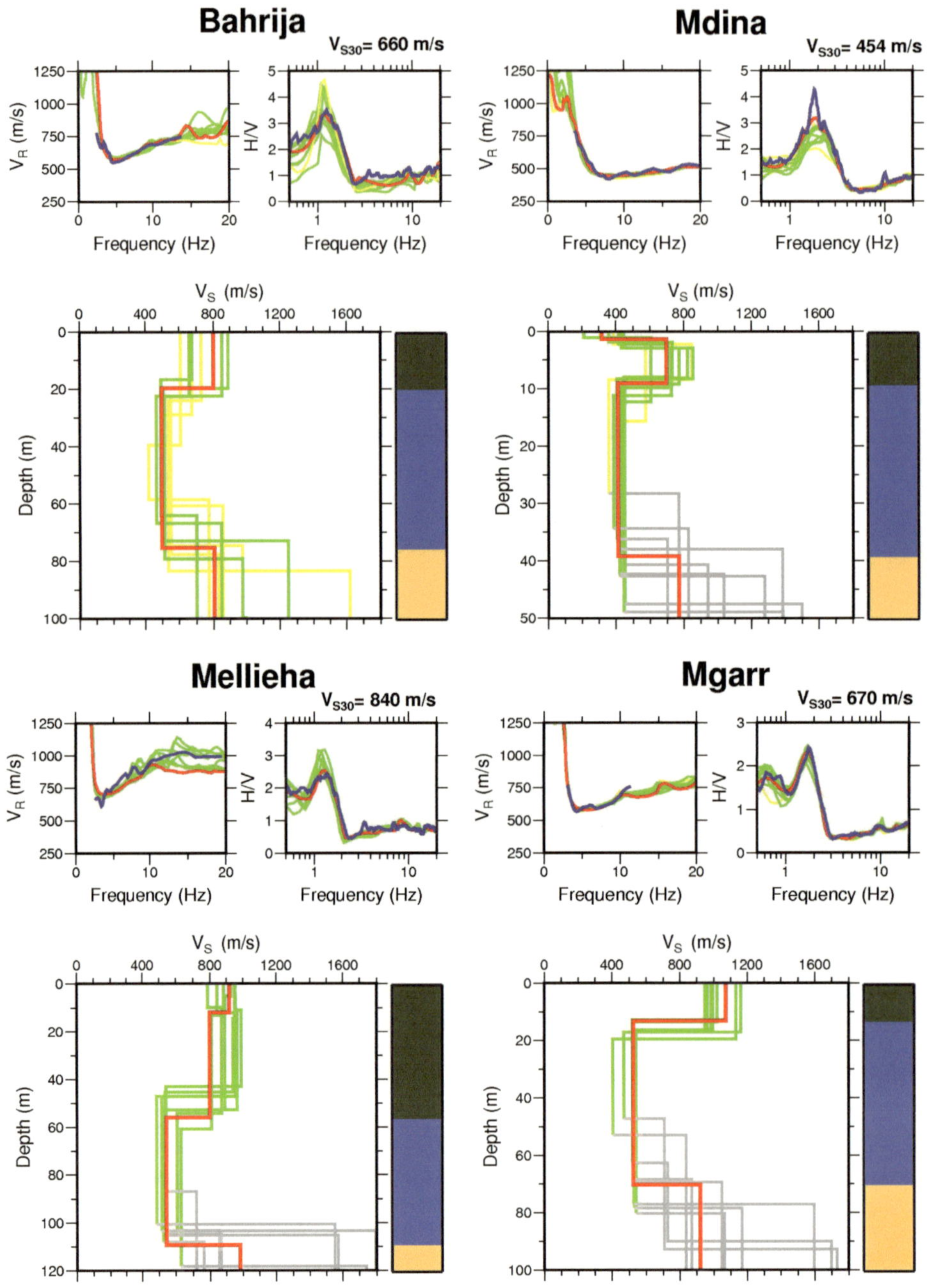

Figure 4.
*The joint inversion results and stratigraphic interpretation (lower panel) for Bahrija, Mdina, Mellieha and Mgarr sites. For each site, the best profiles from each of the 10 inversions are shown, with the red profile representing the one with the lowest misfit. The profiles in green are those characterized by a misfit which is within 50% of the best model's misfit value; the yellow ones are characterized by a misfit greater than 150% of the best model's misfit value. The GL layers are displayed in grey since the values are not reliably constrained by the data. Shown in the upper panel for each site are (from left to right) the effective dispersion and H/V curves. The blue curve is the experimental curve, the red curve shows the best-fitting theoretical curve while the rest (green and yellow) correspond to the other nine profiles. The calculated VS30 for each site is displayed in the top right corner. The colours used in the stratigraphic interpretation correspond to the colours in the geological map (**Figure 1**) [11].*

In general a good match between the theoretical and experimental effective dispersion curves and H/V peak can be observed in all cases. A significantly important feature is that all the final 10 profiles for each site, are in agreement on both the

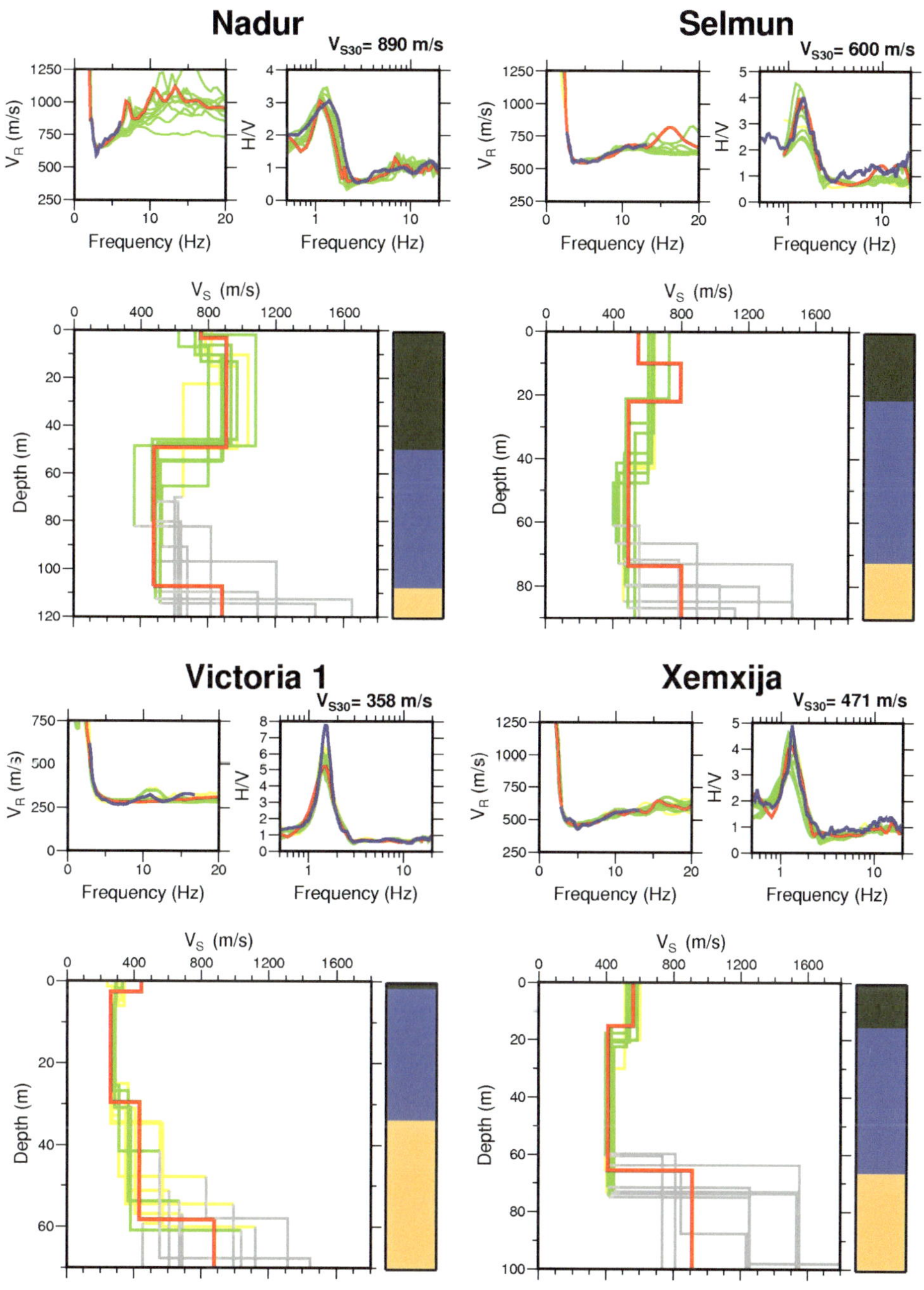

Figure 5.
The joint inversion results for Nadur, Selmun, Victoria 1 and Xemxija sites.

position and S-wave velocity of the low-velocity layer (the Blue Clay layer). Keeping in mind that broad exploration ranges were set in the parametrisations, such an agreement highlights the robustness of the inversion and the sensitivity of the curves to the presence and properties of the low-velocity layer. In addition, this justifies the use of global search methods, such as the GA, which are able to retrieve reasonable profiles without the need of an initial profile close to the solution.

This consistency between the models for a particular site diminishes in the prediction of the velocity of the UCL and more so of the GL layer, where, for example, values between 700 and 1800/s were obtained for the latter. This inconsistency can be attributed to different facts such as the available array conditions, especially length and resonance frequency of geophones which limit the observable depth and the soft BC layer acting as a high-pass filter.

Lithotype	V_S range (m/s)
UCL	550–1100
BC	350–600
GL	700–1400

Table 1.
The V_S ranges for each lithotype obtained from all the studied sites.

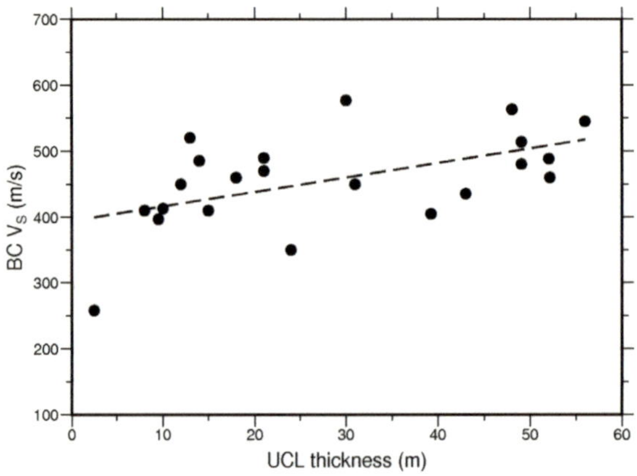

Figure 6.
A graph showing the variation of the BC shear-wave velocity with increasing UCL thickness.

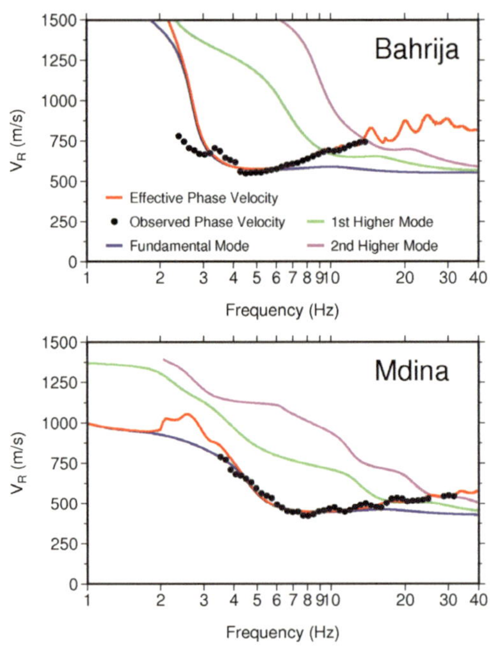

Figure 7.
Comparison of the observed Rayleigh-wave phase velocities (black dots) with the theoretical effective phase velocities and the first three Rayleigh-wave modes for the Bahrija and Mdina sites [11].

The obtained ranges for V_S and thickness for each layer from all the 20 tested sites are shown in **Table 1**. The V_S profiles reveal a variation in the shear-wave velocities of the geological layers at the different sites. Variations in the UCL shear-wave velocities are expected given the fact that the UCL exhibits considerable variation over the islands, ranging from very compact to highly fractured. The V_S in clay varies between 350 and 600 m/s which contrasts with values obtained at sites with outcropping BC layer (between 300 and 400 m/s) [12, 14]. **Figure 6** shows the variation of BC shear-wave velocity with thickness of the overlying UCL considering all the 20 profiles. A trend is clearly visible whereby the higher the thickness of the UCL layer, the higher the shear-wave velocity of the BC layer. This phenomenon can presumably be related to the overburden of the hard UCL layer on the BC, increasing the compactness of the particles, and thus the V_S of the layer.

Finally, in **Figure 7** we show the theoretical individual Rayleigh-wave dispersion curves up to the second higher mode for the best fit models compared with the observed effective dispersion curve for the Bahrija and Mdina models. The computed theoretical effective dispersion curve is also plotted and it can be observed that it fits very well with the observed data. These plots confirm that the effective Rayleigh mode is indeed the superposition of different modes with the higher modes playing an important role in the frequency range when this curve shows an inversely dispersive character [11].

4. Site-specific response analysis

Numerical site-specific response analysis was carried out using the equivalent-linear earthquake site response analysis programme SHAKE2000 [26]. The SHAKE2000 software computes the propagation of shear waves incident vertically on a package of horizontal layers, in which the wave-field in each layer is composed of upward and downward moving waves, whose amplitudes are dependent on the reflectivity/transmission matrices. The programme requires the following three main inputs:

- the soil layer properties (namely the V_S profile);
- the modulus reduction and damping curves for each material;
- the ground motion time-history including the layer number to which the input will be applied.

The V_S profiles obtained from the ambient noise measurements (**Figures 4**, 5 and **8**) were used as input for the soil layer properties for each site. The GL layer was chosen as the bedrock reference layer given that its velocity is generally more than 800 m/s. The modulus reduction and damping curves were chosen from the set of available curves within the package itself after consulting with local geotechnical experts. The chosen curves are displayed in **Figure 9**.

As for ground motion time-history, it is recommended that a suite of records is chosen which are compatible with the national seismic hazard parameters. From the probabilistic seismic hazard analysis conducted by [28], a plausible value for the mean peak ground acceleration (PGA) on rock sites corresponding to a 475-year return period is 0.08g. [28] also note that from the study of historical seismicity and seismotectonic background it is indicated that the seismic hazard of the Maltese islands is related to both moderate magnitude events (M = 5.0–6.0) at short distances (d = 10–40 km) as well as high magnitude events (M = 6.5–8.0) at distances larger than 90 km. In this chapter, the far-field scenario of high magnitude events at

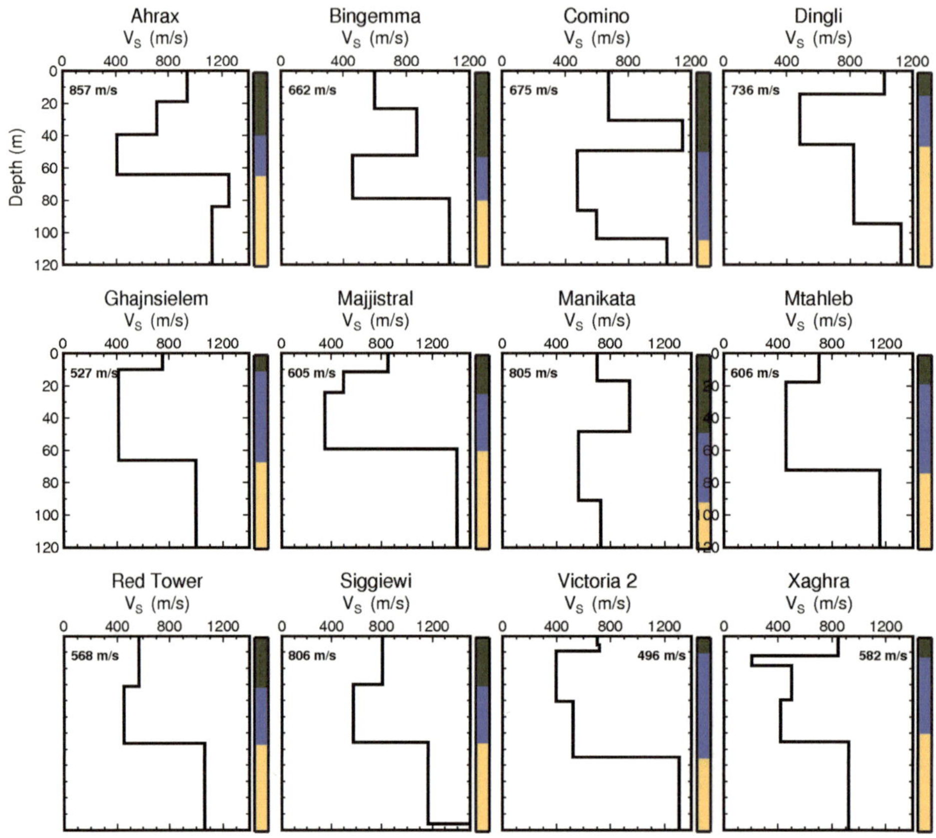

Figure 8.
The best V_S profiles obtained for the 12 sites not shown in ***Figures 4*** *and 5. The procedure of obtaining these profiles is the same as described in Section 3. The coloured bar represents a stratigraphical interpretation using the same colours as in* ***Figures 4*** *and 5.*

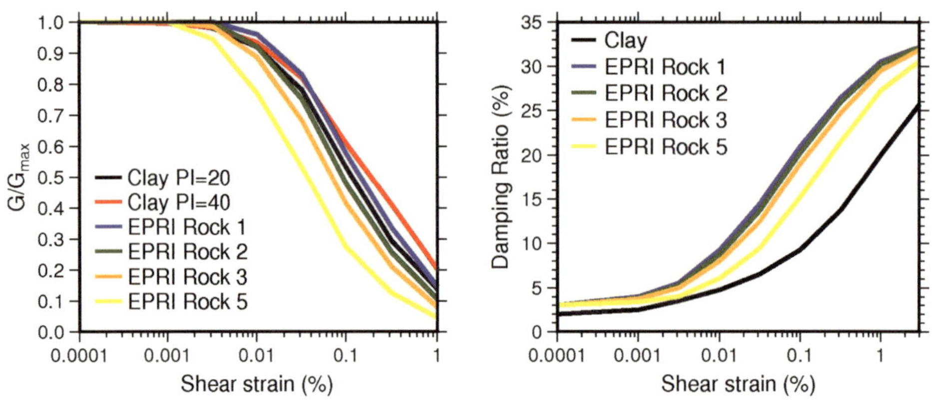

Figure 9.
The modulus reduction (left) and damping (right) curves used for the different layers in the simulation [27].

distances larger than 90 km will be presented considering that the worst case scenario earthquake which has hit the island was the 1693 M7.4 Sicily earthquake.

Since no real data of such earthquakes were ever recorded on the Maltese islands, the "spectrum matching" technique was used to find a suite of seven real accelerograms whose average spectra matched closely a chosen target spectrum. The REXEL code [29], which integrates the European Strong Motion Database, was utilised for this purpose. The EC8 Type 1 spectrum, anchored at a PGA of 0.08 *g* was used as a reference spectrum and the lower and upper limit were set to 90% and 130% of the reference curve respectively (as recommended in the EC8). The search was conducted for magnitudes between 6.5 and 8.0 and distances between 60 and 200 km [28]. Only earthquake records at stations installed on class A sites, according to the EC8, were chosen. The possibility of scaling was allowed with a mean scaling factor not exceeding 5.

Plots of the scaled spectra of the chosen earthquakes together with the EC8 target spectra are presented in **Figure 10**. **Table 2** displays the information about the chosen suite of records.

All the required parameters were inputted in SHAKE2000 and the simulations were run. The software outputs various parameters however we will be focussing mainly on response spectra, PGA values and the theoretical transfer function.

4.1 The response spectra

The 5% damped elastic response spectra were grouped according to the corresponding EC8 site class (based on the V_{S30}) and are displayed in **Figure 11**. Considering the class A sites, it can be observed that all spectra demonstrate a higher response than the EC8 curve at periods longer than 0.5 s with Ahrax and

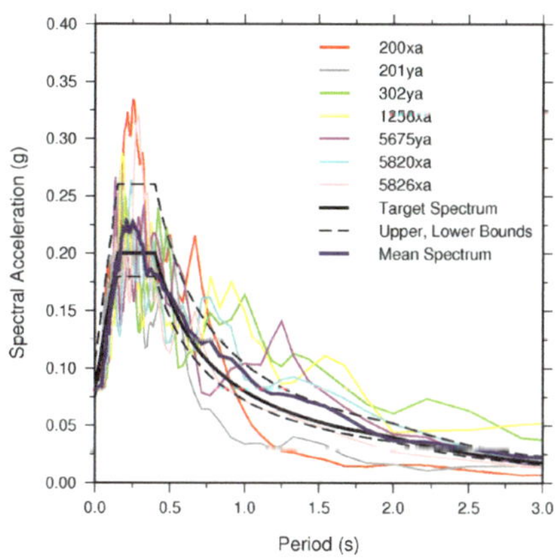

Figure 10.
The response spectra of the compatible combination of acceleration time-histories for the considered scenario. Event information is given in **Table 2** *[27].*

Code	Earthquake Name	Date	MW	Epicentral Distance (km)	Fault Mechanism	Scale Factor
200xa	Montenegro	15/04/1979	6.9	65	Thrust	0.36
201ya	Montenegro	15/04/1979	6.9	105	Thrust	1.07
302ya	Campano	23/11/1990	6.9	92	Normal	5.08
1256xa	Izmit	17/08/1999	7.6	92	Strike-slip	2.27
5675xa	Montenegro	15/04/1979	6.9	180	Thrust	3.67
5820ya	Strofades	18/11/1997	6.6	136	Oblique	0.87
5826xa	Strofades	18/11/1997	6.6	90	Oblique	1.21

Table 2.
Details of the chosen acceleration-time histories [27].

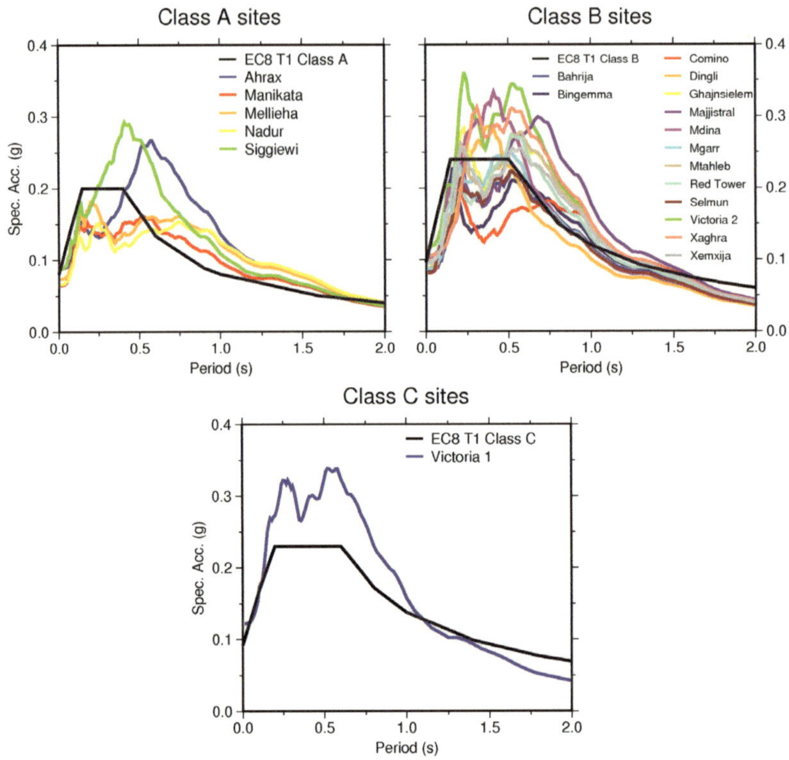

Figure 11.
The response spectra obtained at each site compared to the EC8 recommended site spectra.

Siggiewi showing a well-defined peak reaching up to 150% of the EC8 target spectrum for site class A. On the contrary, at shorter periods (less than 0.2 s), the majority of the sites lie below the EC8 target spectrum. The PGA (spectral acceleration at $T = 0$ s) is also lower than the EC8 target spectrum at almost all the sites. This behaviour is also similar for many of the Class B sites. As regards the class C site (Victoria 1, in Gozo), it is clear that the resulting spectrum is above the EC8 spectrum for a wide period range.

The behaviour highlighted here questions the V_{S30} criterion for site classification and its applicability in such geological settings especially in the context of the design of taller buildings that respond to higher periods of ground shaking.

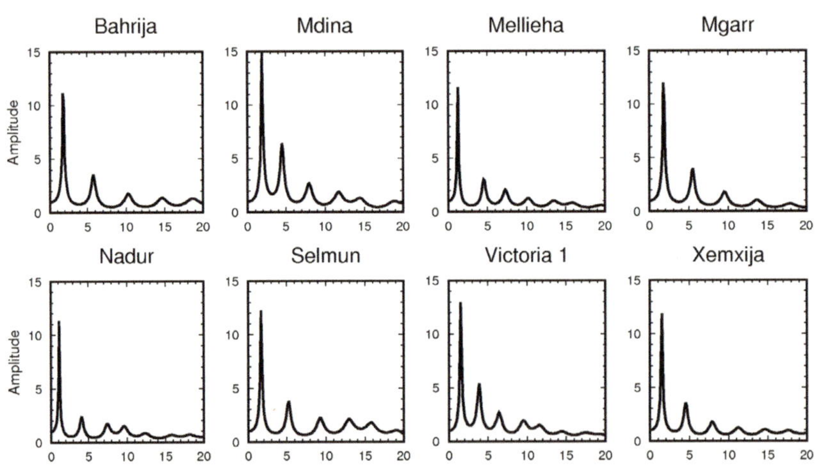

Figure 12.
The transfer functions of the eight sites chosen in Section 3.

Site	H/V peak Frequency (Hz)	Fundamental frequency (Hz)	Predominant period (s)/ Predominant frequency
Ahrax	1.5	1.63	0.58/1.72
Bahrija	1.25	1.75	0.52/1.92
Bingemma	1.47	1.5	0.16/6.25
Comino	0.94	1.25	0.14/7.14
Dingli	1.66	1.5	0.38/2.63
Ghajnsielem	1.34	1.5	0.23/4.35
Majjistral	1.22	1.38	0.68/1.47
Manikata	1.38	1.75	0.52/1.92
Mdina	1.84	1.88	0.41/2.44
Mellieha	1.38	1.25	0.21/4.76
Mgarr	1.75	1.75	0.52/1.92
Mtahleb	1.34	1.75	0.57/1.75
Nadur	1.38	1.13	0.74/1.35
Red Tower	1.41	1.63	0.52/1.92
Selmun	1.47	1.63	0.52/1.92
Siggiewi	1.50	2.13	0.41/2.44
Victoria 1	1.47	1.5	0.57/1.75
Victoria 2	1.41	1.63	0.23/4.35
Xaghra	1.31	1.5	0.31/3.23
Xemxija	1.31	1.5	0.23/4.35

Table 3.
The experimental H/V peak frequency, the fundamental frequency and the predominant period obtained for each study site [27].

Figure 12 shows the resulting transfer functions for the eight chosen sites in Section 3. The fundamental frequency obtained from the 1D transfer function (peak) and the H/V peak frequency obtained for all the sites are tabulated in **Table 3**. The predominant period, which is the period at which the response spectra (**Figure 11**) is the highest, is also tabulated. The fundamental frequencies obtained are in the range 1–2 Hz except for Siggiewi, for which a value slightly higher than 2 was obtained. These are in agreement with the values obtained using the H/V method. The resulting predominant periods range between 0.15 s and 0.75 s. Both the fundamental and predominant periods obtained fall in the range of resonance frequencies of typical 2–10 storey buildings [30, 31], which are becoming increasingly common in the northern part of the islands where the clay is present. Consequently, these buildings might suffer significant damage when this scenario is considered.

4.2 The amplification factors

In order to obtain a comparative measure of the amplification characteristics of different sites from the response spectra, the amplification factors F_{PGA}, FA and FV were calculated, as suggested by [32]. The amplification factors are defined as follows:

$$F_{PGA} = \frac{PGA_{output}}{PGA_{input}};$$

$$FA = \frac{SA_{output}}{SA_{input}};$$

$$FV = \frac{SV_{output}}{SV_{input}}$$

where the input values are obtained from the mean spectrum in **Figure 10** while the output values are obtained from the mean output spectra shown in **Figure 11**. SA and SV are obtained as follows:

$$SA = \frac{\int_{0.5TA}^{1.5TA} S_a \, dT}{TA}$$

where TA is the period at which the spectral acceleration (S_a) is maximum (also called the predominant period in Section 4.1 above); and

$$SV = \frac{\int_{0.8TV}^{1.2TV} S_v \, dT}{0.4TV}$$

with TV representing the period at which the velocity response spectrum (S_v) is maximum.

It can be noted that TA is generally smaller than TV, and that the integral in SA is generally determined over a range that includes shorter periods. In fact [33] deduce that FA is more relevant at shorter periods (< 0.5 s) while FV may be related to the behaviour at longer periods, and thus more relevant in the case of taller buildings. The values of F_{PGA}, FA and FV are mapped in **Figure 13**.

The F_{PGA} and FA values are in the majority of the cases both larger than 1 which implies that amplification is expected. However in a few cases values less than 1 were obtained, indicating that these sites are capable of deamplifying the input ground motion. F_{PGA} values are less than 1 at sites which are characterised by a UCL thickness greater than 48 m and are thus classified as class A according to the EC8

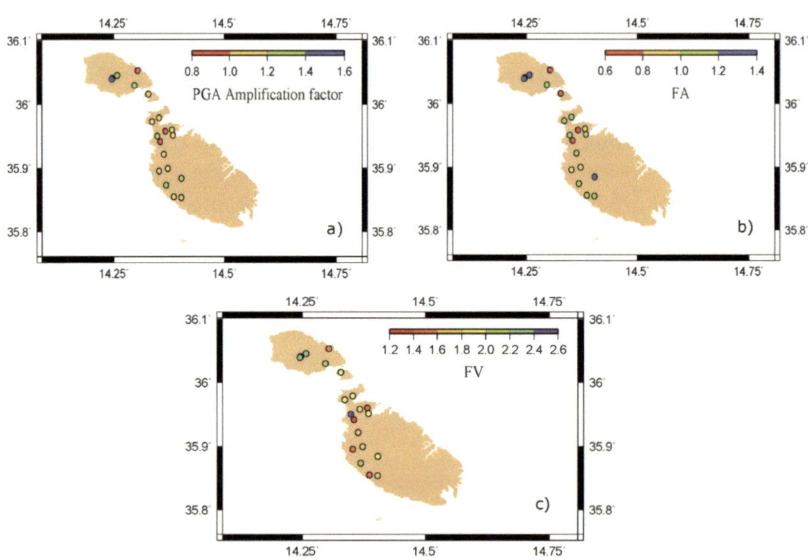

Figure 13.
Maps showing the three amplification factors at the studied sites: (a) F_{PGA}; (b) FA and (c) FV [27].

site classification. It is worth noting however that even though the Bingemma site is characterised by 52 m of UCL, the PGA is still slightly amplified. This could be due to the fact that the UCL is 'layered' into two strata with the upper layer having an average V_S of 600 m/s. In contrast, the highest F_{PGA} values were obtained for those sites where a thin layer of UCL limestone (less than 10 m) outcrops. This is also similar for the amplification factor FA.

In contrast, the amplification factor FV is greater than 1 at all sites, indicating that amplification is expected at longer periods. This is also in agreement with [33] where high FV amplification values were obtained for sites characterised by a low-velocity layer. In general, the same relationship was obtained for FA and FV: sites with thicker UCL exhibited lower amplification values. However, other characteristics of the V_S profiles also play an important role. For example, a higher FV value was obtained at the Majjistral site, which is characterised by a UCL thickness of 24 m, compared to sites which have a thinner UCL layer (e.g. Mdina which has a UCL thickness of 7 m). This could be interpreted as being due to the high impedance contrast (around 4) between the BC and GL layer at the Majjistral site.

To investigate possible correlations between the amplification factors and various parameters of the V_S profiles, the three amplification factors were plotted against: the thickness and V_S of the UCL and BC, the impedance contrast between the BC and underlying GL layer and the V_{S30}. The results are shown in **Figure 14**. The BC thickness exhibits no clear trend with any of the amplification factors. On the other hand, the FV values are seen to increase significantly with decreasing UCL thickness as well as with decreasing UCL and BC shear-wave velocities. These trends are also clear for F_{PGA} and FA, particularly with respect to BC shear-wave velocity. A clear trend can also be observed between FV and the impedance contrast, whereby the FV increases with impedance contrast. Lastly, the three amplification factors can be seen to decrease with an increase in V_{S30}. These observations highlight the role that the different properties which constitute the V_S profile play in site amplification and the difficulty in classifying the sites in rigid groups according to the properties of their upper 30 m of subsoil.

4.3 The in/adequacy of V_{S30}, $V_{Sbedrock}$ and site classification of building codes

Many authors [34, 35] have argued that the V_{S30} is not an adequate proxy to characterize the amplification potential of a site, especially in the type of geological situation being considered here. Since the low-velocity layer is found at a depth which is usually not considered in the V_{S30} calculation, these authors have suggested the use of the travel-time average shear-wave velocity down to the bedrock ($V_{Sbedrock}$) as an alternative.

To assess the reliability of the V_{S30} and $V_{Sbedrock}$ parameters and site classification schemes as proxies for site effects, a number of different shear-wave velocity profiles, all having the same V_{S30} (670 m/s +/− 15 m/s) and $V_{Sbedrock}$ (625 m/s +/− 15 m/s) were randomly constructed. According to the V_{S30} values, these profiles classify as EC8 Class B sites. Within the profiles, the shear-wave velocity and thickness of each layer (UCL, BC, UCL) were constrained to be within the ranges of values measured in the study, shown in **Table 1**. The numerical analysis was conducted again for each of these profiles and the resulting spectra for 4 such profiles, together with the Type 1 EC8 spectra for the different site classes, are presented in **Figure 15**.

Significant differences can be seen between the different response spectra at a wide period range. In particular, the PGA varies from 0.68 g to 0.1 g and the maximum spectral acceleration varies from 0.2 g to almost 0.35 g. As regards the EC8 design spectra, profile 2 is the only profile whose response is comparable with the EC8 site class B design spectrum. The spectrum of profile 4 can be compared

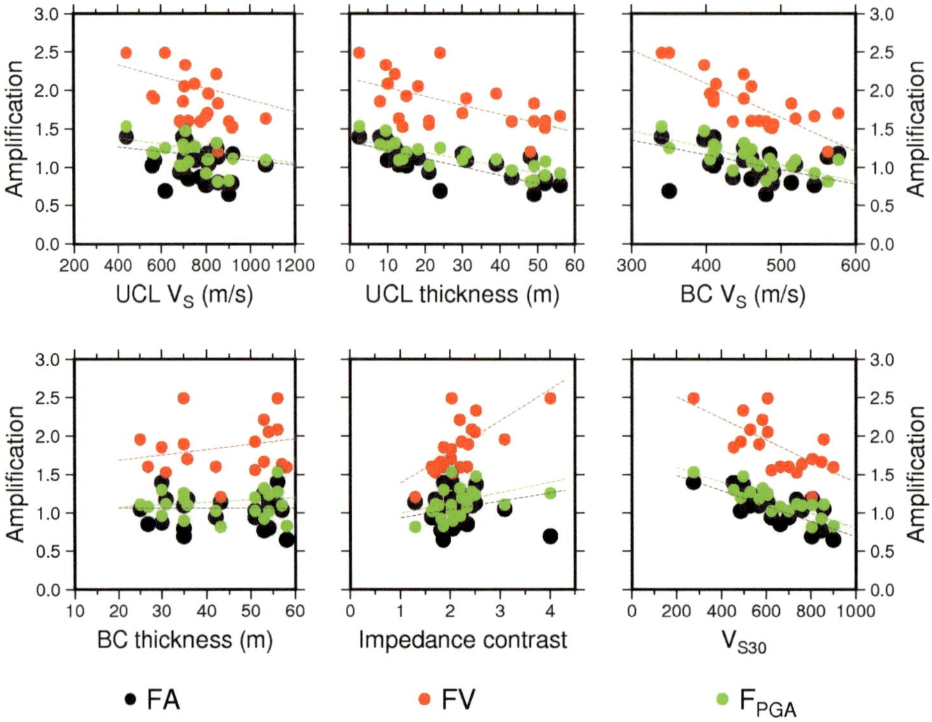

Figure 14.
Graphs showing the variation of the amplification factors with VS profile characteristics [27].

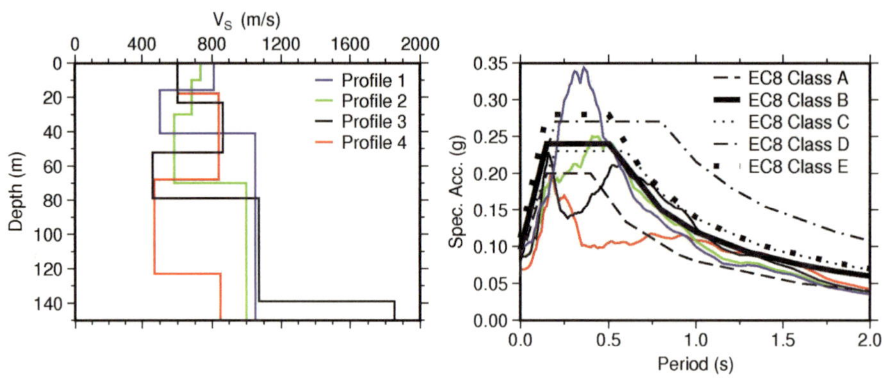

Figure 15.
Left: The hypothetical V_S profiles used for the test. Right: The resulting 5% damped response spectra of the used profiles and the EC8 design spectra.

with the site class A spectrum, profile 3 with site class C, while profile 1 can be seen to exceed the response of a class E site. These results imply that both the V_{S30} and $V_{Sbedrock}$ proxies are not the ideal parameters to use for site response approximations and neither is the use of rigid classes represented by one design spectrum. On the contrary, this continues to support the idea that site-specific response analysis is required, especially for sites characterised by a buried low-velocity layer.

5. Conclusion

Low-velocity layers are known to have the ability to amplify and lengthen the duration of earthquake ground motion. However, their effect when found buried between two geologically harder layers is not well documented. The fact that the Maltese islands are characterized by a thick buried low-velocity layer in almost half of their span provides a good opportunity to study the effect of this type of stratigraphic sequence in a deeper manner. One of the main motivations for this study was the long-standing lack of knowledge within the local scientific and engineering community of how the present, highly densified building stock would fare in the case of a major earthquake in the region. However, the results from this study also provide an important contribution to the international seismological and earthquake engineering community.

A comprehensive investigation was carried out by obtaining V_S profiles at various sites around the islands which were then used as an input to the equivalent-linear site response analysis programme SHAKE2000. The H/V, ESAC and genetic inversion algorithm were used for obtaining V_S profiles at 20 sites and they have been shown to perform very well, particularly in resolving both the presence and the characteristics of a low-velocity layer in the stratigraphy.

The V_S profiles together with dynamic soil properties and design acceleration time histories were then used as inputs to SHAKE2000 to provide the acceleration response spectra at each site. Significant differences were obtained between the EC8 design spectra and some of the site response spectra at various periods. In particular, the response spectrum of the majority of the sites significantly exceeds the EC8 spectra at the plateau, and longer periods of the design spectrum for the site class based on V_{S30}. The predominant period and fundamental frequencies obtained from the transfer function coincide with resonance frequencies of typical 2–10 storey buildings, which are becoming increasingly common in areas where such stratigraphic sequence is present.

From the calculated amplification factors (F_{PGA}, FA and FV), it has been observed that sites with a thin capping of UCL above the clay layer exhibit high amplification factors. However, it was also noted that the other properties of the V_S profiles such as the impedance contrast also contribute to high FV amplification values.

Finally, the inadequacy of using V_{S30} or $V_{Sbedrock}$ to generalize the behaviour of different sites using one design response spectrum was shown by creating layered structures having the same V_{S30} and $V_{Sbedrock}$ but a different V_S profile. The resulting response spectra varied even by a factor of 2 or more at certain periods.

Whereas a site with outcropping clay is normally perceived as vulnerable within the local construction industry, sites on outcropping UCL may often be regarded as "rock sites" without adequate consideration of the effect of the underlying clay. The results here highlight the importance of carrying out site-specific response investigations in areas with buried low-velocity layers, and the need for care in the use of V_{S30} as a proxy for site amplification. The use of non-invasive ambient noise measurements allows a more cost-effective way of obtaining knowledge about the deeper V_S structure, which may influence earthquake response at the site.

Acknowledgements

This work was supported by the Endeavour Scholarship scheme financed out of Government of Malta national fund and also formed part of the SIMIT project (Integrated Italy-Malta Cross-Border System of Civil Protection) (B1-2.19/11) part-financed by the European Union under the ItaliaMalta Cross-Border Cooperation Programme, 2007–2013. The authors are grateful to Dr. D. Albarello and Dr. E.Lunedei for the use of the ESAC and joint inversion codes.

Author details

Daniela Farrugia*, Pauline Galea and Sebastiano D'Amico
University of Malta, Msida, Malta

*Address all correspondence to: daniela.farrugia@um.edu.mt

References

[1] Panzera F, Rigano R, Lombardo G, Cara F, Di Giulio G, Rovelli a. The role of alternating outcrops of sediments and basaltic lavas on seismic urban scenario: the study case of Catania, Italy. Bulletin of Earthquake Engineering. 2010;9(2): 411–439.

[2] Pilz M, Parolai S, Leyton F, Campos J, Zschau J. A comparison of site response techniques using earthquake data and ambient seismic noise analysis in the large urban areas of Santiago de Chile. Geophysical Journal International. 2009;178(2):713–728.

[3] Ansal A, Kurtuluş A, Tönük G. Seismic microzonation and earthquake damage scenarios for urban areas. Soil Dynamics and Earthquake Engineering. 2010;30(11):1319–1328.

[4] Tönük G, Ansal A, Kurtuluş A, Çetiner B. Site specific response analysis for performance based design earthquake characteristics. Bulletin of Earthquake Engineering. 2014;12(3): 1091–1105.

[5] Pedley, H. M., House, M. R., & Waugh, B. The geology of the Pelagian block: the Maltese Islands. The Ocean basins and margins. 1978;417–33.

[6] Zammit-Maempel, G.,. An Outline of Maltese Geology. Progress Press, Malta; 1977.

[7] Pedley M, Clarke MH, Galea P. Limestone Isles in a Crystal Sea. The geology of the Maltese islands. 2002.

[8] Gatt, P. Carbonate facies, depositional sequences and tectonostratigraphy of the Palaeogene Malta Platform [Doctoral dissertation]. Durham University; 2012.

[9] Magri O. A Geological and Geomorphological review of the Maltese Islands with special reference to the coastal zone. Territoris. 2006;6:7–26.

[10] Bard P-Y. Guidelines for the implementation of the H/V spectral ratio technique on ambient vibrations measurements, processing and interpretation. SESAME European research project WP-12-Deliverable D2312, European Commission – Research General Directorate Project No EVG1-CT-2000-00026 SESAME [Internet]. 2004;(December). Available from: ftp://ftp.geo.uib.no/pub/seismo/SOFTWARE/SESAME/USER-GUIDELINES/SESAME-HV-User-Guidelines.pdf

[11] Farrugia F., Paolucci E., D'Amico S., Galea P. Inversion of surface-wave data for subsurface shear-wave velocity profiles characterised by a thick buried low-velocity layer. Geophysical Journal International. 2016;206:1221–31.

[12] Vella A, Galea P, D'Amico S. Site frequency response characterisation of the Maltese islands based on ambient noise H/V ratios. Engineering Geology. 2013 Aug;163:89–100.

[13] Pace S. Shallow Shear Wave Velocity Structure in the Maltese Islands. Master's Thesis. 2012;

[14] Panzera F, D'Amico S, Galea P, Lombardo G, Gallipoli MR, Pace S. Geophysical measurements for site response investigation: preliminary results on the island of Malta. Bollettino di Geofisica Teorica ed Applicata. 2013; 54:111–128.

[15] Galea P, D'Amico S, Farrugia D. Dynamic characteristics of an active coastal spreading area using ambient noise measurements–Anchor Bay, Malta. Geophysical Journal International. 2014 Sep;199(2):1166–1175.

[16] Panzera F, Lombardo G. Seismic property characterization of lithotypes

cropping out in the Siracusa urban area, Italy. Engineering Geology. 2012;153: 12–24.

[17] Castellaro S, Mulargia F. The effect of velocity inversions on H/V. Pure and Applied Geophysics. 2009 Apr;166(4): 567–592.

[18] Di Giacomo D. Analysis and Modeling of HVSR in the Presence of a Velocity Inversion: The Case of Venosa, Italy. Bulletin of the Seismological Society of America. 2005 Dec;95(6): 2364–2372.

[19] Ohori M, Nobata A, Wakamatsu K. A Comparison of ESAC and FK Methods of Estimating Phase Velocity. Bulletin of the Seismological Society of America. 2002;92(6):2323–2332.

[20] Okada H. The Microtremor Survey Method. Tulsa, Oklahoma: Geophysical Monograph Series No. 12, Society of Exploration Geophysicists; 2003.

[21] Parolai S, Richwalski SM, Milkereit C, Fäh D. S-wave Velocity Profiles for Earthquake Engineering Purposes for the Cologne Area (Germany). Bulletin of Earthquake Engineering. 2006 Feb;4:65–94.

[22] Albarello D, Cesi C, Eulilli V, Guerrini F, Lunedei E, Paolucci E, et al. The contribution of the ambient vibration prospecting in seismic microzoning: an example from the area damaged by the April 6, 2009 L'Aquila (Italy) earthquake. Bollettino di Geofisica Teorica ed Applicata. 2011;52 (September):513–538.

[23] Tokimatsu K. Geotechnical site characterization using surface waves. Earthquake Geotechnical Engineering, Ishihara (ed). 1997;1333–1368.

[24] Arai H, Tokimatsu K. S-wave velocity profiling by joint inversion of microtremor dispersion curve and Horizontal-to-Vertical (H/V) spectrum. Bulletin of the Seismological Society of America. 2005 Oct;95(5):1766–1778.

[25] Picozzi M, Albarello D. Combining genetic and linearized algorithms for a two-step joint inversion of Rayleigh wave dispersion and H/V spectral ratio curves. Geophysical Journal International. 2007 Apr;169:189–200.

[26] Ordóñez GA. SHAKE2000 for the 1-D Analysis of Geotechnical Earthquake Engineering Problems User ' s Manual.

[27] Farrugia, D., Galea, P., D'Amico, S. Modelling and assessment of earthquake ground response in areas characterised by a thick buried low-velocity layer. Natural Hazards. 2020;

[28] Panzera F, Longo E, Langer H, Lombardo G, Branca S. Site classification scheme for a complex geologic area : The study case of Mt . Etna. 2015;(November).

[29] Iervolino I, Galasso C, Cosenza E. REXEL: Computer aided record selection for code-based seismic structural analysis. Bulletin of Earthquake Engineering. 2010;8(2): 339–362.

[30] Galea, P., Micallef, T., Muscat, R., & D'Amico, S. Resonance frequency characteristics of buildings in Malta and Gozo using ambient vibrations. In: Georisks in the Mediterranean and their mitigation. 2015.

[31] Panzera, F., D'Amico, S., Lombardo, G., & Longo, E. Evaluation of building fundamental periods and effects of local geology on ground motion parameters in the Siracusa area, Italy. Journal of Seismology. 2016;20(3):1001–19.

[32] Working Group MS. Guidelines for seismic Microzonation. [Internet]. 2015. Available from: http://www.protezionecivile.gov.it/httpdocs/cms/attach_extra/GuidelinesForSeismicMicrozonation.pdf?

[33] Compagnoni M., Pergalani F., and Boncio P. Microzonation study in the Paganica-San Gregorio area affected by the April 6 , 2009 L'Aquila earthquake (Central Italy) and implications for the reconstruction. Bulletin of Earthquake Engineering. 2011;9:181–98.

[34] Pitilakis K, Riga E, Anastasiadis A. New code site classification, amplification factors and normalized response spectra based on a worldwide ground-motion database. Bulletin of Earthquake Engineering. 2013;11(4): 925–966.

[35] Gallipoli MR, Mucciarelli M. Comparison of Site Classification from VS30, VS10, and HVSR in Italy. Bulletin of the Seismological Society of America. 2009 Feb;99(1):340–351.

Chapter 7

Elucidation of Seismic Soil Liquefaction Significant Factors

Mahmood Ahmad, Xiaowei Tang, Feezan Ahmad, Marijana Hadzima-Nyarko, Ahsan Nawaz and Asim Farooq

Abstract

The paper develops a framework to analyze the interactions among seismic soil liquefaction significant factors using the interpretive structural model (ISM) approach based on cone penetration test. To identify the contextual relationships among the significant factors, systematic literature review approach was used bearing in mind the selection principle. Since multiple factors influence seismic soil liquefaction, determining all factors in soil liquefaction would be extremely difficult, as even a few seismic soil liquefaction factors are not easy to deal with. This study highlighted two main characteristics of seismic soil liquefaction factors. First, the seismic soil liquefaction factors–peak ground acceleration F_2 ($a_{\max}$), equivalent clean sand penetration resistance F_5 (q_{c1Ncs}), and thickness of soil layer F_{11} (T_s) influenced soil liquefaction directly and were located at level 2 (top level) in the ISM model, meaning they require additional seismic soil liquefaction factors except thickness of soil layer F_{11} (T_s) to collaboratively impact on soil liquefaction potential. The multilevel hierarchy reveals that depth of soil deposit F_{10} (D_s) is formed the base of ISM hierarchy. Secondly, Matrice d'impacts croisés multiplication appliqués à un classement (MICMAC) analysis has been employed for evaluating these identified factors in accordance with driving power and dependence power. Factors with a higher driving power should be given special consideration. Autonomous soil liquefaction factors have no reliance on other soil liquefaction factors and interfere less. In order to identify the significant factors that affect seismic soil liquefaction susceptibility, the model built in this study clearly illustrates the complex relationships between factors and demonstrates the direct and indirect relationships.

Keywords: Soil liquefaction, Interpretive structural modeling, MICMAC, Cone penetration test

1. Introduction

Seismic soil liquefaction is one of the most complicated geotechnical earthquake engineering problems due to the variability and complexity of site conditions, soil parameters and seismic parameters. All those parameters having a number of factors that cause liquefaction, all of which are of varying importance. Estimating accurate and effective soil liquefaction risks, required identification and benchmarking of the most influential factors that control soil liquefaction need to be comprehensively examined. Limited research has been conducted in the past to identify important parameters of soil liquefaction. Dalvi et al. [1] used the Analytic Hierarchy Process and entropy methods to identify important parameters among 16 factors of soil liquefaction. Zhu [2] analyzed fifteen influencing factors of soil liquefaction by mathematical statistics method. Tang et al. [3] and Ahmad et al. [4] identified significant soil liquefaction factors by employing bibliometric and systematic literature review techniques based on standard penetration test respectively through interpretive structural modeling (ISM) approach. Most of these studies considered the quantification rather than the qualitative information of soil liquefaction factors from scientific publications.

Seismic parameter, soil parameter and site conditions contain variety of factors that trigger liquefaction and discussed in detail in Section 3. As literature review search is the first step in the ISM technique to identify the important factors and their underlying relationships. Therefore, a systematic literature review (SLR) approach is used for this purpose which is described by Okoli and Schabram [5] and Tranfield et al. [6] is used. Warfield developed the ISM method between 1971 and 1974 [7], and it is based on the pair-wise comparison theory. ISM has seen some progress in terms of applications and techniques over the years [8]. Michel Godet and François Bourse introduced the Matrice d'impacts croisés multiplication appliqués à un classement (MICMAC) method. The creation of a graph that classifies factors based on driving power and dependency power is called MICMAC.

In this chapter, ISM and MICMAC methodologies are used to establish and analyze the structural hierarchical relationship and to examine the strength of the relationship between seismic soil liquefaction significant factors based on their driving power and dependence power.

2. Methodology

2.1 Interpretive structural modeling: a qualitative technique

ISM methodology, as interpretive in judgment, can be used as a systematic means of recognizing the contextual relationships between the elements associated with an issue to be examined [9]. The ISM approach has been effectively utilized in diversified set of problems, for instance, risk management in supply chains [10] and energy conversation [11]. ISM can be illustrated in the following steps for the present study, as suggested by Sushil [8]:

Step 1: Identification of factors related to the problem or issue through literature review etc.

Step 2: Using domain information, fix contextual relationships between defined factors (e.g. V–row factor influences the column factor; A–column factor influences the row factor; O–no relationship between the row and column factors; or X–both direction relations from row to column and column to row factors).

Step 3: Construct a structural self-interaction matrix (SSIM) based on pair-wise comparison between factors of system which denotes direct relationship between two factors.

Step 4: SSIM is converted to initial reachability matrix, by replacing 1 or 0 for the original symbols–V, A, X and O as per the rules for transformation (see **Table 1**).

Step 5: The transitivity of initial reachability matrix is checked in order to develop the final reachability matrix. The transitive relationships mean that if variable "x" is associated with variable "y" and variable "y" is associated with variable "z", then variable "x" is certainly associated to variable "z".

If the (*i*, *j*) entry in the SSIM is	Entry in the initial reachability matrix	
	(*i*, *j*)	(*j*, *i*)
V	1	0
A	0	1
X	1	1
O	0	0

Table 1.
Rules for transformation.

Step 6: The reachability and antecedent sets of factors are developed from the final reachability matrix. The reachability set for a particular factor includes the factor itself and other factors which it may help to achieve, and antecedent set includes factor itself and other factors that can help in achieving it. Subsequently, the intersection of these sets is found for the entire factors. The factor for which reachability and intersection sets are identical is listed in the first level. This factor is then separated from other factors for the next iteration process. Repeat the same level of iteration process until all levels of each factor are established.

Step 7: Remove the transitivity links and draw a directed graph (digraph) from the final reachability matrix.

Step 8: Convert the digraph into an ISM-based hierarchical model by replacing the nodes with statements.

Step 9: The conceptual discrepancy of model is verified and improved for necessary modifications and corrections.

2.2 MICMAC analysis

The creation of a graph that classifies factors based on driving and dependency power is a part of the MICMAC study. To arrive at the study's findings and conclusions, MICMAC analysis is used to identify the factors and validate the interpretive structural model factors.

Factors are divided into four clusters based on their driving power and dependency power in MICMAC analysis. The clusters are: Cluster I: Autonomous factors —those that are relatively cut off from the rest of the system and have little or no dependency on others; Cluster II: Dependent factors—cluster II factors are primarily dependent of other factors; Cluster III: Linkage factors—the connecting factors that are unstable and have strong driving power and strong dependence power; and Cluster IV: Independent factors—these factors have weak influence from others factors and have to be paid maximum attention owing to the strong driving power.

3. Application to the case of illustration

3.1 Interpretive structural model of seismic soil liquefaction significant factors

In the ISM technique, a first endeavor is made to ascertain the significant seismic of soil liquefaction factors from the literature using systematic literature review (SLR) approach which is recommended by Okoli and Schabram [5]. The SLR is a systematic, explicit, and reproducible method for identifying, evaluating, and synthesizing the existing body of completed and recorded work published by researchers, scholars, and practitioners [12].

There are three groups of parameters that govern the soil liquefaction phenomenon, according to published research papers, namely seismic parameters, site conditions, and soil parameters [13–18]. Each of these contains a wide range of factors that characterize liquefaction, to a varying degree of significance. The details of these parameters are given below.

3.1.1 Seismic parameter

The vulnerability of any cohesionless soil to liquefaction during an earthquake depends on the magnitude and number of cycles of stresses or strains caused by the seismic excitation. These in turn are correlated to the intensity, duration of ground shaking and predominant frequency. The degree of soil liquefaction varies with the different earthquake magnitude. Based on on-site observations and a simple parametric study, Green and Bommer [19] have concluded that a small earthquake with a moment magnitude of 4.5 will trigger liquefaction in highly susceptible soil deposits. However, for soil profiles suitable for building structures, the minimum earthquake magnitude is about 5 that cause liquefaction. Tesfamariam and Liu [20] considered the Stark and Olson [21] earthquake liquefaction datasets and intuited that with increase in M and $a_{\max}$, the likelihood of liquefaction increases. Peak ground acceleration (PGA) is a function of earthquake magnitude, site to fault distance, fault type and soil type as per Boore et al. [22] and usually used to quantify the ground motion intensity.

Pirhadi et al. [14] used closest distance to rupture surface which is among the other seismic parameters such as earthquake magnitude and peak ground acceleration as an influence factor and concluded that among the seismic parameters earthquake magnitude, peak ground acceleration and closest distance to rupture surface illustrate lesser effects on liquefaction triggering as compared to the cumulative absolute velocity. It is generally agreed, that earthquake magnitude, peak ground acceleration, and closest distance to rupture surface are the three major factors that affect the seismic intensity at the site.

3.1.2 Soil parameter

Liquefaction is usually observed in shallow, loose, saturated cohesionless soils subjected to strong ground motions. In case of in-situ cone penetration test, soil behavior type index is used to classify soils based on fines content presented by Robertson and Wride [23]. The liquefaction susceptibility depends on soil type, where fine-size particles are easier to liquefaction than coarse particles.

The type of soil that is more prone to liquefaction is one in which deformation resistance is mobilized by particle friction. When other factors like grain shape, uniformity coefficient, and relative density are held constant, the frictional resistance of cohesion less soils decreases as grain size decreases. Gravelly soils mobilize more strength during shearing and dissipate excess pore pressures more quickly than sandy soils. There are some case histories [24–26] that show liquefaction in loose gravelly soils during severe ground shaking or when the gravel layer is confined by an impervious layer.

The strength of soil liquefaction may vary depending on the fines content. Several studies have found that fines content has a significant impact on soil susceptibility to liquefaction [24–26]. Soil liquefaction potential increases as fines content exceeds 30%. When fines content exceeds 50%, however, the soil's liquefaction potential is reduced [27].

Zhou et al. [27] concluded that the cone tip resistance (q_c) factor is sensitive among the predictor variables in CPT in-situ test method, which provides meaningful guiding significance for the subsequent prediction of seismic liquefaction potential. Furthermore Ahmad et al. [18] concluded that cone tip resistance (q_c) has a considerable influence on liquefaction triggering. Furthermore, Ahmad et al. [28] used the equivalent clean sand penetration resistance (q_{c1Ncs}) to decrease uncertainty and has found the strongest influence on liquefaction potential.

3.1.3 Site condition

It is widely known that the increase in the vertical effective stress increases the bearing capacity and shear strength of soil, and consequently increases the shear stress required to cause liquefaction and decreases the potential for liquefaction. Many researchers have reported that saturated sands deeper than 15 to 18 m are not probably to liquefy [29]. These depths are in general agreement to Kishada [30], who states that a saturated sandy soil is not liquefiable if the value of the vertical effective stress exceeds 190 kN/m^2. It is reported that an increase in the overburden pressure the occurrence of liquefaction decreases [31, 32]. Tesfamariam and Liu [20] considered the Stark and Olson [21] earthquake liquefaction datasets and, intuited that, with a decrease in vertical effective stress, the likelihood of soil liquefaction increases. As vertical effective stress is conditioned on total vertical stress therefore accordingly, total and vertical effective stresses are included in the proposed model as governing factors.

In order to induce extensive damage at ground surface level due to liquefaction, the liquefied soil layer must be sufficient thick thereby resulting uplift pressure and amount of water expelled from the liquefied layer can result in ground damage such as sand boiling and fissuring (Ishihara [26]; Dobry [33]). If the liquefied sand layer is thin and deposited within the soil profile, the presence of a non-liquefiable surface layer may prevent the effects of the at-depth liquefaction from reaching the surface. Ishihara [26] established a standard that specifies a threshold value for the thickness of a non-liquefiable surface layer to avoid ground damage due to liquefaction.

It was intuited in the survey report prepared by Japan society of Civil Engineers that the big-sized earthquake liquefied the sand layer when the thickness is more than 3.0 m. When the thickness of the liquefied layer is very thin, the presence of a non-liquefiable surface layer may prevent the effects of the in-depth liquefaction from reaching the surface.

The resistance of soil to liquefaction is weakened as groundwater levels rise. The effect on soil liquefaction potential increases as groundwater levels rise above 2 m [34]. The water table regime must be minimized as one of the design criteria against seismic soil liquefaction [35]. The vertical effective stress is closely related to the depth of soil deposit. The vertical effective stress increases as the depth of the soil deposit increases. Increased vertical stress has been shown to improve the soil's bearing capacity and shear strength, reducing the risk of liquefaction. Even liquefaction from very loose sand is almost impossible for over 15 m of overburden, according to Florin and Ivanov [36], and Satyam [37] concluded the same for the preliminary assessment of the soil liquefaction potential in a seismically active region.

The significant factors of seismic soil liquefaction that are identified through SLR approach are presented in **Table 2**.

Field experts' examined and analyzed the preliminary list and they believed that the soil liquefaction factors retrieved from the literature were important for expanding exploratory research by developing structural self-interaction matrix for interpretive structural modeling. The set of liquefaction factors identified in **Table 2** for seismic soil liquefaction potential was used to develop the model which represented the correlation between eleven seismic soil liquefaction factors. In the ISM model, for the development of the structural self-interaction matrix (SSIM),

Factor (F_i) code	Significant factor
F_1	Earthquake magnitude, (M)
F_2	Peak ground acceleration, ($a_{\max}$)
F_3	Closest distance to rupture surface, (R/r_{rup})
F_4	Fines content, (FC)
F_5	Equivalent clean sand penetration resistance, (q_{c1Ncs})
F_6	Soil behavior type index, (I_c)
F_7	Vertical effective stress, (σ'_v)
F_8	Total vertical stress, (σ_v)
F_9	Groundwater table, (D_w)
F_{10}	Depth of soil deposit, (D_s)
F_{11}	Thickness of soil layer, (T_s)

Table 2.
List of significant factors of seismic soil liquefaction.

F_1	F_2	F_3	F_4	F_5	F_6	F_7	F_8	F_9	F_{10}	F_{11}	F_{12}	F_i
	V	O	O	O	O	O	O	O	O	O	V	F_1
		A	O	O	A	O	O	O	O	O	V	F_2
			O	O	O	O	O	O	O	O	V	F_3
				V	V	O	O	O	O	O	V	F_4
					A	A	O	A	A	O	V	F_5
						A	A	O	O	O	V	F_6
							A	A	A	O	V	F_7
								O	A	O	V	F_8
									O	O	V	F_9
										O	V	F_{10}
											V	F_{11}
												F_{12}

Note: V–row factor influences the column factor; A–column factor influences the row factor; O–no relationship between the row and column factors.

Table 3.
Structural self-interaction matrix for seismic soil liquefaction factors.

pair-wise comparison were made by the correlation criteria and four symbols V, A, X, or O were used (see **Table 3**). For example, earthquake magnitude F_1 (M)–row factor influences the peak ground acceleration, F_2 ($a_{\max}$)– column factor so the symbol used is V. Groundwater table depth F_9 (D_w)–column factor influences the vertical effective stress, F_7 (σ'_v)–row factor so the symbol used is A. Earthquake magnitude F_1 (M)–row factor has no relation with the thickness of soil layer F_{11} (T_s)–column factor so the symbol used is O. Field experts' made consensus on the pair-wise comparison and the results are shown in **Table 3**.

SSIM is converted to a binary matrix called the initial reachability matrix by replacing the original symbols V, A, X, and O with 1 or 0 (**Table 4**) as per the rule illustrated in **Table 1**. When pair of the same factor, i.e., F_1 (M) with F_1 (M) is formed, it is represented by 1. The concept of transitivity is introduced in **Table 4**

F_i	F_1	F_2	F_3	F_4	F_5	F_6	F_7	F_8	F_9	F_{10}	F_{11}	F_{12}
F_1	1	1	0	0	0	0	0	0	0	0	0	1
F_2	0	1	0	0	0	0	0	0	0	0	0	1
F_3	0	1	1	0	0	0	0	0	0	0	0	1
F_4	0	0	0	1	1	1	0	0	0	0	0	1
F_5	0	0	0	0	1	0	0	0	0	0	0	1
F_6	0	1	0	0	1	1	0	0	0	0	0	1
F_7	0	0	0	0	1	1	1	0	0	0	0	1
F_8	0	0	0	0	0	1	1	1	0	0	0	1
F_9	0	0	0	0	1	0	1	0	1	0	0	1
F_{10}	0	0	0	0	1	0	1	1	0	1	0	1
F_{11}	0	0	0	0	0	0	0	0	0	0	1	1
F_{12}	0	0	0	0	0	0	0	0	0	0	0	1

Table 4.
Initial reachability matrix.

F_i	F_1	F_2	F_3	F_4	F_5	F_6	F_7	F_8	F_9	F_{10}	F_{11}	F_{12}	Dri.	Rank
F_1	1	1	0	0	0	0	0	0	0	0	0	1	3	III
F_2	0	1	0	0	0	0	0	0	0	0	0	1	2	II
F_3	0	1	1	0	0	0	0	0	0	0	0	1	3	III
F_4	0	1*	0	1	1	1	0	0	0	0	0	1	5	V
F_5	0	0	0	0	1	0	0	0	0	0	0	1	2	II
F_6	0	1	0	0	1	1	0	0	0	0	0	1	4	IV
F_7	0	1*	0	0	1	1	1	0	0	0	0	1	5	V
F_8	0	1*	0	0	1*	1	1	1	0	0	0	1	6	VI
F_9	0	0	0	0	1	1*	1	0	1	0	0	1	5	V
F_{10}	0	0	0	0	1	1*	1	1	0	1	0	1	6	VI
F_{11}	0	0	0	0	0	0	0	0	0	0	1	1	2	II
F_{12}	0	0	0	0	0	0	0	0	0	0	0	1	1	I
Dep.	1	7	1	1	7	6	4	2	1	1	1	12	44/44	
Rank	VI	II	VI	VI	II	III	IV	V	VI	VI	VI	I		VI/VI

Note: 1 indicates the values after applying transitivity; Dep. represents dependence power; Dri. represents driving power.*

Table 5.
Final reachability matrix.

when the initial reachability matrix has been obtained and is presented in final reachability matrix (**Table 5**), wherein entries marked (*) show the transitivity. For example, in **Table 4**, the initial reachability matrix shows that F_4 (*FC*) is related to F_6 (I_c), and F_6 (I_c) is related to F_2 (a_{max}), then the interaction F_4 (*FC*) and F_2 (a_{max}) having 0 value is transformed into 1* in **Table 5**. The reachability sets are determined from the factor itself and other factors which have influence in the horizontal direction, while the antecedent sets consist of the factor itself and other factors which have influence in the vertical direction for each significant soil liquefaction factor. For example, in the case of F_1 (M) in the final reachability matrix (**Table 5**), the reachability set will be all factors with values of 1 or 1* in the row intersections with F_1 (M): F_1 (M), F_2 (a_{max}), and F_{12} (LP). The antecedent set is all factors with values of 1 or 1* in the column intersections with F_1 (M): F_1 (M) only. When the intersection and reachability sets appear in the same intersection and reachability columns, then the corresponding factor is confirmed at a level (e.g. F_{12} (LP) as level 1) and the factor in that level is separated out, e.g. F_{12} (LP) from the other factors for the next level-iteration process.

The same level-iteration process is repeated until the level of each seismic soil liquefaction factor is established. Level partitioning of the soil liquefaction factors is accomplished in six iterations and shown in **Tables 6–11**. The ISM model is developed on the level partitions basis from **Tables 6–11**.

Factor (F_i)	Reachability set $R(F_i)$	Antecedent set $A(F_i)$	Intersection set $R(F_i) \cap A(F_i)$	Level L_i
F_1	F_1,F_2,F_{12}	F_1	F_1	
F_2	F_2,F_{12}	$F_1,F_2,F_3,F_4,F_6,F_7,F_8$	F_2	
F_3	F_2,F_3,F_{12}	F_3	F_3	
F_4	F_2,F_4,F_5,F_6,F_{12}	F_4	F_4	
F_5	F_5,F_{12}	$F_4,F_5,F_6,F_7,F_8,F_9,F_{10}$	F_5	
F_6	F_2,F_5,F_6,F_{12}	$F_4,F_6,F_7,F_8,F_9,F_{10}$	F_6	
F_7	F_2,F_5,F_6,F_7,F_{12}	F_7,F_8,F_9,F_{10}	F_7	
F_8	$F_2,F_5,F_6,F_7,F_8,F_{12}$	F_8,F_{10}	F_8	
F_9	F_5,F_6,F_7,F_9,F_{12}	F_9	F_9	
F_{10}	$F_5,F_6,F_7,F_8,F_{10},F_{12}$	F_{10}	F_{10}	
F_{11}	F_{11},F_{12}	F_{11}	F_{11}	
F_{12}	F_{12}	$F_1,F_2,F_3,F_4,F_5,F_6,F_7,F_8,F_9,F_{10},F_{11},F_{12}$	F_{12}	L_1

Table 6.
Level partition—Iteration 1.

Factor (F_i)	Reachability set $R(F_i)$	Antecedent set $A(F_i)$	Intersection set $R(F_i) \cap A(F_i)$	Level L_i
F_1	F_1,F_2	F_1	F_1	
F_2	F_2	$F_1,F_2,F_3,F_4,F_6,F_7,F_8$	F_2	L_2
F_3	F_2,F_3	F_3	F_3	
F_4	F_2,F_4,F_5,F_6	F_4	F_4	
F_5	F_5	$F_4,F_5,F_6,F_7,F_8,F_9,F_{10}$	F_5	L_2
F_6	F_2,F_5,F_6	$F_4,F_6,F_7,F_8,F_9,F_{10}$	F_6	
F_7	F_2,F_5,F_6,F_7	F_7,F_8,F_9,F_{10}	F_7	
F_8	F_2,F_5,F_6,F_7,F_8	F_8,F_{10}	F_8	
F_9	F_5,F_6,F_7,F_9	F_9	F_9	
F_{10}	F_5,F_6,F_7,F_8,F_{10}	F_{10}	F_{10}	
F_{11}	F_{11}	F_{11}	F_{11}	L_2

Table 7.
Level partition—Iteration 2.

Factor (F_i)	Reachability set $R(F_i)$	Antecedent set $A(F_i)$	Intersection set $R(F_i) \cap A(F_i)$	Level L_i
F_1	F_1	F_1	F_1	L_3
F_3	F_3	F_3	F_3	L_3
F_4	F_4,F_6	F_4	F_4	
F_6	F_6	$F_4,F_6,F_7,F_8,F_9,F_{10}$	F_6	L_3
F_7	F_6,F_7	F_7,F_8,F_9,F_{10}	F_7	
F_8	F_6,F_7,F_8	F_8,F_{10}	F_8	
F_9	F_6,F_7,F_9	F_9	F_9	
F_{10}	F_6,F_7,F_8,F_{10}	F_{10}	F_{10}	

Table 8.
Level partition—Iteration 3.

Factor (F_i)	Reachability set $R(F_i)$	Antecedent set $A(F_i)$	Intersection set $R(F_i) \cap A(F_i)$	Level L_i
F_4	F_4	F_4	F_4	L_4
F_7	F_7	F_7,F_8,F_9,F_{10}	F_7	L_4
F_8	F_7,F_8	F_8,F_{10}	F_8	
F_9	F_7,F_9	F_9	F_9	
F_{10}	F_7,F_8,F_{10}	F_{10}	F_{10}	

Table 9.
Level partition—Iteration 4.

Factor (F_i)	Reachability set $R(F_i)$	Antecedent set $A(F_i)$	Intersection set $R(F_i) \cap A(F_i)$	Level L_i
F_8	F_8	F_8,F_{10}	F_8	L_5
F_9	F_9	F_9	F_9	L_5
F_{10}	F_8,F_{10}	F_{10}	F_{10}	

Table 10.
Level partition—Iteration 5.

Factor (F_i)	Reachability set $R(F_i)$	Antecedent set $A(F_i)$	Intersection set $R(F_i) \cap A(F_i)$	Level L_i
F_{10}	F_{10}	F_{10}	F_{10}	L_6

Table 11.
Level partition—Iteration 6.

The digraph is developed from final reachability matrix by removing the transitivity links and on the level partitions basis from **Tables 6–11**. The digraph is converted in to an ISM-based hierarchical model by replacing nodes with statements. Each seismic soil liquefaction factor is positioned as per the consequent level and the relationships of the soil liquefaction factors are fixed from the bottom (level 6) to the top of the model (level 1). The seismic soil liquefaction factors are connected by arrows from the bottom of the model (higher-level) to the top of model (lower-level). The multilevel hierarchy model developed with identified relations between the significant factors of seismic soil liquefaction potential (LP) is shown in **Figure 1**.

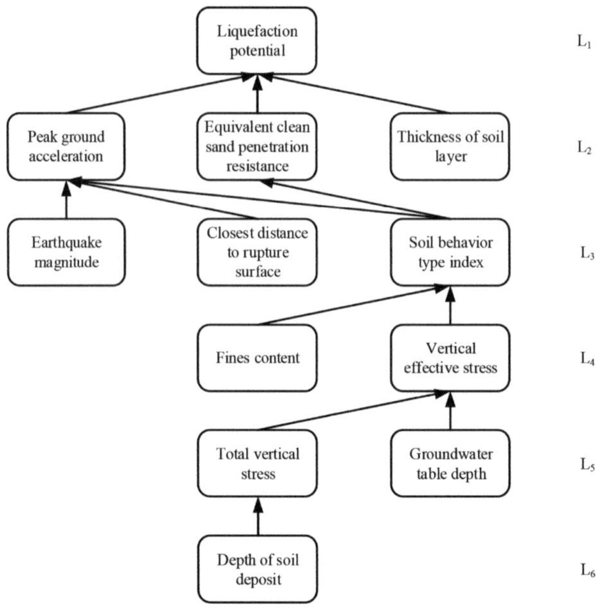

Figure 1.
Model depicting the relationships between seismic soil liquefaction significant factors based on ISM technique.

3.2 MICMAC analysis: classification of CPT-based seismic soil liquefaction significant factors

The driving power and dependency power of each variable was measured using the final reachability matrix to analyze the strength of the relationship between seismic soil liquefaction significant factors. Driving power is characterized as an activity that propels other activities, while dependency power is defined as an activity that is driven by other activities. The driving power and dependency power are determined from the final reachability matrix by adding the sum of all '1's in that factor's corresponding row and column.

This is considered as an input to build a graph to classify the factors into four clusters i.e., Autonomous, Dependent, Linkage, and Independent factors (see **Figure 2**). Autonomous factors (first cluster) have weak driving power and weak dependence power. Dependent factors (second cluster) have weak driving power and strong dependence power. Linkage factors (third cluster) have strong driving power and strong dependence power. In the dependent factors (fourth cluster) acquires strong driving power and weak dependence power. The soil liquefaction factors have been categorized based on these aforementioned clusters. The four clusters of soil liquefaction factors are:

3.2.1 Cluster I: autonomous factors

Cluster I, represents autonomous factors and consists of soil liquefaction factors which have weak driving power and dependence power. This cluster has six seismic soil liquefaction factors (55%). Cluster I factors are relatively disconnected from the system. Autonomous factors in cluster I are earthquake magnitude F_1 (M), closest distance to rupture surface F_3 (R/r_{rup}), fines content F_4 (FC), vertical effective stress F_7 (σ'_v), groundwater table F_9 (D_w), and thickness of soil layer F_{11} (T_s).

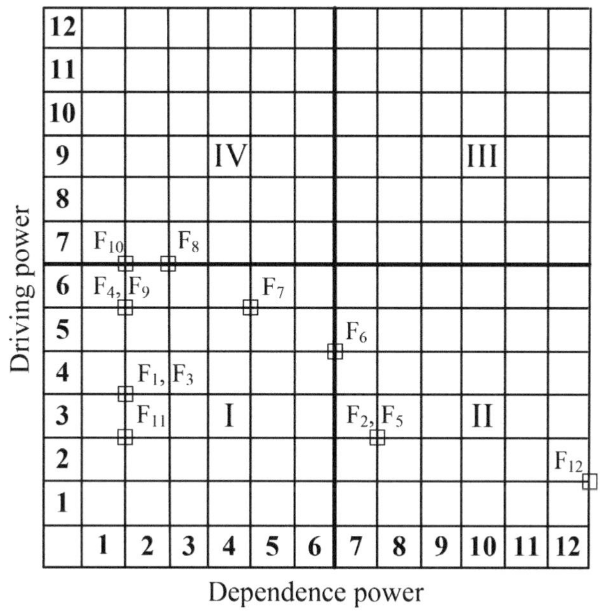

Figure 2.
MICMAC Analysis of seismic soil liquefaction factors.

3.2.2 Cluster II: dependence factors

Dependence factors have a strong dependence power and weak driving power. This dependence cluster has two seismic soil liquefaction factors (18%) except liquefaction potential that including peak ground acceleration F_2 (a_{max}) and equivalent clean sand penetration resistance F_5 (q_{c1Ncs}), while, the liquefaction potential F_{12} (LP) falls in this cluster is not an influence factor of earthquake liquefaction, but a discriminate index. It is just proved that the driving power is poor and needs to rely on other factors to discernment liquefaction. In the ISM model, these factors form the top levels which need other soil liquefaction factors that collectively act to influence soil liquefaction.

3.2.3 Cluster III: linkage factors

Linkage factors have a strong driving power as well as strong dependence power. The factors affect each other and directly affect the liquefaction system. Therefore, the factors in this cluster are unstable. No factor in this model fall into this cluster, which indicates that the liquefaction influencing factors in this model are relatively stable.

3.2.4 Cluster IV: independent/driving factors

In this cluster, factors have strong driving power but weak dependence power. It is often the most critical factors of the system and also the essential factors. No factor in this model fall in this cluster.

A special case can be observed on factors that are depth of soil deposit F_{10}, (D_s) and total vertical stress F_8, (σ_v), on the middle between independent and autonomous factors whereas soil behavior type index F_6, (I_c), on the middle between dependence and autonomous factors. Factors that are depth of soil deposit F_{10}, (D_s) and total vertical stress F_8, (σ_v), are lower on dependence but higher on driving power, are located between two clusters, I and IV. Similarly, soil behavior type index F_6, (I_c), factor is intermediate on dependence but lower driving power, but is located between two clusters, I and II. These factors need attention owing to establish and provide a more accurate and caution way of selecting significant factors for seismic soil liquefaction potential and its induced-hazards risk lassessment modeling.

4. Discussion and conclusions

The intention of this research study is the identification and benchmarking the seismic soil liquefaction factors of seismic soil liquefaction and the understanding of their relationship. ISM-based hierarchical model has been developed to examine the cone penetration test based significant factors of seismic soil liquefaction potential. The ISM model presents the relationships between seismic soil liquefaction factors and their benchmarking position from higher to lower-level significant factors in hierarchy. The results provide a more accurate and caution way for establishment of seismic soil liquefaction potential and liquefaction-induced hazards risk assessment models. Seismic soil liquefaction factors located on top level hierarchy are greatly influenced by the interconnection of left-over factors. It is evident from the ISM model that the factor depth of soil deposit, F_{10} (D_s) at level 6, forms the base of the ISM hierarchy and has high driving power and low dependence power, whereas peak ground acceleration F_2 ($a_{\max}$), equivalent clean sand penetration resistance F_5 (q_{c1Ncs}), and thickness of soil layer F_{11} (T_s), in the second level directly influence liquefaction potential. The other soil liquefaction potential factors are earthquake magnitude F_1 (M), closest distance to rupture surface F_3 (R/r_{rup}), and soil behavior type index F_6 (I_c) at level 3, fines content F_4 (FC) and vertical effective stress F_7 (σ'_v) at level 4, total vertical stress F_8 (σ_v) and groundwater table depth F_9 (D_w) at level 5, as per the outcomes of the ISM hierarchical model and are classified as indirect factors that affect soil liquefaction.

By performing MICMAC analysis, the dependence-driving diagram is plotted which offers information about the relative significance and the interdependencies among various factors of seismic soil liquefaction. It is found in this study, that there exists no independent and linkage factor. Among the 11 factors studied, 2 factors are falling in dependent quadrant in the dependence-driving diagram and it is recognized that these particular factors will depend on other factors. Further, 6 factors fall under the autonomous quadrant. Rest of the 3 factors, in which 2 of them i.e., depth of soil deposit F_{10}, (D_s) and total vertical stress F_8, (σ_v), are located between two clusters, I and IV owing to intermediate diving power. Similarly, soil behavior type index F_6, (I_c), is located between two clusters, I and II owing to intermediate dependence power. These factors i.e., depth of soil deposit, total vertical stress and soil behavior type index need attention owing to provide a more accurate and caution way for further establishment of seismic soil liquefaction potential and liquefaction-induced hazards risk assessment models.

Acknowledgements

The work presented in this paper was part of the research sponsored by the Key Program of National Natural Science Foundation of China under Grant No. 51639002 and the National Key Research and Development Plan of China under Grant No. 2018YFC1505300-5.3.

Conflict of interest

The authors declare no conflict of interest.

Author details

Mahmood Ahmad[1,2], Xiaowei Tang[1]*, Feezan Ahmad[1], Marijana Hadzima-Nyarko[3], Ahsan Nawaz[4] and Asim Farooq[5]

1 State Key Laboratory of Coastal and Offshore Engineering, Dalian University of Technology, Dalian, China

2 Department of Civil Engineering, University of Engineering and Technology Peshawar (Bannu Campus), Bannu, Pakistan

3 Faculty of Civil Engineering and Architecture Osijek, Josip Juraj Strossmayer University of Osijek, Osijek, Croatia

4 Institute of Construction Project Management, College of Civil Engineering and Architecture, Zhejiang University, Hangzhou, China

5 Department of Transportation Engineering, Pak-Austria Fachhochschule: Institute of Applied Sciences and Technology, Haripur, Pakistan

*Address all correspondence to: tangxw@dlut.edu.cn

References

[1] Dalvi, A.N.; Pathak, S.R.; Rajhans, N. R. Entropy analysis for identifying significant parameters for seismic soil liquefaction. *Geomechanics and Geoengineering* **2014**, *9*, 1-8.

[2] Zhu, S. Statistical analysis of factors causing liquefaction of sand during Tangshan Earthquake. *Seismology and Geology* **1980**, *2*, 79-80.

[3] Tang, X.-W.; Hu, J.-L.; Qiu, J.-N. Identifying significant influence factors of seismic soil liquefaction and analyzing their structural relationship. *KSCE Journal of Civil Engineering* **2016**, *20*, 2655-2663.

[4] Ahmad, M.; Tang, X.-W.; Qiu, J.-N.; Ahmad, F. Interpretive Structural Modeling and MICMAC Analysis for Identifying and Benchmarking Significant Factors of Seismic Soil Liquefaction. *Applied Sciences* **2019**, *9*, 233.

[5] Okoli, C.; Schabram, K. A guide to conducting a systematic literature review of information systems research. Working Papers on Information Systems **2010**, *10*, 1-51.

[6] Tranfield, D.; Denyer, D.; Smart, P. Towards a methodology for developing evidence-informed management knowledge by means of systematic review. *Br. J. Manage.* **2003**, *14*, 207-222.

[7] Olsen, S. *Group Planning and Problem-Solving Methods in Engineering Management*; John Wiley & Sons, Inc., 605 3rd Ave., New York, NY 10158: 1982.

[8] Sushil, S. Interpreting the interpretive structural model. *Global Journal of Flexible Systems Management* **2012**, *13*, 87-106.

[9] Warfield, J.N. Developing interconnection matrices in structural modeling. *IEEE Transactions on Systems, Man, and Cybernetics* **1974**, *SMC-4*, 81-87.

[10] Pfohl, H.-C.; Gallus, P.; Thomas, D. Interpretive structural modeling of supply chain risks. *International Journal of Physical Distribution & Logistics Management* **2011**, *41*, 839-859.

[11] Saxena, J.; Vrat, P. Impact of indirect relationships in classification of variables—a micmac analysis for energy conservation. *Systems Research* **1990**, *7*, 245-253.

[12] Fink, A. *Conducting research literature reviews: From the internet to paper*; Sage publications: 2019.

[13] Seed, H.B. Ground motions and soil liquefaction during earthquakes. *Earthquake engineering research insititue* **1982**.

[14] Pirhadi, N.; Tang, X.; Yang, Q.; Kang, F. A New Equation to Evaluate Liquefaction Triggering Using the Response Surface Method and Parametric Sensitivity Analysis. *Sustainability* **2019**, *11*, 112.

[15] Ahmad, M.; Tang, X.-W.; Qiu, J.-N.; Ahmad, F.; Gu, W.-J. A step forward towards a comprehensive framework for assessing liquefaction land damage vulnerability: Exploration from historical data. *Frontiers of Structural and Civil Engineering* **2020**, *14*, 1476-1491.

[16] Hu, J.-L.; Tang, X.-W.; Qiu, J.-N. Assessment of seismic liquefaction potential based on Bayesian network constructed from domain knowledge and history data. *Soil Dyn. Earthquake Eng.* **2016**, *89*, 49-60.

[17] Ahmad, M.; Tang, X.-W.; Ahmad, F.; Jamal, A. Assessment of soil liquefaction potential in Kamra, Pakistan. *Sustainability* **2018**, *10*, 4223.

[18] Ahmad, M.; Tang, X.-W.; Qiu, J.-N.; Gu, W.-J.; Ahmad, F. A hybrid approach for evaluating CPT-based seismic soil

liquefaction potential using Bayesian belief networks. *Journal of Central South University* **2020**, *27*, 500-516.

[19] Green, R.A.; Bommer, J.J. What is the smallest earthquake magnitude that needs to be considered in assessing liquefaction hazard? *Earthquake Spectra* **2019**, *35*, 1441-1464.

[20] Tesfamariam, S.; Liu, Z. Seismic risk analysis using Bayesian belief networks. In *Handbook of seismic risk analysis and management of civil infrastructure systems*, Elsevier: 2013; pp. 175-208.

[21] Stark, T.D.; Olson, S.M. Liquefaction resistance using CPT and field case histories. *Journal of geotechnical engineering* **1995**, *121*, 856-869.

[22] Boore, D.M.; Joyner, W.B.; Fumal, T.E. Equations for estimating horizontal response spectra and peak acceleration from western North American earthquakes: A summary of recent work. *Seismol. Res. Lett.* **1997**, *68*, 128-153.

[23] Robertson, P.; Wride, C. Evaluating cyclic liquefaction potential using the cone penetration test. *Canadian geotechnical journal* **1998**, *35*, 442-459.

[24] Seed, H.B. Landslides during earthquakes due to liquefaction. *Journal of Soil Mechanics & Foundations Div* **1968**.

[25] Andrus, R.; Stokoe, K.H.; Roesset, J. Liquefaction of gravelly soil at Pence Ranch during the 1983 Borah Peak, Idaho earthquake. *International Journal of Rock Mechanics and Mining Sciences Abstracts* **1993**, *30*, 98A.

[26] Ishihara, K. Stability of natural deposits during earthquakes. Proceeding of the 11th conference. In Proceedings of Soil Mechanics and Foundation Engineering, San Francisco.

[27] Zhou, J.; Li, E.; Wang, M.; Chen, X.; Shi, X.; Jiang, L. Feasibility of stochastic gradient boosting approach for evaluating seismic liquefaction potential based on SPT and CPT case histories. *J. Perform. Constr. Facil.* **2019**, *33*, 04019024.

[28] Ahmad, M.; Tang, X.-W.; Qiu, J.-N.; Ahmad, F. Evaluating Seismic Soil Liquefaction Potential Using Bayesian Belief Network and C4. 5 Decision Tree Approaches. *Applied Sciences* **2019**, *9*, 4226.

[29] *Liquefaction potential of cohesionless soils. 2007. GDP-9, Revision# 2, State of New York, Department of Transportation*; Procedure, Geotechnical Design, Geotechnical Engineering Bureau: 2007.

[30] Kishada, H. A note on liquefaction of hydraulic fill during the Tokachi-Oki earthquake. In Proceedings of Second Seminar on Soil Behavior and Ground Response During Earthquakes, University of California, Berkeley, CA.

[31] Boulanger, R.W. High overburden stress effects in liquefaction analyses. *Journal of Geotechnical and Geoenvironmental Engineering* **2003**, *129*, 1071-1082.

[32] Seed, R.B.; Cetin, K.O.; Moss, R.E.; Kammerer, A.M.; Wu, J.; Pestana, J.M.; Riemer, M.F.; Sancio, R.B.; Bray, J.D.; Kayen, R.E. Recent advances in soil liquefaction engineering: a unified and consistent framework. In Proceedings of Proceedings of the 26th Annual ASCE Los Angeles Geotechnical Spring Seminar: Long Beach, CA.

[33] Dobry, R. Some Basic Aspects of Soil Liquefaction during Earthquakes a. *Ann. N. Y. Acad. Sci.* **1989**, *558*, 172-182.

[34] Hannich, D.; Hoetzl, H.; Ehret, D.; Huber, G.; Danchiv, A.; Bretotean, M. Liquefaction probability in Bucharest and influencing factors. In Proceedings of Proceedings of the International

Symposium on Strong Vrancea Earthquake and Risk Mitigation, Bucharest, Romania; pp. 4-6.

[35] Fioravante, V.; Giretti, D.; Abate, G.; Aversa, S.; Boldini, D.; Capilleri, P. P.; Cavallaro, A.; Chamlagain, D.; Crespellani, T.; Dezi, F. Earthquake geotechnical engineering aspects of the 2012 Emilia-Romagna earthquake (Italy). In Proceedings of Case histories in geotechnical engineering, Chicago; pp. 1-34.

[36] Florin, V.; Ivanov, P. Liquefaction of saturad sandy soils. In Proceedings of Proc. 5th Int. Soil Mech. Foundat. Eng., Paris; p. 111.

[37] Satyam, N. Review on liquefaction hazard assessment. In *Advances in Geotechnical Earthquake Engineering—Soil Liquefaction and Seismic Safety of Dams and Monuments*, 2012; pp. 63-82.

Section 3

Building Foundations

Chapter 8

The Dynamic Behaviour of Pile Foundations in Seismically Liquefiable Soils: Failure Mechanisms, Analysis, Re-Qualification

Rohollah Rostami, Slobodan B. Mickovski, Nicholas Hytiris and Subhamoy Bhattacharya

Abstract

This chapter presents a concise overview of the mechanics of failure, analysis and requalification procedures of pile foundations in liquefiable soils during earthquakes. The aim is to build a strong conceptual and technical interpretation in order to gain insight into the mechanisms governing the failure of structures in liquefaction and specify effective requalification techniques. In this regard, several most common failure mechanisms of piles during seismic liquefaction such as bending (flexural), buckling instability and dynamic failure of the pile are introduced. Furthermore, the dynamic response commentary is provided by critically reviewing experimental investigations carried out using a shaking table and centrifuge modelling procedures. The emphasis is placed on delineating the concept of seismic design loads and important aspects of the dynamic behaviour of piles in liquefiable soils. In this context, using Winkler foundation approach with the proposed p–y curves and finite-element analyses in conjunction with numerical analysis methods, are outlined. Moreover, the feasibility of successful remediation techniques for earthquake resistance is briefly reviewed in light of the pile behaviour and failure. Finally, practical recommendations for achieving enhanced resistance of the seismic response of pile foundation in liquefiable soil are provided.

Keywords: liquefaction, dynamic behaviour, pile, failure mechanisms, requalification

1. Introduction (Characterisation of liquefaction behaviour)

The liquefaction of loose, saturated sands, particularly cohesionless soils is caused by earthquake shaking or cyclic (monotonically increasing) undrained loading. The early work in liquefaction soil in the laboratory apparently emerged from the experience of the Fukui earthquake in 1948 in Japan [1]. It was regarded as a milestone from researchers since its devastating failures were prevalent following the major earthquakes in Niigata, Japan and Alaska, USA, in 1964 [2–4].

Soil liquefaction has been responsible for extremely damaged structures and foundation piles of bridges and buildings and has resulted in severe loss of strength and stiffness of saturated cohesionless soil. Liquefaction was first introduced by Hazen [5] that he used to describe the 1918 collapse of Calaveras Dam in California [6]. Typically, liquefaction occurs when a deposit of loose saturated sand layers are subject to shaking during a seismic event, the progressive build-up of excess pore water pressure and the stiffness of the liquefied soil drops to a value of near-zero. The reduction in strength and stiffness of liquefied soil often leads to permanent deformation in sloping grounds, commonly termed as lateral spreading (a few centimetres or metres) or flow failure (hundreds of meters). Flow failures have been observed in a number of hydraulically filled earth dams, constructed tailings dams and in coastal and/or offshore areas [6]. This hydraulic problem was observed as secondary and progressive liquefaction surrounding the majority of slides as a result of the generation of excess pore pressure and the upward flow of water almost immediately prior to the initiation of the slide flow [6]. This also may lead to formation of sand boils, which have been illustrated by Ishihara [7] in terms of the relative thicknesses of liquefiable and overlying non-liquefiable layers in case history data from the 1976 Tangshan and 1983 Nihonkai earthquakes. In this respect, Huang and Yu [8] also classified the liquefaction related damage to soils and foundations during earthquakes in the first part of the twenty-first century in: ground subsidence, lateral spread, and damage induced by buoyancy (uplift).

Laboratory studies carried out to investigate the liquefaction susceptibility and conventionally evaluate the undrained behaviour of sandy soils under monotonic shearing. This tendency is generally expressed in terms of void ratio (e) and relative density (Dr). In this state, sand is flowing under constant shear stress at constant effective minor principal stress and at constant volume [6, 9–11]. Poulos [11] included the requirement of constant velocity, the "steady-state of deformation," and the relationship between the steady-state effective stress and the void ratio i.e. the "steady-state" line. The response for very loose sand shows fully contractive behaviour is reached at large strains, as delineated in **Figure 1**. This behaviour reported as "spontaneous liquefaction" and also known as flow. In in the case of sand with slightly higher density, the strain softening is followed by the strain hardening and the sand recovers its strength and restores stability. This type of behaviour was first called "limited liquefaction" by Castro [9] and known as limited flow.

In medium-dense and dense sands exhibiting dilative behaviour, ever-increasing shear stress is needed to induce shear strain and eventually obtain the steady state

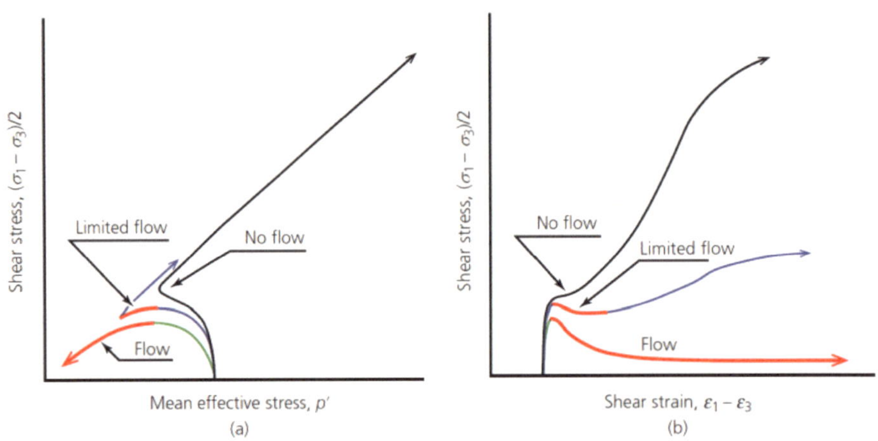

Figure 1.
Monotonic behaviour of different sands: (a) effective stress path; (b) shear stress–shear strain relation [12].

of deformation (no flow) [12]. Kramer et al. [6] indicated that at relative densities greater than those corresponding to the steady state line (slightly greater than 44%), the soil exhibit dilative behaviour, with no potential for liquefaction.

Figure 2a and **b** illustrate a typical cyclic loading test on loose sand (2a) and dense sand (2b) using a torsional shear apparatus. The upper graphs describe the time history of cyclic shear stress ratio applied to constant-amplitude cyclic loading, while the lower graphs show the development of shear strain and the generation of excess pore pressure with time. In the early stages of loading the effective stress path moves to the left and the shear strain is negligible. However, as the loading progresses, the pore pressure builds up until the stress path begins to cross the phase transformation line identified by Ishihara [7] and eventually reaches a value equal to the initial confining pressure, which is called cyclic mobility. It can be seen that in both sand the effective confining stress decreases, but the shear strain increases in a slower manner for dense sand.

The liquefaction potential of the soil is generally estimated by comparing the anticipated earthquake loading and its inherent liquefaction resistance. This comparison is most commonly base on cyclic shear stress amplitude usually normalised by initial vertical effective stress and known as a cyclic stress ratio (CSR) for loading and a cyclic resistance ratio (CRR) for resistance. The potential for liquefaction is expressed in terms of a factor of safety against liquefaction, FL = CRR/CSR. If the FL > 1.0 soil profile can be safe against liquefaction. Standard penetration test (SPT) and cone penetration test (CPT) are the two empirical methods that use to obtain the cyclic resistance to liquefaction. The susceptibility of soil deposits to liquefaction is determined by a combination of various factors such as soil properties, geological conditions and ground motion characteristics. The soil's CRR is also affected by the duration of shaking (which is correlated to the earthquake magnitude scaling factor, MSF) and effective overburden stress (which is expressed through a Kσ factor). The evaluating of the soil liquefaction potential based on the SPT and CPT values are well explained in [12].

The seismic response of a soil profile is strongly influenced by the effective stress of an earthquake ground motions. The nature of ground motions at sites containing potentially liquefiable soils can affect the potential damage to pile foundations. The

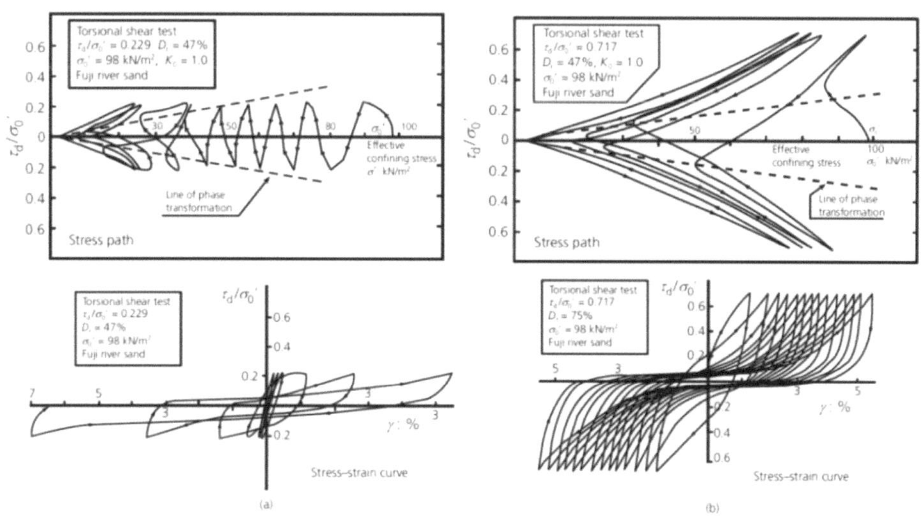

Figure 2.
Stress path behaviour and stress–strain curve of (2a) loose sand and (2b) dense sand from the cyclic torsion shear test (Ishihara, [7]) (D_r: Relative density of the soil, K_0: Coefficient of earth pressure, γ : Shear strain, τ_d / σ_0: Shear stress ratio).

relevant characteristics of ground motions are the frequency content, amplitude, and duration, and they can provide insight into the effects of ground motion duration on liquefaction hazards. The initial characteristic site period for a simple layer (ground of thickness H) with constant initial shear wave velocity, V_{so}, is given by Tso = 4 H/ V_{so}. The velocity is related to frequency f and wavelength λ by $v = f\lambda$. The relation between v_s and the SPT N value indicates the soil type [12]. The effects of liquefaction and generation positive of pore pressure leads to decrease in the effective stresses and the shear modulus of the soil. As a result, there is a reduction in soil stiffness which, in turn, increases prevalence of low frequency motions. The accelerations recorded of the Wildlife liquefaction array in the 1987 Superstition Hills earthquake (NS component) at the ground surface and at 7.5 m depth and the associated excess pore water pressure measured at 2.9 m depth are shown in **Figure 3**.

In each case, a clear and gradual stiffness degradation associated with an increase in the pore water pressure can be observed. Acceleration time history of the surface record showing clear visual evidence of the high-frequency portion of the motion from the beginning of the record (about 18 s); which is consistent with a series of isolated high-frequency pulses of acceleration (see numbers) [12]. Kramer et al. [14] reported that these pulses have amplitudes smaller than those of the pulses that occur prior to liquefaction, but in some cases, the peak ground acceleration of the entire motion is produced by a strong dilation pulse occurring near, or even after, the initiation of liquefaction. Hall et al. [15] examined the transient vibration characteristics of two 2 × 2 pile-group models based on the wavelet. They found that liquefaction causes a decrease in structural frequency, whose reduction depends on the rate of excess pore pressure build-up, whereby high rates ("fast liquefaction") lead to greater reduction, i.e. up to 51%.

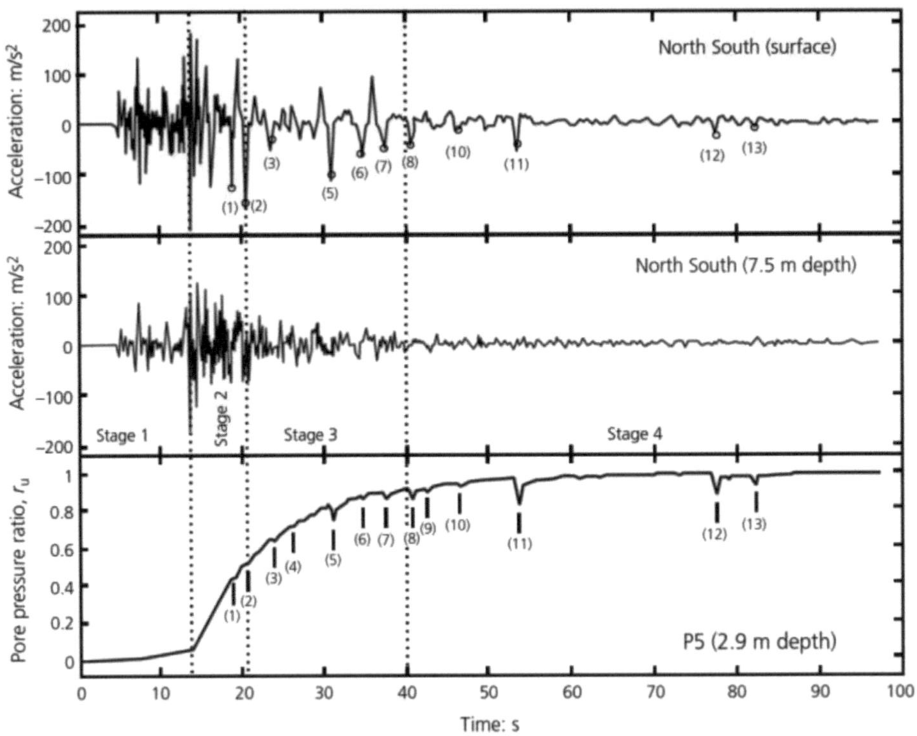

Figure 3.
Acceleration time histories time histories and associated pore water pressure of north–south components at wildlife site during the 1987 Superstition Hills earthquake (Zeghal and Elgamal, [13]) (numbers show high-frequency pulses of acceleration).

Ozener et al. [16] used of Stockwell spectrograms and indicated that as a result of the changes in stiffness; the response of a liquefied soil is often markedly different before and after triggering of liquefaction.

The liquefaction potential index (LPI) of Iwasaki et al. [17] is an integral number of factors of safety (FS) values weighted by depths of soil layers, which has increasingly been used for assessing the severity of liquefaction hazards [17, 18]. This is expressed by Eq. (1):

$$LPI = \int_{0}^{20} F.w(z)dz \tag{1}$$

Where z is the depth (0–20 m) and w(z) = weighting function (10–0.5z) and F is a function of the liquefaction resistance. Sites with LPI > 15 have increasing susceptibility to liquefaction and potential for severe damage. The risk of liquefaction tends to decrease with depth while if LPI < 5, the effects are minor due to increasing effective stress.

2. Pile failure

Piles are a particular type of deep foundation generally constructed to support heavily loaded structures to transfer the loads from superstructures to the deeper layers of soil, relying on both skin friction and tip resistance [19]. Piles are also used in seismic-prone zones comprising loose to medium-dense sandy soil or soft clay. However, during earthquake shaking when the soil around the pile loses much of its stiffness and strength due to liquefaction, the pile will act like a long laterally unsupported column and could buckle under the high axial load from the superstructure, affecting the foundations. Collapse and damage of pile-supported structures due to liquefaction have been observed after many major earthquakes [2–4]. The observations from many historic cases indicated that the failure of foundations occurred at unexpected locations (see **Figure 4**).

During earthquakes, the response of pile-supported structures to liquefiable soils would depend of the stiffness of the pile foundation type, the response of the soil surrounding the pile, and the soil-pile interaction effects. The analysis of this response requires accurate characterisation of the interaction effects include the inertial loading exerted by the superstructure and the kinematic loading induced by the soil surrounding the pile. **Figure 5** illustrates four critical stages of loading on the piles during a seismic liquefaction-induced event.

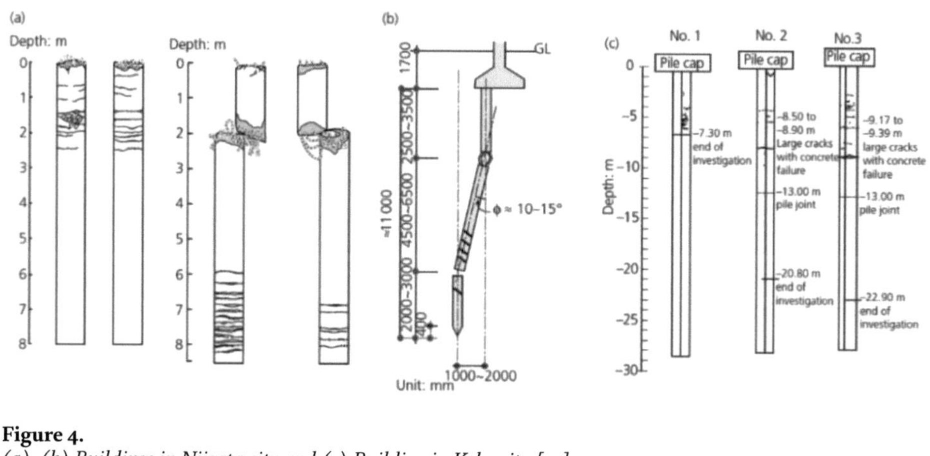

Figure 4.
(a), (b) Buildings in Niigata city and (c) Building in Kobe city [12].

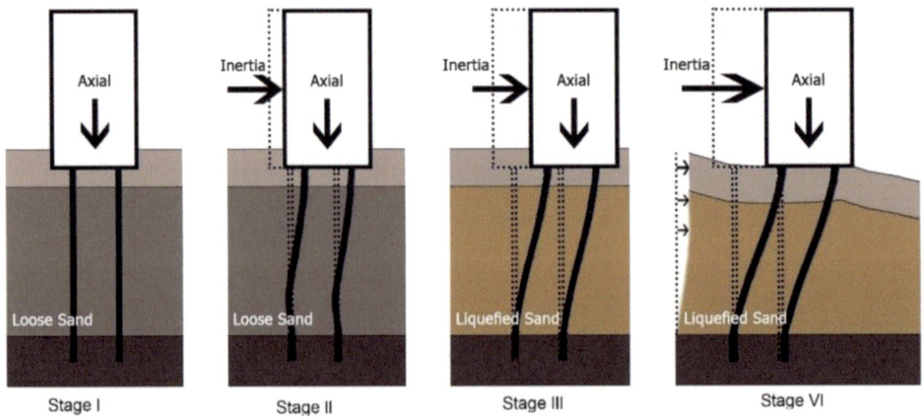

Figure 5.
Different stages of loading and failure mechanism of pile during earthquake [20].

Before, or just at the onset of the earthquake, the axial load on the piles can be estimated based on static equilibrium. Upon commencement of the seismic vibration, and before the excess pore water pressure build-up, this axial compressive load may increase/decrease further due to the inertial effect of the superstructure (due to oscillation of superstructure) and the kinematic effects of the soil flow past the foundation (due to ground movement). This change in loading can be transient (during the vibration, due to the dynamic effects of the soil mass) and residual (after the vibration, due to soil flow, often known as "lateral spreading" [21]). However, at this stage, with pore water pressure built up (at full liquefaction, the excess pore water pressures reach the overburden vertical effective stress), the soil loses its strength and stiffness, and the pile acts as an unsupported slender columns over the liquefied depth. Most of the efforts have been made to greatly improve understanding of pile failure mechanism due to liquefaction [20–24]. Two plausible mechanisms of pile failure: bending (due to inertia of the superstructure and/or kinematic loads due to lateral soil pressure) and buckling (due to axial load), have been studied in detail separately. However, dynamic failure (bending–buckling interaction) of a pile foundation may also occur in a seismically liquefiable soil deposits and lead to failure of the structure.

2.1 Bending failure

The bending failure mechanism due to liquefaction-induced lateral spreading of soil is often demonstrated as one of the predominant causes of pile foundation failures during earthquakes [21–24]. The bending failure mechanism can occur when soil liquefies and loses much of its stiffness, causing the piles to act as unsupported slender columns. Piles can exhibit bending failure as a result of one or both of two mechanisms. First, during seismic shaking the lateral flow of soils at a particular depth induces additional forces on piles and simultaneously the bending moment is generated in the pile due to the summation of inertia and kinematic loads. Second, at the end of the shaking, the bending moment is expected only due to kinematic flow as a result of the full dissipation of pore pressure [25]. The bending behaviour of a pile depends on the bending strength (e.g. yielding of the pile materials) and the flexural stiffness (changes in geometry of the moment-resisting pile section) [26]. Most of the current design methods, such as JRA [27], NEHRP [28], IS:1893 [29], and Eurocode 8 [30] focus on bending strength of the pile to avoid bending failure due to lateral loads (combination of inertia and/or lateral spreading). When pile-supported structures are embedded deep enough to move with the non-liquefiable layer, the displacement at the pile cap is equal to the ground displacement, and this condition yields maximum bending moments different from the free-end condition. Hwang et al. [31] evaluated the influence of liquefied ground flow on the pile behaviour. It was found that by increasing the slope angle of the liquefying ground, the shear force, the bending moment, and the lateral displacement of the pile increased. For the pile head condition, the bending moment also increased with depth. However, for the fixed pile head condition, the maximum bending moment of free head was about 1.5 times greater than that under the fixed pile head condition. The head supports for numerical analysis will be explained in Section 3.3.2.

2.2 Buckling failure

The second mechanism is the buckling instability under the interaction of axial and lateral loads, and piles acting as beam-columns under both axial and lateral loading [32–35]. Bhattacharya [35] argued this failure mechanism and suggested piles become laterally unsupported in the liquefiable zone during strong shaking which the axial load applied on pile and the soil around the pile liquefies loses of its stiffness and strength. Next, the piles act as unsupported long slender columns, and soils cannot support the corresponding action. Buckling failure depends on the geometrical properties of the member (i.e. slenderness ratio). The buckling mechanism is in the length of in touch with liquefied soil. The lateral loads for structural elements, due to slope movement increase lateral spread displacement demands, which in can cause plastic hinge to form and reducing the buckling load.

Extensive research has been carried out on the buckling instability of pile in liquefied soils. One early method for the stability of beams on elastic foundations proposed by Hetenyi [36] may be the base of the buckling analysis of pile foundations. The lateral loads, due to inertia or lateral spreading, could increase the lateral deflection of pile and thus reduce the buckling load [35]. On the other hand, there will always be confining pressure around the pile even if the soil has fully liquefied, and it could provide some lateral support to the pile and increase the buckling load [37]. As observed by Bhattacharya et al. [24], Knappett and Madabhushi [32], and Zhang et al. [38], buckling failure of the end-bearing pile normally occurs when the soil is fully liquefied. And pile buckling in partially liquefied soil would require a higher buckling load than that in the fully liquefied soil. In other words, when predicting the critical buckling load of pile in liquefiable soils, only the soil that has fully been liquefied needs to be considered. Zhang et al. [38] found that the critical buckling load of piles in liquefied soils increases with the increase of soil relative density and flexural rigidity of the pile and decreases with the increase of initial geometric imperfections of the pile and pier height. Shanker et al. [39] proposed an analytic method to predict the critical buckling load of pile under partial to full loss of lateral support.

2.3 Dynamic failure (bending–buckling)

A collapse of pile-supported structures in liquefiable deposits may occur under the combined action of lateral load and axial load. Bhattacharya et al. [37] included the dynamics failure on the combined axial and lateral loads on a pile foundation. In this mechanism, piles are subjected to both axial and lateral loads during seismic shaking and piles act like beam-column members (**Figure 6**).

As a result of this combination (axial- and lateral-loading) on piles during a seismic liquefaction-induced event, the influence of the axial load, P, in piles causes a loss of lateral stiffness (y is the lateral displacement) until the axial load

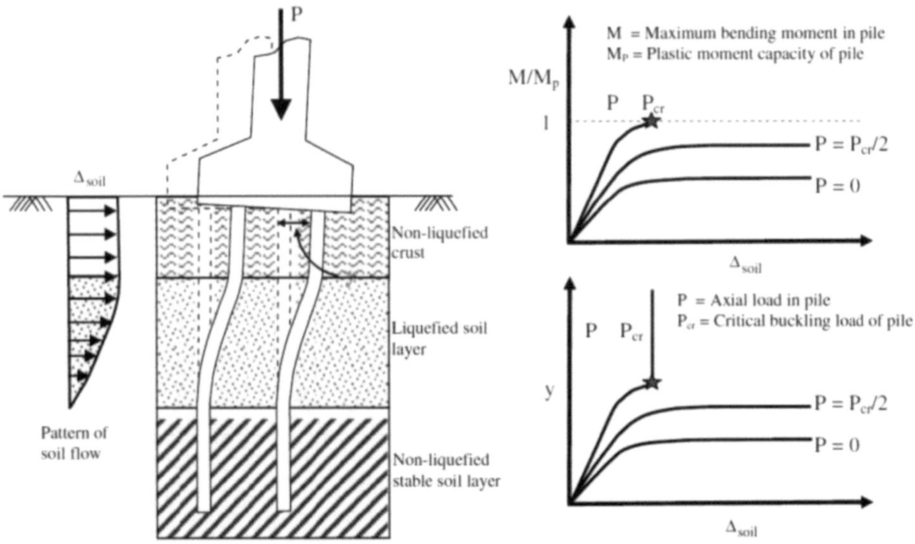

Figure 6.
Schematic of the effect of bending–buckling interaction on the response of pile foundation [40].

approaches the critical value ($P=P_{cr}$). The loss of lateral stiffness in association with the axial load (i.e., pile deflection), Δ, is dominated by the excessive moment caused by the P-Δ effect (see Section 3.3.1). Subsequently, the large deflection of the beam may then induce plasticity in the beam resulting in an early failure. The same failure point of the pile is also possible when bending moment reaches Mp and pile continues to deflect without any additional loading.

Dash et al. [40] investigated the importance of bending–buckling interaction in seismic design of piles in liquefiable soils using numerical techniques. They concluded that if a pile is designed for bending and buckling criteria separately and safe for these individual design criteria, it may fail due to their combined effect. Recently, this is also suggested by Zhang et al. [41] to consider the buckling mechanism together with the effect of lateral load. It is hence important for the designers to consider a possible boundary for safe design to avoid failure of the pile.

3. Dynamic behaviour of pile

A variety of design procedures have been adopted by design guidelines and codes for assessing the behaviour of piles in liquefiable ground. Dynamic loading during earthquake be superimposed onto the working loads of the piles. Predicting seismic response of pile foundations in liquefied soil layers is much more complex due to uncertainties in the mechanisms involved in soil–pile-superstructure interaction (different dynamic loads, the stiffness and shear strength of the surrounding soil and pore water pressure generation). In practice, different design procedures have been used for the seismic design of pile-supported structures. The Japanese Highway Code of Practice (JRA) [27], for example, advises the practicing engineers to consider both of the loading conditions mentioned above. However, it suggests a separate bending failure check for the effects of kinematic and inertial forces. Similarly, BS EN ISO 2008 [30] advises pile design against bending due to inertial and kinematic forces arising from the deformation of the surrounding soil. In the event of liquefaction, Eurocode 8 also suggests that "the side resistance of soil layers that are susceptible to liquefaction or to substantial strength degradation shall be ignored". The NEHRP [28], on the other hand, focuses on the bending strength of the piles by treating them as laterally loaded beams and assuming that the lateral load due to inertia and soil movement causes bending failure.

Since the mid-1960s, significant research has been conducted to understand the dynamic behaviour of pile foundations in liquefiable soils using various experimental techniques as well as various numerical modelling methods. These investigations can be divided into three categories: field observations (case histories), laboratory tests (dynamic centrifuge experiments, shaking table tests and full-scale field tests), and numerical modelling (Winkler analyses with linear-elastic or hysteretic soil behaviour, finite-element analyses). These will be discussed in detail in the following subsections.

3.1 Field observations (case histories)

This section provides a brief review of case histories. This can help to appropriately understand the phenomena involved and to identify important aspects of pile-soil-interaction behaviour. These case histories are primarily from the past 50 to 60 years, which describes the observed of some of the damaged piled foundations from the literature (see **Table 1**).

Iwasaki [51] reported the results of investigations on seismic damages to highway bridges during major eight earthquakes in Japan (occurred in 1923 to 1983). Their observation described that many of reinforced concrete buildings, highway bridges and other structures sustained considerable damage due to liquefaction of sandy soils (e.g., Showa Bridge 1964, Yuriage Bridge 1978, Shizunai Bridge 1982, Gomyoko Bridge 1983). The Showa bridge collapse has been a case history of interest in many publications and it was as an iconic example of the detrimental effects of liquefaction-induced lateral spreading on the ground. Hamada [2] argued that a more plausible explanation could be offered based on the ground displacements suffered due to liquefaction induced lateral spreading. In this respect, the JRA code [27] tried to formalise this research and presented methods of estimating the loading due to lateral spreading ground on pile foundations. This problem was revisited by Yoshida et al. [52] and they collated a number of eye-witness accounts to establish the timing of the bridge collapse as well as the lateral spreading of the river banks. It was suggested that lateral spreading of the surrounding ground started after the bridge had collapsed. Madabhushi and Bhattacharya [21] reanalysed the bridge and showed that lateral spreading hypothesis could not explain the failure of the bridge. A similar explanation was reported by Kerciku et al. [53]. As a final remark, Bhattacharya et al. [12, 54] and Mohanty et al. [55] suggested that the Showa Bridge could have collapsed because of bending, buckling, and combined action of bending of pile foundations.

The Niigata Family Court House building was a four-storey building constructed on concrete pile foundations. Hamada [2] suggested that one pile suffered relatively modest damage, as it did not penetrate into the deeper, non-liquefied ground. Madabhushi et al. [56] concluded that the laterally spreading ground around the piles caused the observed distress in these piles.

Further example on the probability of identifying collapse mechanisms is the Kandla Port one of the largest ports in India, located in the western state of Gujarat. Following the Bhuj earthquake of 2001, there was some damage to the port facilities [49]. Dash et al. [57] used conventional analysis of a single pile or a pile group to predict collapse. They concluded that the foundation mats over the non-liquefied crust shared a considerable amount of load of the superstructure and resisted the complete collapse of the building.

Case history	Earthquake event	Magnitude (M_L)	Pile type	Pile diameter (m)	Pile length (m)	Pile length in liquefiable soil (m)	Pile performance	References
10 storey-Hokuriku building	1964 Niigata	7.5	Reinforced concrete	0.4	12	5	Cracking on piles	Hamada [2]
Showa bridge	1964 Niigata	7.5	Steel tube	0.6	25	19	Collapse due to buckling of pile foundations	Hamada [2]
Landing Road bridge	1987 Edgecumbe	6.3	Square Reinforced concrete	0.4	9	4	Minor cracking at pile heads	Berrill et al. [42]
10-story Hotel east of State Highway	1989 Loma	7.0	Prestressed concrete piles	0.36	12	5	Piles performed well.	Adib et al. [43]
Buildings on Port Island	1995 Kobe	6.9	Reinforced concrete	0.45	9 9	4 4	Severe damage at liquefiable/non-liquefiable interface	Fujii et al. [44]
Building near the Higashi Kobe bridge	1995 Kobe	6.9	Prestressed concrete pile	0.45	10	6	Failures at about 1-m below the reclaimed fill, many cracks between 3 and 6 m below ground surface	Fujii et al. [44]
14 storey building in American park	1995 Kobe	6.9	Reinforced concrete	2.5	33	12.2	Damage most common at pile heads.	Tokimatsu et al. [45]
Hanshin expressway	1995 Kobe	6.9	Reinforced concrete	1.5	41	15	Heavy damage including collapse	Ishihara [46]
NHK and NFCH buildings	1964 Niigata	7.5	Reinforced concrete	0.35	16	8	Severe damage at interfaces between liquefied and non-liquefied soils.	Hamada [2]
Buildings on Fukaehama	1995 Kobe	6.9	High-strength concrete piles,	0.35–0.6	20	16	Pile failures at the interface between liquefied and non-liquefied layer.	Tokimatsu and Asaka [47]

Case history	Earthquake event	Magnitude (M_L)	Pile type	Pile diameter (m)	Pile length (m)	Pile length in liquefiable soil (m)	Pile performance	References
Building on Higashinada-ku	1995 Kobe	6.9	Prestressed concrete piles	0.4	12	5	Cracks on piles at near the pile head, in the middle and the bottom of the liquefied layer.	Tokimatsu et al. [48]
Harbour Master's building, Kandla	2001 Bhuj	7.7	Reinforced concrete	0.4	25	15	Piles performed poor.	Madabhushi et al. [49]
Miaoziping Bridge	2008 Wenchuan	7.9	Reinforced concrete	—	100	45	One approach bridge span collapsed.	Kawashima et al., [50]

Table 1.
Summary of case histories on pile foundation performance in past earthquakes (adapted from Bhattacharya et al. [33]).

Bhattacharya et al. [33] collated 15 case histories of pile earthquake performance and classified them according to their Euler buckling load when the soil was fully liquefied. In most of the cases where the axial load in the pile was 50% or more of the buckling loads, the foundation suffered significant damage.

Based on these observations, the failure of pile foundations occurred in both laterally spreading ground and in level ground where no lateral spreading would be anticipated. The cracks observed were near the bottom and at interfaces between liquefied and non-liquefied layers and often at the pile head. Additionally, severe damage had also formed at the boundaries of the liquefiable and non-liquefiable layers and at various depths between. The plastic hinge formation occurred at the boundaries of the liquefiable and non-liquefiable layers and at various depths.

3.2 Laboratory testing

Laboratory studies are also critical to elucidate the failure mechanisms and the behaviour of soil–pile interaction in liquefiable soils and its relevant fundamental parameters such as relative density, confining pressure, shear strength, frequency content and amplitude, damping ratio, pile bending moment, and deformed shape of the soil profile. Therefore, while the aim of the work presented in this section is to review and provide well-interpreted field of the dynamic response of piles by different physical model tests that can be used to evaluate analytical procedures and design methods. Many studies have investigated the seismic response of pile, soil and superstructure using shake-table experiments [58–63], dynamic centrifuge experiments [19, 64, 65], and full-scale field tests that utilise blast-induced liquefaction [66, 67]. The requirements for a model container for carrying out seismic soil-structure interactions (SSI) at 1-g (shaking table) and N-g (geotechnical centrifuge at N times earth's gravity) are well introduced in [68].

3.2.1 Shaking table tests

Iwasaki et al., [69] used a shaking table to estimate the liquefaction potential by using fundamental properties of the soil. Meymand [70] carried out a set of soil-pile interaction tests using the large shaking table operated by U.C. Berkeley. It was reported that damping for the single piles computed from 10 to 20%. Yao et al., [71] highlighted that the transient state prior to soil liquefaction was important in the design of piles due to dynamic earth pressure showed peak response in this state. Tokimatsu et al. [72] investigated pile under the combination of inertial and kinematic forces and reported that the pile foundation response depends on the time period of the ground as well as the superstructure. Cubrinovski et al. [73] discussed the behaviour of pile foundations under lateral spreading. Chau et al. [74] suggested that seismic pounding between the laterally compressed soil and the pile near the pile cap level can be one of the probable causes of pile damages. Motamed & Towhata [75] carried out a series of 1 g small-scale shake table model tests on a 3 × 3 pile group located behind a sheet-pile quay wall. It reported that fixed-end mitigating sheet pile can reduce the bending moment of pile. This is depended on the pile position within the group [76]. Gao et al. [61] studied the dynamic interactive behaviour of soil–pile foundations in liquefying ground under different shaking frequency and amplitude. They reported that the frequency of motion does not have a significant effect on the pile and soil response; however, these responses depend on the shaking amplitude. Besides, Lombardi & Bhattacharya [34] concluded that natural frequency of pile foundation decreases due to liquefaction; also they found that the damping ratio may increase due to liquefaction in excess of 20% (**Figure 7**).

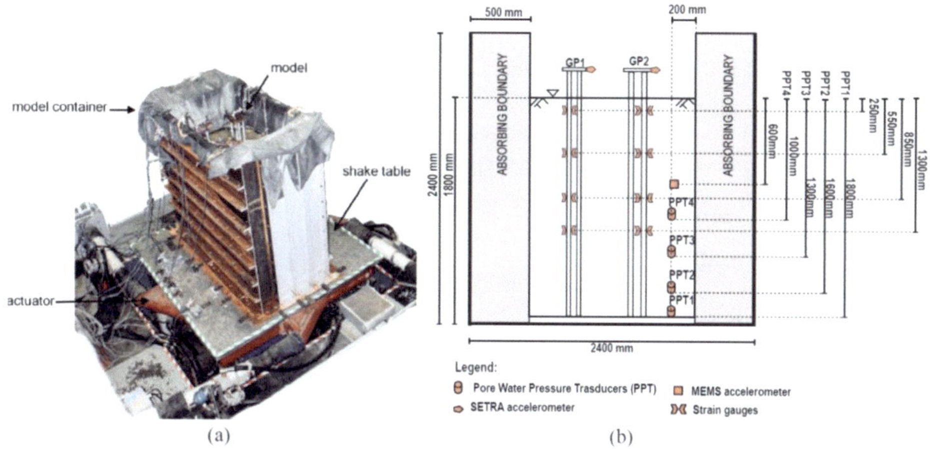

Figure 7.
(a) The shaking table test and (b) elevation view of test [34].

A similar observation was reported by Tang and Ling [62] and Tang et al. [63], which conducted a shaking table experiment to investigate the dynamic behaviour of a reinforced-concrete (RC) elevated cap pile foundation during (and prior to) soil liquefaction. These works indicated decreasing the frequency and increasing the amplitude of earthquake excitation. Next to this, Chen et al. [77] suggested that the seismic response of the soil and structure depends on input motion with richer low frequency components. On the other hand, Su et al. [58] document thicker pile having higher displacement. Likewise, the work performed by Liu et al. [78] was demonstrated that pile group bending moment was able to increase dramatically as the diameter increases. Four large laminar-box shaking table experiments used by Ebeido et al. [59] to examine pile response due to the mechanism of liquefaction-induced lateral spreading. They concluded that the highest pile lateral loads occurred at the initial stages of lateral deformation. Zhang et al. [41] reported that a pile collapsed due to buckling instability, which happened after the soil fully liquefied.

3.2.2 Centrifuge tests

Similar to the shaking table test, centrifuge test enables to address liquefaction, lateral spreading, and their effect on pile foundations by the simulation of gravity-induced stress conditions in scale models of soil structures at N times earth's gravity. Conceptually, this technique consists of the linear dimensions in the model soil by a factor 1/N and the confining stress is identical by a factor of unity. Thus, scale models law to simulate the behaviour of full-scale earth structures by reduced scale and provide data applicable to full-scale problems [68].

Several experimental studies have investigated of piles subject to liquefaction by using centrifuge method testing. McVay et al. [79] analysed the behaviour of the laterally loaded on pile group in sand with different pile group models. They reported that by changing the size of the group, there was no change in the group's lateral resistance, but only was a function of row position. Wilson [80] and Wilson et al. [81] performed dynamic centrifuge tests on the soil-pile interaction (**Figure 8**), which was directly obtained from the observed p-y response through back analysis of a single pile. This analyse presented the first dynamic characterisation of p–y behaviour of pile foundations in liquefying sand.

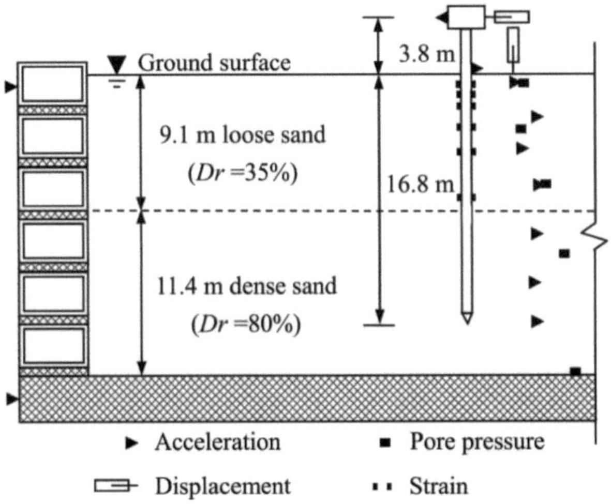

Figure 8.
Layout of the centrifuge test setup by Wilson [80].

Abdoun et al. [82] estimated the peak subgrade reaction values in liquefiable sand from centrifuge tests and concluded that the largest free head pile bending moment occurred at the boundary between the liquefied and non-liquefied strata. Similarly, a large displacement due to the liquefaction of the backfill soil was observed between the rubble mound and the bearing stratum, which produced a large bending moment at the top of the pile [83]. Brandenberg et al. [84] conducted various aspects of bending failure mechanism. In contrast, Bhattacharya et al. [85] proposed buckling instability failure mechanism as a new theory of pile failure mechanism verified by dynamic centrifuge tests. Another experiment concluded that an increase in excess pore pressure around and beneath end-bearing piles might induce the instability failure caused by liquefaction [86]. A similar observation was reported to clarify the buckling instability failure mechanism by Knappett and Madabhushi [32]. Recently, Garala et al. [19] conducted unexplored aspects of kinematic pile bending.

3.3 Numerical modelling

The numerical simulation tools have been prominent for analysing liquefaction problems in the light of potential disadvantages of physical models used in experimental simulation. This section presents different numerical platforms used in modelling of pile foundations under dynamic loading and their capabilities and limitations. Review of the recent relevant works delineates the important aspects of the seismic analysis of piles in liquefiable soils.

Numerical modelling can be divided into three categories: Beam on nonlinear Winkler Foundation (BNWF) approach with the proposed p–y curves, two-dimensional numerical modelling and the full three-dimensionality of model. Due to computationally complex and time-consuming of two- and three-dimensional numerical modelling, most of the researchers prefer to use the pseudo-static analyse is based on Winkler method for the seismic analysis of pile foundations. Winkler models are approximately capable of predicting maximum lateral displacement and maximum bending moment of pile foundations in liquefied soils. However, it is not able to simulate the prototype model accurately because it is difficult to estimate the accurate values for the springs and dashpots coefficients, which considerably change over time, especially during strong shaking.

3.3.1 p–y curves

The beam-on-nonlinear-Winkler-foundation (BNWF) method (also known as p-y method or Winkler method) is widely used in the modelling of soil–pile–structure interaction due to its simplicity in modelling and computational efficiency. This method is based on the hypothesis that the reaction exerted by the soil at a given depth on the pile shaft is proportional to the relative pile–soil lateral deflection. In the BNWF method, soil-pile interactions are modelled by a set of nonlinear soil springs, whereas the horizontal responses are analysed using p-y spring, for simulating shaft friction controlling vertical loading characteristics by t-z spring and end-bearing at the bottom of the piles responses are represented via the q-z spring (**Figure 9**). Each spring can be defined by means of a non-linear relationship between the soil reaction (per unit length of the pile) p and the corresponding relative soil–pile displacement.

This method is based on the beam on elastic foundation approach of Hetényi [36] and Winkler [88]. The p–y curves have been used to model the reaction of the foundation with consideration of inertial effects and seismic soil–pile interaction. Guidelines for the p – y curves as prescribed by current codes of practice are based on the works of Matlock [89] for soft clay, Reese [90] for cohesionless soils and Cox et al. [91] and O'Neill and Murchison [92] for sand, published by API [93] and DNV [94]. A different approach based on the assumption that the liquefied soil behaves like soft clay applied to account for the effects of liquefaction on the p–y curves, which is known as "residual strength" method [95]. However, applications of these curves were developed from a number of field tests with relatively few inherent limitations. Therefore, numerous works have been carried out to evaluate p–y curves for laterally loaded piles in liquefiable soils, such as Dobry et al. [96], Yasuda et al., [97], Sivathayalan and Vaid [98] and Rollins et al. [99]. Subsequently, the p–y method was extended to liquefiable soils by applying a "p multiplier" [96, 100], which is a reduction factor (m_p). Combining the force- and displacement-based methods,

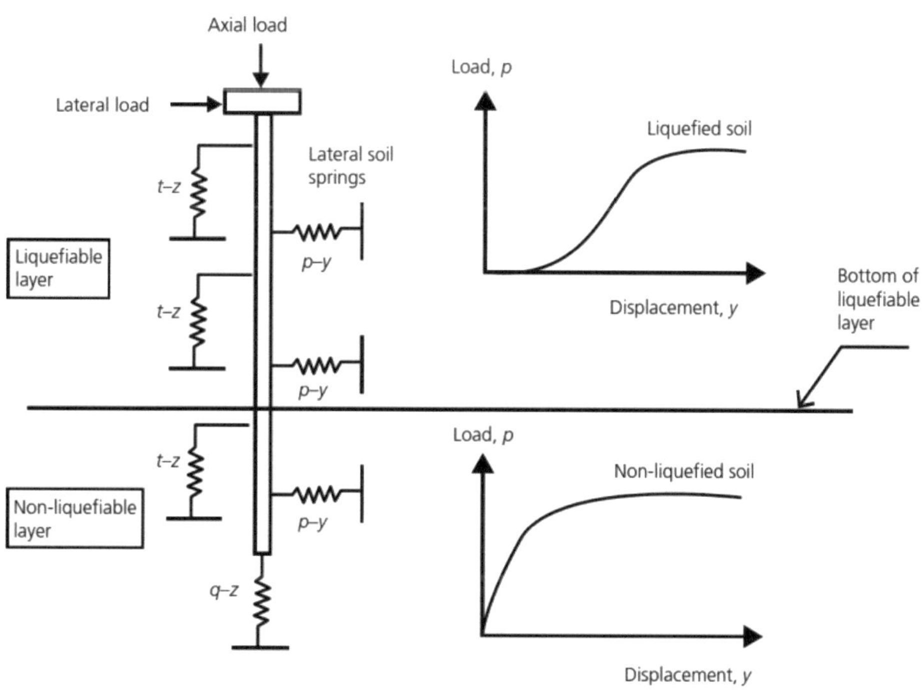

Figure 9.
Method of pile analysis using p-y curves, Dash et al. [87].

Cubrinovski et al. [100] proposed to use limit pressures for non-liquefied crust layers and linear springs with a "stiffness degradation factor" (known as the p-multiplier) for liquefied layers during liquefaction-induced lateral spreading. Several analyses of the full-scale tests [81, 101–103] conducted and observed the actual shape of post-cyclic stress–strain response of liquefied soils. They suggested an S-curve shape of the "p–y" curve for liquefied soil. Similarly, the post-liquefaction behaviour of sands observed in element tests by [104, 105]. Lombardi et al. [106] and Dash et al. [107] adopted a new set of p–y curves that can be obtained by modifying the conventional p–y curves (for non-liquefied soils) in such a way that replicates the strain hardening behaviour with practically-zero stiffness at low strain.

3.3.2 Two- and three-dimensional numerical modelling

To gain further insight into the field of simulation of the soil–pile interaction, the general design process is presented here. Finn and Fujita [108] developed and recently reviewed [109] an approximate method for nonlinear, three-dimensional analysis of pile foundations using PILE-3D. Constitutive models for simulating the nonlinear behaviour of pile in liquefiable soil have been proposed and typically implemented through the finite-element using two- or three-dimensional numerical modelling. Some of the more popular computer programs used are FLAC (Itasca), DIANA (DIANA Analysis) and PLAXIS. Many researchers have been using the Open System for Earthquake Engineering Simulation (OpenSees) (Mazzoni et al., [110]) as a ground response analysis tool. However, the main challenge of numerical modelling remains the simultaneous numerical prediction of accelerations, generation and redistribution of excess pore pressures, and the resulting of deformations [111]. Yu et al. [112] suggested that the finite element method (FEM) based on solid mechanics can accurately simulate the soil behaviour (for the initial stage) and the smoothed particle hydrodynamics (SPH) method in the framework of fluid dynamics is more suitable (for the flow stage). In this regard, Finite difference numerical models were developed using ABAQUS/Explicit, SAP2000, FLAC 3D, SANISAND and PDMY02 models or FEM using DBLEAVES code. The bounding surface constitutive model simple anisotropic sand (SANISAND) developed by Dafalias and Manzari [113] and implemented in OpenSees by Ghofrani and Arduino [114], was utilised to represent the behaviour of the liquefiable soil layer. Pressure-Dependent multiyield surface constitutive model version 02 (PDMY02) developed and implemented in OpenSees by Elgamal et al. [115] and Yang et al. [116] to simulate soil behaviour. Ramirez et al. [111] compared the predictive capabilities of PDMY02 and SANISAND platforms. Furthermore, Drucker-Prager [117] and Mohr-Coulomb plasticity [118] models are also soil constitutive modelling approaches in 3D analyses of soil-foundation systems.

Some of the recent numerical studies on pile foundations in liquefiable soil are listed in **Table 2**.

A fully coupled formulation (u–P or u–P–U) has been used to analyse soil displacements and pore water pressures [136]. The u–P formulation (called EightNodeBrick–u–P element in OpenSees framework) is the simplification of the u-p-U approach, which captures the movements of the soil skeleton (u) and the change of the pore pressure (P). More detail about description, formulation and implementation of this theory can be found in [120–122]. **Figure 10** is an example of three numerical modelling of pile in liquefiable of soil.

BNWF analyses of piles and pile groups have been exclusively employed for modelling case histories. Dash et al. [57] created 3D non-linear model of Tower of Kandla Port by using a finite element program SAP2000. A good agreement between the analytical and field observations analysis was reported. Similarly, Dash et al. [40] investigated the bending-buckling mechanism by exploring the Showa

References	Analysis type	Test type	Soil type	Pile configuration	Pile type	Pile head condition
Cubrinovski et al. [119]	3D DIANA	Shaking table	Sand	3 × 3	Stainless steel	Fixed
Cheng and Jeremic′ [120]	3D (u-p-U) FEM	Full scale	Sand	Single	Aluminium	Free
Dash et al. [57]	3D (p-y) SAP2000	Full scale	Clay soil	3 × 4	Concrete	Fixed
McGann et al. [117]	3D (p-y) FEM	Full scale	Sand	Single	Reinforced Concrete	Free and Fixed
Rahmani and Pak [121]	3D (p-u) OpenSees	Centrifuge	Sand	Single	Concrete	Free and Fixed
Wang et al. [122]	3D (p-u) OpenSees	Centrifuge	Sand	Single	Concrete	Fixed
Valsamis et al. [123]	p-y NASTRAN	Full scale	Sand	Single	Concrete	Fixed
Bhowmik et al. [124]	3D Abaqus	Full scale	Sandy clay	Single	steel	Fixed
Wang and Orense [125]	2D (p-y) OpenSees	Centrifuge	Sand	Single	Steel pipe	Fixed
Finn [109]	PILE3-D	Full scale	Sand	4 × 4	Concrete	Free and Fixed
Sextos et al. [126]	3D (p-y) SAP2000	Full scale	silty sands	2 × 2, 4 × 4	Reinforced concrete	Fixed
Hamayoon et al. [127]	3D FEM	Shaking table	Sand	3 × 3	Aluminium pipes	Fixed
Li and Motamed [128]	2D (p-y) OpenSees	Shaking table	Silica sand	3 × 3	Steel	Fixed
Lombardi and Bhattacharya [129]	2D (p-y) SAP2000	Shaking table	Sand	Single, 2 × 2	Aluminium	Free and Fixed
Rostami et al. [118]	3D (p-y) Abaqus	Shaking table	Sand	Single	Reinforced concrete	Free and Fixed
Zhang et al. [130]	3D (p-u) OpenSees	Centrifuge	Sand	Single	Concrete pipe pile	Free
Wang et al. [131]	2D (p-y) OpenSees	Centrifuge	Sand	3 × 2	Concrete	Fixed
López Jiménez et al. [132]	3D Flac SANISAND	Full scale	Sand	3 × 4	Concrete	Fixed
Li et al. [133]	3D Flac	Full scale	Sand	3 × 3	XCC pile	Fixed
Zhang et al. [41]	2D (p-y) OpenSees	Shaking table	Sand	Single	Aluminium	Free
Kazemi Esfeh and Kaynia [134]	3D Flac SANISAND	Centrifuge	Ottawa sand	OWT	Monopile	Fixed
Rajeswari and Sarkar [135]	3D (p-u) OpenSees	Full scale	Nevada sand	Single	Concrete	Fixed

Table 2.
Summary of recent numerical studies on pile foundation in liquefiable soil.

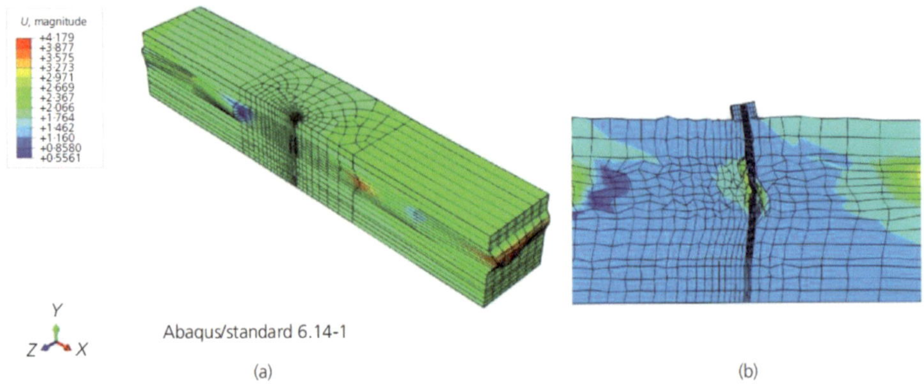

Figure 10.
An example of three numerical modelling of pile in liquefiable of soil [118]. (a) Deformed shape of model of unimproved soil with 3 m thickness of liquefiable soil; (b) pile deformation.

bridge pile failure in 1964 Niigata earthquake. In this context, McGann et al. [117] proposed a simplified procedure for the analysis of single piles subject to lateral spreading based on a parametric study. Moreover, the values of degradation factors of p-y curves in liquefiable soils computed in 3D FEM using OpenSees [137].

Two-dimensional models have been used to study soil-structure interaction in the majority of the numerical analyses using OpenSees. Haldar and Babu [138] examined the failure mechanism in piles and observed the failure mode was greatly dependent on the depth of the liquefiable soil layer. Zhang and Hutchinson [139] proposed a strategy by integrating the calculated plastic curvature at all integration points along the pile shaft. It was reported that the plastic hinge length of piles extending through liquefiable layers is about 1.4 times larger than that of non-liquefiable conditions. However, the location of plastic hinges can be affected by a variety of factors, such as material properties, pile length and thickness of the liquefied soil layer [118]. Wang and Orense [125] used a 2DBNWF finite element model implemented via Open Sees to analyse the response of raked pile foundations in liquefying ground. Bhowmik et al. [124] investigated the nonlinear behaviour of single hollow pile in layered soil subjected to varying levels of horizontal dynamic load. It found that separation of pile from the surrounding soil considerably affects the resonance frequency and amplitude of the pile foundations. Finn [109] compered different factors that can take into the behaviour of pile foundations during earthquakes in both liquefiable and non-liquefiable soils. Through a 2D nonlinear dynamic finite element (FE) modelling, Li and Motamed [128] presented a large-scale shake table test. It demonstrated that the FE model was able to reproduce the shaking table model behaviour with reasonable accuracy.

Three-dimensional analysis has become more common in the analysis of a full behaviour soil–pile–superstructure system with the greatest potential for accurately and certainly using either solid elements or beam-column elements. Zhang and Liu [140] performed a total of 90 3D finite element analyses using ABAQUS/Explicit to investigate the seismic response of different pile-raft-superstructure systems constructed on soft clay subjected to far-field ground motions. Zhang et al. [130] reported a good agreement between the numerical and the experimental data. Jiménez et al. [132] analysed the effects of this interaction, numerical models with a 3-storey reinforced concrete building using Flac 3D. Esfeh and Kaynia [134] used the software FLAC3D and the SANISAND constitutive model to conduct the nonlinear dynamic analyses for Offshore Wind Turbines. It was found that SANISAND model is capable of simulating the pore pressure generation in the free-field as observed in a recent centrifuge test. Recently, Manzari et al. [141] compered 11 sets of Type-B numerical simulations with the results of a selected set of centrifuge tests conducted in the LEAP- 2017 project. They obtained a good match trends in a number of numerical simulations with their experimental results.

4. Remediation schemes

Various ground improvement techniques have been developed for remediation of piled foundations in liquefiable soils over the past few decades. New techniques are introduced either to prevent liquefaction or to minimise the resulting settlements. Piled foundations of existing buildings are often difficult to access for retrofitting and, in addition, any procedure must ensure that the superstructure is not damaged during remediation [142]. Remediation of liquefiable soils for pile foundations needs to meet the several design performances required [143]. First, the most appropriate method for remediation should be selected for a specific portion or area (e.g. ground improvement). Next, the effective of the remedial measure should be appropriately determined to eliminate liquefaction and the associated ground deformations (lateral spreading and settlement). Moreover, the economic viability of the scheme should be evaluated to reduce or avoid potential structural damage caused by liquefied soil. The most common remediation techniques for pile foundations founded in liquefiable soils are summarised in **Figure 11**.

Installation of drains (e.g., using stone columns, sand compaction piles, prefabricated vertical drains (PVDs)) can prevent or delay liquefaction by enhancing dissipation of excess pore pressures and preventing void redistribution and the formation of a water lens below a low permeability crust [144–147]. However, deviatoric deformation and volumetric strains due to localised drainage during shaking significantly influenced the effectiveness of drains [148].

A number of densification techniques (e.g., using deep dynamic compaction, vibro-compaction, compaction piles) have been widely studied, because these techniques are relatively simple and practical, and the resulting remediation success can be easily verified by using in-situ penetration techniques [149–152]. However, Rayamajhi et al. [153, 154] reported that the densification and drainage techniques of improvement are often ineffective while the soil-cement columns were relatively

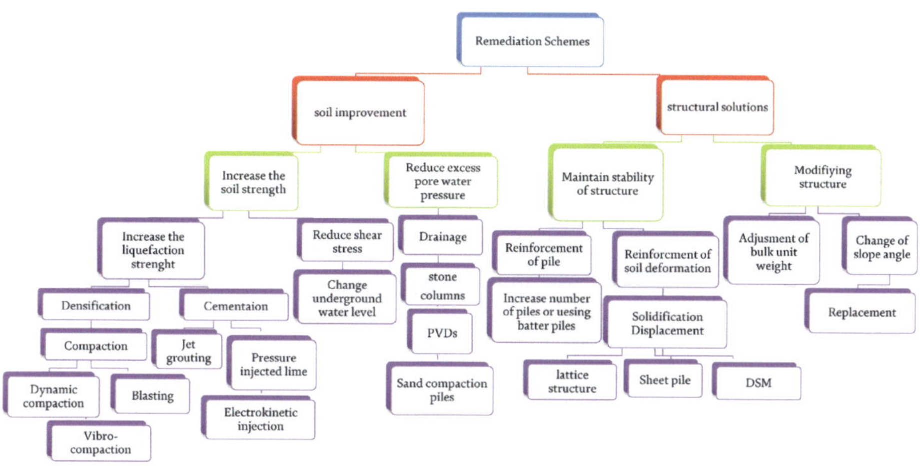

Figure 11.
Summary of the most common remediation techniques.

ineffective in reducing the potential for liquefaction triggering in saturated silty soils. This method also may not reduce permanent and transient tilt [148]. In this regard, Olarte et al. [155] compared drainage, densification and reinforcement with in-ground structural walls techniques. It was reported that both drainage and densification can reduce excess pore pressures and permanent foundation settlement and the performance of the reinforcement wall depended on the properties of the earthquake motion.

The soil stiffness of the liquefiable layer must be chosen carefully for a reliable analysis because it significantly affects the pile response.

Solidification methods (e.g., using deep soil mixing, jet grouting, walls, deep soil-mixed (DSM), sheet piles or lattice-shaped walls) are promising ground improvement methods which have proven to be effective in stabilising potentially liquefiable soil at several sites during earthquakes not only in controlling lateral spread but also in preventing liquefaction [20, 156–159]. This method was confirmed by a three-dimensional finite difference model using FLAC3D [157]. Moreover, the foundation ground of the 14-storey Meriken Park Hotel was improved using the deep cement mixing (DCM) method and it had good performance and survived the Kobe earthquake without damage [160]. In addition, authors [20] proposed a seismic requalification methodology of a pile-supported structure based on numerical simulations. It was recommended to use cementation or/and lattice structure techniques for reducing liquefaction hazard.

5. Summary

Procedures for identifying pile failure mechanisms due to liquefaction have been developed by reviewing of case histories data, experimental and numerical techniques. Examination of dynamic behaviour of pile foundations in seismically liquefiable soils led to the following practical recommendations:

- In order to consider a possible boundary for safe design and avoid failure of the pile, it is suggested to consider the buckling mechanism together with the effect of lateral load.

- Lessons from case histories reported that plastic hinge formation occurred at various locations which cannot be predicted with certainty. However, it was emphasised that, pile foundations could have collapsed because of bending, buckling, and combined action of bending of pile foundations.

- The review of results from various physical modelling techniques indicated that the dynamic behaviour of pile foundations depends not only on the site-specific characteristics (e.g., frequency of input motion, the amplitude, etc.), but also on the dynamic (modal) characteristics of the system (fixity of the pile head, the pile tip, pile position within the group, thickness of liquefying soil layer, etc.), which considerably affect piles performance in liquefied grounds.

- Overall, numerical modelling is capable of representing the most important aspects of pile failure mechanisms. The results obtained from different studies reported a good match in a number of numerical simulations compared to their experimental observations. In this respect, 3D analysis, in particular, favourably performed in the pursuit of identifying pile failure mechanisms.

- Various remediation techniques are available to mitigate the pile foundations on deposits of liquefiable sand. Nevertheless, despite being convenient for this purpose, it should be noted that considering one or more methods combined can provide economical solutions for liquefaction remediation problems. It would be rational to consider the remediation methods that have been implemented at many sites or tested/modelled by few large-scale earthquakes, which performed well. The above reasoning recommends using a combination of solidification methods.

Author details

Rohollah Rostami[1*], Slobodan B. Mickovski[1], Nicholas Hytiris[1] and Subhamoy Bhattacharya[2]

1 Built Environment Asset Management Centre, Glasgow Caledonian University, Glasgow, UK

2 University of Surrey, Guildford, UK

*Address all correspondence to: rohollah.rostami@gcu.ac.uk

References

[1] Ishihara, K., 1993. Liquefaction and flow failure during earthquakes. Geotechnique, 43(3), pp.351-451.

[2] Ross G, Seed HB, Migliacio R. Performance of highway bridge foundations in the great Alaska earthquake of 1964. Committee on the Alaskan Earthquake of the Division of Earth Sciences National Research Council. The Great AlaskaEarthquake of 1964. Washington, DC: National Academy of Sciences; 1973.

[3] Hamada M. Large ground deformations and their effects on lifelines: 1964 Niigata Earthquake. Case studies of liquefaction and lifelines performance during past earthquake. Technical report NCEER-92-0001. Buffalo: National Centre for Earthquake Engineering Research; 1992.

[4] Youd TL, Bartlett SF. Case histories of lateral spreads from the 1964 Alaskan earthquake. In: Proceedings of the third Japan–US workshop on Earth resistant design of life line facilities and counter measures for soils liquefaction. In: NCEER, vol.91-0001,February1,1989.

[5] Hazen, A. (1920). "Hydraulic fill dams." Trans., ASCE, Vol. LXXXIII, 1713-1745.

[6] Kramer, S.L. and Seed, H.B., 1988. Initiation of soil liquefaction under static loading conditions. Journal of Geotechnical Engineering, 114(4), pp.412-430.

[7] Ishihara K (1985) Stability of natural deposits during earthquakes. Proceedings 11th International Conference on Soil Mechanics and Foundation Engineering, vol 1, pp 321-376.

[8] Huang, Y. and Yu, M., 2013. Review of soil liquefaction characteristics during major earthquakes of the twenty-first century. Natural hazards, 65(3), pp.2375-2384.

[9] Castro, G. (1969). "Liquefaction of sands." Harvard Soil Mechanics Series, No. 81, Harvard University Press, Cambridge, Mass.

[10] Castro, G., and Poulos, S. J. (1977). "Factors affecting liquefaction and cyclic mobility." J. Geotech. Engrg. Div., ASCE, 103(GT6), 501-506.

[11] Poulos, S. J. (1981). "The steady state of deformation." J. Geotech. Engrg. Div., ASCE, 107(GT5), 553-562.

[12] Bhattacharya, S., Orense, R.P. and Lombardi, D., 2019. Seismic Design of Foundations: Concepts and applications. ICE Publishing.

[13] Zeghal, M. and Elgamal, A.W., 1994. Analysis of site liquefaction using earthquake records. Journal of geotechnical engineering, 120(6), pp.996-1017.

[14] Kramer, S.L., Asl, B.A., Ozener, P. and Sideras, S.S., 2015. Effects of liquefaction on ground surface motions. In Perspectives on earthquake geotechnical engineering (pp. 285-309). Springer, Cham.

[15] Hall, F.E., Lombardi, D. and Bhattacharya, S., 2018. Identification of transient vibration characteristics of pile-group models during liquefaction using wavelet transform. Engineering Structures, 171, pp.712-729.

[16] Özener, P.T., Greenfield, M.W., Sideras, S.S. and Kramer, S.L., 2020. Identification of time of liquefaction triggering. Soil Dynamics and Earthquake Engineering, 128, p.105895.

[17] Iwasaki, T., Tokida, K.I., Tatsuoka, F., Watanabe, S., Yasuda, S. and Sato, H., 1982, June. Microzonation

for soil liquefaction potential using simplified methods. In Proceedings of the 3rd international conference on microzonation, Seattle (Vol. 3, No. 2, pp. 1310-1330).

[18] Chung, J. and David Rogers, J., 2017. Deterministic and probabilistic assessment of liquefaction hazards using the liquefaction potential index and liquefaction reduction number. Journal of Geotechnical and Geoenvironmental Engineering, 143(10), p.04017073.

[19] Garala, T.K., Madabhushi, G.S. and Di Laora, R., 2020. Experimental investigation of kinematic pile bending in layered soils using dynamic centrifuge modelling. Géotechnique, pp.1-16.

[20] Rostami, R., Bhattacharya, S., Hytiris, N. and Mickovski, S.B., 2018. Seismic risk management of piles in liquefiable soils stabilised with cementation or lattice structures. Geotechnical Research, 6(2), pp.130-143.

[21] Bhattacharya, S. and Madabhushi, S.P.G., 2008. A critical review of methods for pile design in seismically liquefiable soils. Bulletin of Earthquake Engineering, 6(3), pp.407-446.

[22] Dobry, R., Abdoun, T., O'Rourke, T.D. and Goh, S.H., 2003. Single piles in lateral spreads: Field bending moment evaluation. Journal of Geotechnical and Geoenvironmental Engineering, 129(10), pp.879-889.

[23] Bhattacharya, S., Madabhushi, S.P.G. and Bolton, M.D., 2004. An alternative mechanism of pile failure in liquefiable deposits during earthquakes. Geotechnique, 54(3), pp.203-213.

[24] Bhattacharya, S., Adhikari, S. and Alexander, N.A., 2009. A simplified method for unified buckling and free vibration analysis of pile-supported structures in seismically liquefiable soils. Soil Dynamics and Earthquake Engineering, 29(8), pp.1220-1235.

[25] Haldar, S. and Babu, G.S., 2010. Failure mechanisms of pile foundations in liquefiable soil: Parametric study. International Journal of Geomechanics, 10(2), pp.74-84.

[26] Kramer, S.L. and Heavey, E.J., 1988. Analysis of laterally loaded piles with nonlinear bending behavior. Transportation Research Record, (1169).

[27] JRA (Japanese Road Association) (2002) Specification for Highway Bridges: Part V: Seismic Design. JRA, Tokyo, Japan.

[28] Fema (Federal Emergency Management Agency) (2000) NEHRP Recommended Provisions and Commentary for Seismic Regulations for New Buildings and Other Structures. Fema, Washington, DC, USA, Fema 369.

[29] IS-1893. Part 1: Criteria for Earthquake Resistant Design of Structures. Bureau of Indian Standard: New Delhi, 2000.

[30] EN, B., 2004. Eurocode 8: Design of structures for earthquake resistance-Part 5: Foundations, retaining structures and geotechnical aspects.

[31] Hwang, J.I., Kim, C.Y., Chung, C.K. and Kim, M.M., 2006. Viscous fluid characteristics of liquefied soils and behavior of piles subjected to flow of liquefied soils. Soil Dynamics and Earthquake Engineering, 26(2-4), pp.313-323.

[32] Knappett, J.A. and Madabhushi, S.P.G., 2009. Influence of axial load on lateral pile response in liquefiable soils. Part I: physical modelling. Geotechnique, 59(7), pp.571-581.

[33] Bhattacharya, S., Blakeborough, A. and Dash, S., 2008, November. Learning

from collapse of piles in liquefiable soils. In Proceedings of the Institution of Civil Engineers-Civil Engineering (Vol. 161, No. 6, pp. 54-60). Thomas Telford Ltd.

[34] Lombardi, D. and Bhattacharya, S., 2014. Modal analysis of pile-supported structures during seismic liquefaction. Earthquake Engineering & Structural Dynamics, 43(1), pp.119-138.

[35] Bhattacharya, S. and Goda, K., 2013. Probabilistic buckling analysis of axially loaded piles in liquefiable soils. Soil Dynamics and Earthquake Engineering, 45, pp.13-24.

[36] Hetényi, M., 1971. Beams on elastic foundation: theory with applications in the fields of civil and mechanical engineering (No. BOOK). University of Michigan.

[37] Bhattacharya, S., Adhikari, S. and Alexander, N.A., 2009. A simplified method for unified buckling and free vibration analysis of pile-supported structures in seismically liquefiable soils. Soil Dynamics and Earthquake Engineering, 29(8), pp.1220-1235.

[38] Zhang, X., Tang, L., Ling, X. and Chan, A., 2020. Critical buckling load of pile in liquefied soil. Soil Dynamics and Earthquake Engineering, 135, p.106197.

[39] Shanker K, Basudhar PK, Patra NR. Buckling of piles under liquefied soil conditions. Geotechnical and Geological Engineering. 2007 Jun 1;25(3):303-13.

[40] Dash, S.R., Bhattacharya, S. and Blakeborough, A., 2010. Bending–buckling interaction as a failure mechanism of piles in liquefiable soils. Soil Dynamics and Earthquake Engineering, 30(1-2), pp.32-39.

[41] Zhang, X., Tang, L., Li, X., Ling, X. and Chan, A., 2020. Effect of the combined action of lateral load and axial load on the pile instability in liquefiable soils. Engineering Structures, 205, p.110074.

[42] Berrill, J.B., Christensen, S.A., Keenan, R.P., Okada, W. and Pettinga, J.R., 2001. Case study of lateral spreading forces on a piled foundation. Geotechnique, 51(6), pp.501-517.

[43] Adib, M., Villet, W.C. and Nisar, A., 1995. Prestressed concrete piles under seismic loading: Case history.

[44] Fujii, S., Isemoto, N., Satou, Y., Kaneko, O., Funahara, H., Arai, T. and Tokimatsu, K., 1998. Investigation and analysis of a pile foundation damaged by liquefaction during the 1995 Hyogoken-Nambu earthquake. Soils and Foundations, 38, pp.179-192.

[45] Tokimatsu, K., Mizuno, H. and Kakurai, M., 1996. Building damage associated with geotechnical problems. Soils and foundations, 36, pp.219-234.

[46] Ishihara, K., 1997. Terzaghi oration: Geotechnical aspects of the 1995 Kobe earthquake. Proceedings of ICSMFE, Hamburg, 1997, pp 2047-2073.

[47] Tokimatsu, K. and Asaka, Y., 1998. Effects of liquefaction-induced ground displacements on pile performance in the 1995 Hyogoken-Nambu earthquake. Soils and Foundations, 38, pp.163-177.

[48] Tokimatsu, K., Oh-oka, H., Shamoto, Y., Nakazawa, A. and Asaka, Y., 1997. Failure and deformation modes of piles caused by liquefaction—induced lateral spreading in 1995 Hyogoken—Nambu Earthquake. Geotechnical Engineering in Recovery from Urban Earthquake Disaster, KIG FOR UM, 97.

[49] Madabhushi, S.P.G., Patel, D. and Haigh, S.K., 2005. Geotechnical aspects of the Bhuj earthquake, Chapter 3, EEFIT Report, Institution of Structural Engineers, London.

[50] Kawashima, K., Takahashi, Y., Ge, H., Wu, Z. and Zhang, J., 2009. Reconnaissance report on damage of bridges in 2008 Wenchuan, China, earthquake. Journal of Earthquake Engineering, 13(7), pp.965-996.

[51] Iwasaki, T., 1984. A case history of bridge performance during earthquakes in Japan.

[52] Yoshida, N., Tazoh, T., Wakamatsu, K., Yasuda, S., Towhata, I., Nakazawa, H. and Kiku, H., 2007. Causes of Showa Bridge collapse in the 1964 Niigata earthquake based on eyewitness testimony. Soils and Foundations, 47(6), pp.1075-1087.

[53] Kerciku, A.A., Bhattacharya, S., Lubkowski, Z.A. and Burd, H.J., 2008. Failure of Showa Bridge during 1964 Niigata earthquake: lateral spreading or buckling instability. In Proceedings of the 14th World Conference on Earthquake Engineering.

[54] Bhattacharya, S., Tokimatsu, K., Goda, K., Sarkar, R., Shadlou, M. and Rouholamin, M., 2014. Collapse of Showa Bridge during 1964 Niigata earthquake: A quantitative reappraisal on the failure mechanisms. Soil Dynamics and Earthquake Engineering, 65, pp.55-71.

[55] Mohanty, P., Dutta, S.C. and Bhattacharya, S., 2017. Proposed mechanism for mid-span failure of pile supported river bridges during seismic liquefaction. Soil Dynamics and Earthquake Engineering, 102, pp.41-45.

[56] Madabhushi, G., Knappett, J. and Haigh, S., 2010. Design of pile foundations in liquefiable soils. Imperial College Press.

[57] Dash, S.R., Govindaraju, L. and Bhattacharya, S., 2009. A case study of damages of the Kandla Port and Customs Office tower supported on a mat–pile foundation in liquefied soils under the 2001 Bhuj earthquake. Soil Dynamics and Earthquake Engineering, 29(2), pp.333-346.

[58] Su, L., Tang, L., Ling, X., Liu, C. and Zhang, X., 2016. Pile response to liquefaction-induced lateral spreading: a shake-table investigation. Soil Dynamics and Earthquake Engineering, 82, pp.196-204.

[59] Ebeido, A., Elgamal, A., Tokimatsu, K. and Abe, A., 2019. Pile and Pile-Group Response to Liquefaction-Induced Lateral Spreading in Four Large-Scale Shake-Table Experiments. Journal of Geotechnical and Geoenvironmental Engineering, 145(10), p.04019080.

[60] Li, G. and Motamed, R., 2017. Finite element modeling of soil-pile response subjected to liquefaction-induced lateral spreading in a large-scale shake table experiment. Soil Dynamics and Earthquake Engineering, 92, pp.573-584.

[61] Gao, X., Ling, X.Z., Tang, L. and Xu, P.J., 2011. Soil–pile-bridge structure interaction in liquefying ground using shake table testing. Soil Dynamics and Earthquake Engineering, 31(7), pp.1009-1017.

[62] Tang, L. and Ling, X., 2014. Response of a RC pile group in liquefiable soil: A shake-table investigation. Soil Dynamics and Earthquake Engineering, 67, pp.301-315.

[63] Tang, L., Ling, X., Zhang, X., Su, L., Liu, C. and Li, H., 2015. Response of a RC pile behind quay wall to liquefaction-induced lateral spreading: a shake-table investigation. Soil Dynamics and Earthquake Engineering, 76, pp.69-79.

[64] Brandenberg, S.J., Boulanger, R.W., Kutter, B.L. and Chang, D., 2005. Behavior of pile foundations in laterally spreading ground during

centrifuge tests. Journal of Geotechnical and Geoenvironmental Engineering, 131(11), pp.1378-1391.

[65] Zhang, S., Wei, Y., Cheng, X., Chen, T., Zhang, X. and Li, Z., 2020. Centrifuge modeling of batter pile foundations in laterally spreading soil. Soil Dynamics and Earthquake Engineering, 135, p.106166.

[66] Rollins, K.M., Gerber, T.M., Lane, J.D. and Ashford, S.A., 2005. Lateral resistance of a full-scale pile group in liquefied sand. Journal of Geotechnical and Geoenvironmental Engineering, 131(1), pp.115-125.

[67] Ashford, S.A., Juirnarongrit, T., Sugano, T. and Hamada, M., 2006. Soil–pile response to blast-induced lateral spreading. I: field test. Journal of Geotechnical and Geoenvironmental Engineering, 132(2), pp.152-162.

[68] Bhattacharya, S., Lombardi, D., Dihoru, L., Dietz, M.S., Crewe, A.J. and Taylor, C.A., 2012. Model container design for soil-structure interaction studies. In Role of seismic testing facilities in performance-based earthquake engineering (pp. 135-158). Springer, Dordrecht.

[69] Iwasaki, T., Arakawa, T. and Tokida, K.I., 1984. Simplified procedures for assessing soil liquefaction during earthquakes. International Journal of Soil Dynamics and Earthquake Engineering, 3(1), pp.49-58.

[70] Meymand, P.J., 1998. Shaking table scale model tests of nonlinear soil-pile-superstructure interaction in soft clay.

[71] Yao, S., Kobayashi, K., Yoshida, N. and Matsuo, H., 2004. Interactive behavior of soil–pile-superstructure system in transient state to liquefaction by means of large shake table tests. Soil Dynamics and Earthquake Engineering, 24(5), pp.397-409.

[72] Tokimatsu, K., Suzuki, H. and Sato, M., 2005. Effects of inertial and kinematic interaction on seismic behavior of pile with embedded foundation. Soil Dynamics and Earthquake Engineering, 25(7-10), pp.753-762.

[73] Cubrinovski, M., Ishihara, K. and Poulos, H., 2009. Pseudo-static analysis of piles subjected to lateral spreading. Bulletin of the New Zealand Society for Earthquake Engineering, 42(1), pp.28-38.

[74] Chau, K.T., Shen, C.Y. and Guo, X., 2009. Nonlinear seismic soil–pile–structure interactions: shaking table tests and FEM analyses. Soil Dynamics and Earthquake Engineering, 29(2), pp.300-310.

[75] Motamed, R., Towhata, I., Honda, T., Tabata, K. and Abe, A., 2013. Pile group response to liquefaction-induced lateral spreading: E-Defense large shake table test. Soil Dynamics and Earthquake Engineering, 51, pp.35-46.

[76] Haeri, S.M., Kavand, A., Rahmani, I. and Torabi, H., 2012. Response of a group of piles to liquefaction-induced lateral spreading by large scale shake table testing. Soil Dynamics and Earthquake Engineering, 38, pp.25-45.

[77] Chen, G., Chen, S., Qi, C., Du, X., Wang, Z. and Chen, W., 2015. Shaking table tests on a three-arch type subway station structure in a liquefiable soil. Bulletin of Earthquake Engineering, 13(6), pp.1675-1701.

[78] Liu, C., Tang, L., Ling, X., Deng, L., Su, L. and Zhang, X., 2017. Investigation of liquefaction-induced lateral load on pile group behind quay wall. Soil Dynamics and Earthquake Engineering, 102, pp.56-64.

[79] McVay, M., Zhang, L., Molnit, T. and Lai, P., 1998. Centrifuge testing of large laterally loaded pile groups

in sands. Journal of Geotechnical and Geoenvironmental Engineering, 124(10), pp.1016-1026.

[80] Wilson, D.W., 1998. Soil-pile-superstructure interaction in liquefying sand and soft clay (Doctoral dissertation, University of California, Davis).

[81] Wilson, D.W., Boulanger, R.W. and Kutter, B.L., 2000. Observed seismic lateral resistance of liquefying sand. Journal of Geotechnical and Geoenvironmental Engineering, 126(10), pp.898-906.

[82] Abdoun, T., Dobry, R., O'Rourke, T.D. and Goh, S.H., 2003. Pile response to lateral spreads: centrifuge modeling. Journal of Geotechnical and Geoenvironmental engineering, 129(10), pp.869-878.

[83] Takahashi, A. and Takemura, J., 2005. Liquefaction-induced large displacement of pile-supported wharf. Soil Dynamics and Earthquake Engineering, 25(11), pp.811-825.

[84] Brandenberg, S.J., Boulanger, R.W., Kutter, B.L. and Chang, D., 2005. Behavior of pile foundations in laterally spreading ground during centrifuge tests. Journal of Geotechnical and Geoenvironmental Engineering, 131(11), pp.1378-1391.

[85] Bhattacharya, S. and Bolton, M., 2004, August. Buckling of piles during earthquake liquefaction. In Proc. 13th World conference on Earthquake Engineering (pp. 1-4).

[86] Knappett, J.A. and Madabhushi, S.P., 2008. Liquefaction-induced settlement of pile groups in liquefiable and laterally spreading soils. Journal of geotechnical and geoenvironmental engineering, 134(11), pp.1609-1618.

[87] Dash, S., Rouholamin, M., Lombardi, D. and Bhattacharya, S., 2017. A practical method for construction of py curves for liquefiable soils. Soil Dynamics and Earthquake Engineering, 97, pp.478-481.

[88] Winkler, E., 1867. Die Lehre von der Elasticitaet und Festigkeit: mit besonderer Rücksicht auf ihre Anwendung in der Technik für polytechnische Schulen, Bauakademien, Ingenieue, Maschinenbauer, Architecten, etc (Vol. 1). Dominicus.

[89] Matlock, H., 1970. Correlations for design of laterally loaded piles in soft clay. Offshore technology in civil engineering's hall of fame papers from the early years, pp.77-94.

[90] Reese, L.C., Cox, W.R. and Koop, F.D., 1974. Analysis of laterally loaded piles in sand. Offshore technology in civil engineering hall of fame papers from the early years, pp.95-105.

[91] Cox, W.R., Reese, L.C. and Grubbs, B.R., 1974, January. Field testing of laterally loaded piles in sand. In Offshore Technology Conference. Offshore Technology Conference.

[92] O'Neill, M.W. and Murchison, J.M., 1983. An evaluation of py relationships in sands. A report to the American Petroleum Institute. Houston, TX, USA: University of Houston.

[93] API, Recommended practice for planning, designing and constructing fixed offshore platforms-working stress design. 2A-WSD (RP2A-WSD) 21stEdition, December; 2000.

[94] DNV. Offshore Standard DNV-OS-J101. Design of offshore wind turbine structures. Det Norske Veritas, 2007.

[95] Goh, S., 1999. Limit state model for soil-pile interaction during lateral spread. In Proc. 7th. US-Japan Workshop on Earthquake Resistant Design of

Lifeline Facilities and Countermeasures against Soil Liquefaction, Seattle, WA, 1999.

[96] Dobry, R., Taboada, V., Liu, L. Centrifuge modelling of liquefaction effects during earthquakes. Proc. 1st Intl. Conf. On Earthquake Geotechnical Engineering, IS-Tokyo, 1995, 14-16.

[97] Yasuda, S., 1998. Post liquefaction behavior of several sands. In Proc. Of the 11^< th> European Conference on Earthquake Engineering.

[98] Sivathayalan, S. and Vaid, Y.P., 2004, August. Cyclic resistance and post liquefaction response of undisturbed in-situ sands. In Proceedings of the 13th world conference on earthquake engineering.

[99] Rollins, K.M., Gerber, T.M., Lane, J.D. and Ashford, S.A., 2005. Lateral resistance of a full-scale pile group in liquefied sand. Journal of Geotechnical and Geoenvironmental Engineering, 131(1), pp.115-125.

[100] Cubrinovski, M., Kokusho, T. and Ishihara, K., 2006. Interpretation from large-scale shake table tests on piles undergoing lateral spreading in liquefied soils. Soil Dynamics and Earthquake Engineering, 26(2-4), pp.275-286.

[101] Bhattacharya, S., Bolton, M.D. and Madabhushi, S.P.G., 2005. A reconsideration of the safety of piled bridge foundations in liquefiable soils. Soils and foundations, 45(4), pp.13-25.

[102] Brandenberg, S.J., 2005. Behavior of pile foundations in liquefied and laterally spreading ground. University of California, Davis.

[103] Takahashi, A., Kuwano, Y. and Yano, A., 2002, July. Lateral resistance of buried cylinder in liquefied sand. In Proceedings of the International Conference on physical modelling in geotechnics, ICPMG-02, St. John's, Newfoundland, Canada.

[104] Dash, S.R., 2010. Lateral pile soil interaction in liquefiable soils (Doctoral dissertation, Oxford University).

[105] Lombardi, D., Bhattacharya, S., Hyodo, M. and Kaneko, T., 2014. Undrained behaviour of two silica sands and practical implications for modelling SSI in liquefiable soils. Soil Dynamics and Earthquake Engineering, 66, pp.293-304.

[106] Lombardi, D., Dash, S.R., Bhattacharya, S., Ibraim, E., Muir Wood, D. and Taylor, C.A., 2017. Construction of simplified design p–y curves for liquefied soils. Géotechnique, 67(3), pp.216-227.

[107] Dash, S., Rouholamin, M., Lombardi, D. and Bhattacharya, S., 2017. A practical method for construction of py curves for liquefiable soils. Soil Dynamics and Earthquake Engineering, 97, pp.478-481.

[108] Finn, W.D.L. and Fujita, N., 2002. Piles in liquefiable soils: seismic analysis and design issues. Soil Dynamics and Earthquake Engineering, 22(9-12), pp.731-742.

[109] Finn, W.L., 2015. 1st Ishihara Lecture: An overview of the behavior of pile foundations in liquefiable and non-liquefiable soils during earthquake excitation. Soil Dynamics and Earthquake Engineering, 68, pp.69-77.

[110] Mazzoni, S., McKenna, F., Scott, M.H. and Fenves, G.L., 2006. OpenSees command language manual. Pacific Earthquake Engineering Research (PEER) Center, 264.

[111] Ramirez, J., Barrero, A.R., Chen, L., Dashti, S., Ghofrani, A., Taiebat, M. and Arduino, P., 2018. Site response in a layered liquefiable deposit: evaluation of different numerical tools

and methodologies with centrifuge experimental results. Journal of Geotechnical and Geoenvironmental Engineering, 144(10), p.04018073.

[112] Yu, M., Huang, Y. and Xu, Q., 2014. Numerical Methods for Deformation Analysis of Liquefiable Soils. In Engineering Geology for Society and Territory–Volume 4 (pp. 55-57). Springer, Cham.

[113] Dafalias, Y.F. and Manzari, M.T., 2004. Simple plasticity sand model accounting for fabric change effects. Journal of Engineering mechanics, 130(6), pp.622-634.

[114] Ghofrani, A. and Arduino, P., 2018. Prediction of LEAP centrifuge test results using a pressure-dependent bounding surface constitutive model. Soil Dynamics and Earthquake Engineering, 113, pp.758-770.

[115] Elgamal, A., Yang, Z. and Parra, E., 2002. Computational modeling of cyclic mobility and post-liquefaction site response. Soil Dynamics and Earthquake Engineering, 22(4), pp.259-271.

[116] Yang, Z., Lu, J. and Elgamal, A., 2008. OpenSees soil models and solid-fluid fully coupled elements. User's Manual. Ver, 1, p.27.

[117] McGann, C.R., Arduino, P. and Mackenzie-Helnwein, P., 2011. Applicability of conventional py relations to the analysis of piles in laterally spreading soil. Journal of geotechnical and geoenvironmental engineering, 137(6), pp.557-567.

[118] Rostami, R., Hytiris, N., Bhattacharya, S. and Giblin, M., 2017. Seismic analysis of pile in liquefiable soil and plastic hinge. Geotechnical Research, 4(4), pp.203-213.

[119] Cubrinovski, M., Uzuoka, R., Sugita, H., Tokimatsu, K., Sato, M., Ishihara, K., Tsukamoto, Y. and Kamata, T., 2008. Prediction of pile response to lateral spreading by 3-D soil–water coupled dynamic analysis: Shaking in the direction of ground flow. Soil Dynamics and Earthquake Engineering, 28(6), pp.421-435.

[120] Cheng, Z. and Jeremić, B., 2009. Numerical modeling and simulation of pile in liquefiable soil. Soil Dynamics and Earthquake Engineering, 29(11-12), pp.1405-1416.

[121] Rahmani, A. and Pak, A., 2012. Dynamic behavior of pile foundations under cyclic loading in liquefiable soils. Computers and Geotechnics, 40, pp.114-126.

[122] Wang, R., Fu, P. and Zhang, J.M., 2016. Finite element model for piles in liquefiable ground. Computers and Geotechnics, 72, pp.1-14.

[123] Valsamis, A.I., Bouckovalas, G.D. and Chaloulos, Y.K., 2012. Parametric analysis of single pile response in laterally spreading ground. Soil Dynamics and Earthquake Engineering, 34(1), pp.99-110.

[124] Bhowmik, D., Baidya, D.K. and Dasgupta, S.P., 2013. A numerical and experimental study of hollow steel pile in layered soil subjected to lateral dynamic loading. Soil Dynamics and Earthquake Engineering, 53, pp.119-129.

[125] Wang, S. and Orense, R.P., 2014. Modelling of raked pile foundations in liquefiable ground. Soil Dynamics and Earthquake Engineering, 64, pp.11-23.

[126] Sextos, A.G., Mylonakis, G.E. and Mylona, E.K.V., 2015. Rotational excitation of bridges supported on pile groups in soft or liquefiable soil deposits. Computers & Structures, 155, pp.54-66.

[127] Hamayoon, K., Morikawa, Y., Oka, R. and Zhang, F., 2016. 3D dynamic

finite element analyses and 1 g shaking table tests on seismic performance of existing group-pile foundation in partially improved grounds under dry condition. Soil Dynamics and Earthquake Engineering, 90, pp.196-210.

[128] Li, G. and Motamed, R., 2017. Finite element modeling of soil-pile response subjected to liquefaction-induced lateral spreading in a large-scale shake table experiment. Soil Dynamics and Earthquake Engineering, 92, pp.573-584.

[129] Lombardi, D. and Bhattacharya, S., 2016. Evaluation of seismic performance of pile-supported models in liquefiable soils. Earthquake Engineering & Structural Dynamics, 45(6), pp.1019-1038.

[130] Zhang, X., Tang, L., Ling, X., Chan, A.H.C. and Lu, J., 2018. Using peak ground velocity to characterize the response of soil-pile system in liquefying ground. Engineering Geology, 240, pp.62-73.

[131] Wang, X., Ye, A. and Ji, B., 2019. Fragility-based sensitivity analysis on the seismic performance of pile-group-supported bridges in liquefiable ground undergoing scour potentials. Engineering Structures, 198, p.109427.

[132] Jiménez, G.A.L., Dias, D. and Jenck, O., 2019. Effect of layered liquefiable deposits on the seismic response of soil-foundations-structure systems. Soil Dynamics and Earthquake Engineering, 124, pp.1-15.

[133] Li, W., Stuedlein, A.W., Chen, Y., Liu, H. and Cheng, Z., 2019. Response of pile groups with X and circular cross-sections subject to lateral spreading: 3D numerical simulations. Soil Dynamics and Earthquake Engineering, 126, p.105774.

[134] Esfeh, P.K. and Kaynia, A.M., 2020. Earthquake response of monopiles and caissons for Offshore Wind Turbines founded in liquefiable soil. Soil Dynamics and Earthquake Engineering, 136, p.106213.

[135] Rajeswari, J.S. and Sarkar, R., 2020. Estimation of Transient Forces in Single Pile Embedded in Liquefiable Soil. International Journal of Geomechanics, 20(9), p.06020023.

[136] Zienkiewicz, O.C. and Shiomi, T., 1984. Dynamic behaviour of saturated porous media; the generalized Biot formulation and its numerical solution. International journal for numerical and analytical methods in geomechanics, 8(1), pp.71-96.

[137] Choobbasti, A.J. and Zahmatkesh, A., 2016. Computation of degradation factors of py curves in liquefiable soils for analysis of piles using three-dimensional finite-element model. Soil Dynamics and Earthquake Engineering, 89, pp.61-74.

[138] Haldar, S. and Babu, G.S., 2010. Failure mechanisms of pile foundations in liquefiable soil: Parametric study. International Journal of Geomechanics, 10(2), pp.74-84.

[139] Zhang, J. and Hutchinson, T.C., 2012. Inelastic pile behavior with and without liquefaction effects. Soil Dynamics and Earthquake Engineering, 36, pp.12-19.

[140] Zhang, L. and Liu, H., 2017. Seismic response of clay-pile-raft-superstructure systems subjected to far-field ground motions. Soil Dynamics and Earthquake Engineering, 101, pp.209-224.

[141] Manzari, M.T., El Ghoraiby, M., Zeghal, M., Kutter, B.L., Arduino, P., Barrero, A.R., Bilotta, E., Chen, L., Chen, R., Chiaradonna, A. and Elgamal, A., 2020. LEAP-2017: Comparison of the Type-B numerical simulations with centrifuge test results. In Model

Tests and Numerical Simulations of Liquefaction and Lateral Spreading (pp. 187-218). Springer, Cham.

[142] Mitrani, H. and Madabhushi, S.G., 2013. Geomembrane containment walls for liquefaction remediation. Proceedings of the Institution of Civil Engineers-Ground Improvement, 166(1), pp.9-20.

[143] Iai, S., 2005. Remediation of liquefiable soils for port structures in Japan—analysis, design and performance. Journal of earthquake engineering, 9(spec01), pp.77-103.

[144] Asgari, A., Oliaei, M. and Bagheri, M., 2013. Numerical simulation of improvement of a liquefiable soil layer using stone column and pile-pinning techniques. Soil Dynamics and Earthquake Engineering, 51, pp.77-96.

[145] Howell, R., Rathje, E.M., Kamai, R. and Boulanger, R., 2012. Centrifuge modeling of prefabricated vertical drains for liquefaction remediation. Journal of geotechnical and geoenvironmental engineering, 138(3), pp.262-271.

[146] Tang, L., Cong, S., Ling, X., Lu, J. and Elgamal, A., 2015. Numerical study on ground improvement for liquefaction mitigation using stone columns encased with geosynthetics. Geotextiles and Geomembranes, 43(2), pp.190-195.

[147] Vytiniotis, A., Panagiotidou, A.I. and Whittle, A.J., 2019. Analysis of seismic damage mitigation for a pile-supported wharf structure. Soil Dynamics and Earthquake Engineering, 119, pp.21-35.

[148] Dashti, S., Bray, J.D., Pestana, J.M., Riemer, M. and Wilson, D., 2010. Centrifuge testing to evaluate and mitigate liquefaction-induced building settlement mechanisms. Journal of geotechnical and geoenvironmental engineering, 136(7), pp.918-929.

[149] Baez, J.I. and Martin, G.R., 1993, May. Advances in the design of vibro systems for the improvement of liquefaction resistance. In Symposium Ground Improvement (pp. 1-16).

[150] Balakrishnan, A., and B. L. Kutter. "Settlement, sliding, and liquefaction remediation of layered soil." Journal of geotechnical and geoenvironmental engineering 125, no. 11 (1999): 968-978.

[151] Coelho, P.A.L.F., Haigh, S.K., Gopal Madabhushi, S.P. and O'brien, T.S., 2007. Post-earthquake behaviour of footings employing densification to mitigate liquefaction. Proceedings of the Institution of Civil Engineers-Ground Improvement, 11(1), pp.45-53.

[152] Mitchell, J.K., 2008. Mitigation of liquefaction potential of silty sands. In From research to practice in geotechnical engineering (pp. 433-451).

[153] Rayamajhi, D., Nguyen, T.V., Ashford, S.A., Boulanger, R.W., Lu, J., Elgamal, A. and Shao, L., 2014. Numerical study of shear stress distribution for discrete columns in liquefiable soils. Journal of Geotechnical and Geoenvironmental Engineering, 140(3), p.04013034.

[154] Rayamajhi, D., Tamura, S., Khosravi, M., Boulanger, R.W., Wilson, D.W., Ashford, S.A. and Olgun, C.G., 2015. Dynamic centrifuge tests to evaluate reinforcing mechanisms of soil-cement columns in liquefiable sand. Journal of Geotechnical and Geoenvironmental Engineering, 141(6), p.04015015.

[155] Olarte, J., Paramasivam, B., Dashti, S., Liel, A. and Zannin, J., 2017. Centrifuge modeling of mitigation-soil-foundation-structure interaction on liquefiable ground. Soil Dynamics and Earthquake Engineering, 97, pp.304-323.

[156] Hasheminezhad, A. and Bahadori, H., 2019. Seismic response of shallow

foundations over liquefiable soils improved by deep soil mixing columns. Computers and Geotechnics, 110, pp.251-273.

[157] Nguyen, T.V., Rayamajhi, D., Boulanger, R.W., Ashford, S.A., Lu, J., Elgamal, A. and Shao, L., 2012. Effect of DSM grids on shear stress distribution in liquefiable soil. In GeoCongress 2012: State of the Art and Practice in Geotechnical Engineering (pp. 1948-1957).

[158] Nguyen, T.V., Rayamajhi, D., Boulanger, R.W., Ashford, S.A., Lu, J., Elgamal, A. and Shao, L., 2013. Design of DSM grids for liquefaction remediation. Journal of geotechnical and geoenvironmental engineering, 139(11), pp.1923-1933.

[159] Yamauchi, T., Tezuka, H. and Tsukamoto, Y., 2017. Development of rational soil liquefaction countermeasure consisting of lattice-shaped soil improvement by jet grouting for existing housing estates. In Geotechnical Hazards from Large Earthquakes and Heavy Rainfalls (pp. 49-59). Springer, Tokyo.

[160] Suzuki, Y., Saito, S., Onimaru, S., Kimura, T., Uchida, A. and Okumura, R., 1996. Grid-shaped stabilized ground improved by deep cement mixing method against liquefaction for a building foundation; Shinso kongo shori koho wo mochiita koshijo jiban kairyo ni yoru ekijoka taisakuko.